Methods in Molecular Biology

Series Editor
John M. Walker
School of Life and Medical Sciences
University of Hertfordshire
Hatfield, Hertfordshire, AL10 9AB, UK

For further volumes:
http://www.springer.com/series/7651

Cell Viability Assays

Methods and Protocols

Edited by

Daniel F. Gilbert

*Friedrich-Alexander University (FAU) Erlangen-Nürnberg,
Institute of Medical Biotechnology, Erlangen, Germany*

Oliver Friedrich

*Friedrich-Alexander University (FAU) Erlangen-Nürnberg,
Institute of Medical Biotechnology, Erlangen, Germany*

Humana Press

Editors
Daniel F. Gilbert
Friedrich-Alexander University (FAU) Erlangen-Nürnberg
Institute of Medical Biotechnology
Erlangen, Germany

Oliver Friedrich
Friedrich-Alexander University (FAU) Erlangen-Nürnberg
Institute of Medical Biotechnology
Erlangen, Germany

ISSN 1064-3745 ISSN 1940-6029 (electronic)
Methods in Molecular Biology
ISBN 978-1-4939-6959-3 ISBN 978-1-4939-6960-9 (eBook)
DOI 10.1007/978-1-4939-6960-9

Library of Congress Control Number: 2017936200

Printed on acid-free paper

This Humana Press imprint is published by Springer Nature
The registered company is Springer Science+Business Media LLC
The registered company address is: 233 Spring Street, New York, NY 10013, U.S.A.

Preface

In vitro assessment of cellular viability has become a generic approach in addressing a vast range of biological questions in many areas of biomedical research. The spectrum of available cell viability indicators assessing individual physiological, structural, or functional parameters is large and is continuously increasing with the availability and optimization of new or existing technologies. Depending on the number and diversity of employed fitness indicators, a cell viability assay can generate fitness phenotypes of varying complexity: when a single indicator is used, the information provided on the cellular condition is very limited, potentially resulting in poor dataset concordance, whereas when various indicators are employed, e.g., in a multiplexing approach, combining different methods in one experiment, cellular fitness is reflected more comprehensively, allowing for decreased interassay variability and increased reproducibility of experimental results. While cell-based viability screening is typically carried out using simple and single indicator-based approaches, a paradigm shift toward more advanced methods generating complex cell fitness phenotype readouts is currently taking over as indicated by an increasing availability of protocols describing multiparameter assaying techniques.

This book is intended to provide an overview and to discuss the strengths and pitfalls of commonly used cell fitness indicators. We aim to give an in-depth view of protocols that are used in the classical cell-based viability screening approach and to provide experimental methods for advanced cell viability assaying strategies, including evaluation of e.g. cellular transporter activity, intracellular calcium signaling, electrical network activity, synaptic vesicle recycling or ligand-gated ion channel function. In this volume, we cover biochemical, fluorescence and luminescence-based strategies as well as computational and label-free methodologies for assaying cellular viability by means of e.g. viscoelastic properties, impedance and multiphoton microscopy. The biological samples used in the described approaches cover a broad range of specimen including conventional culture models, stem and primary cells as well as parasites. These chapters address an interdisciplinary audience, including graduate students, postdoctoral fellows, and scientists in all areas of biomedical research. As the concept of this series is meant to shed light into the sometimes tiny "tips and tricks" that decide over the success or flaw of biological experiments, we hope that the chapters will provide useful hints to the community.

Erlangen, Germany

Daniel F. Gilbert
Oliver Friedrich

Contents

Contributors

SEPIDEH ABOLPOUR MOFRAD • *Institute of Medical Biotechnology, Friedrich-Alexander-Universität Erlangen-Nürnberg, Erlangen, Germany*
GAMZE ATES • *Department of In Vitro Toxicology and Dermato-Cosmetology, Faculty of Medicine and Pharmacy, Vrije Universiteit Brussel, Brussels, Belgium*
EMILIO BENFENATI • *Department of Environmental Health Sciences, Laboratory of Environmental Chemistry and Toxicology, IRCCS—Istituto di Ricerche Farmacologiche "Mario Negri", Milan, Italy*
ACHIM BREILING • *DKFZ ZMBH Alliance, Division of Epigenetics, German Cancer Research Center, Heidelberg, Germany*
JASMINE P. BROWN • *Integrated Systems Toxicology Division, NHEERL, US EPA, NC, USA*
JOSEPH BRUTON • *Karolinska Institutet, Stockholm, Sweden*
ANDREAS BUTTGEREIT • *Institute of Medical Biotechnology, Friedrich-Alexander-Universität Erlangen-Nürnberg, Erlangen, Germany*
LEO LI-YING CHAN • *Department of Technology R&D, Nexcelom Bioscience LLC, Lawrence, MA, USA*
DAISHI CHEN • *Translational Cell Biology and Neurooncology Laboratory of the Universitätsklinikum Erlangen (UKER), Friedrich-Alexander University of Erlangen–Nürnberg (FAU), and Department of Neurosurgery of the Universitätsklinikum Erlangen, Universitätsklinikum Erlangen (UKER), Friedrich-Alexander University of Erlangen – Nürnberg (FAU), Erlangen, Germany*
BASTIAN CHRIST • *Translational Center Würzburg "Regenerative Therapies for Oncology and Musculosceletal Diseases", Branch of Fraunhofer Institute for Interfacial Engineering and Biotechnology IGB, Würzburg, Germany*
ALEVTINA CUBUKOVA • *Translational Center Würzburg "Regenerative Therapies for Oncology and Musculosceletal Diseases", Branch of Fraunhofer Institute for Interfacial Engineering and Biotechnology IGB, Würzburg, Germany*
ITAEVIA M. CURRY-CHISOLM • *Integrated Systems Toxicology Division, NHEERL, US EPA, NC, USA*
HAIDER DAKHIL • *Institute of Fluid Mechanics, Friedrich-Alexander-Universität Erlangen-Nürnberg (FAU), Erlangen, Germany; Faculty of Engineering, University of Kufa, Najaf, Iraq*
SOFIA DEMBSKI • *Chair Tissue Engineering and Regenerative Medicine, University Hospital Würzburg, Würzburg, Germany; Fraunhofer Institute for Silicate Research ISC, Würzburg, Germany*
ALAN J. DICKSON • *Faculty of Life Sciences, The University of Manchester, Manchester, UK*
MARK ELVIN • *Faculty of Life Sciences, The University of Manchester, Manchester, UK*
HANNES ENGELHARDT • *Institute of Bioprocess Engineering, Friedrich-Alexander University Erlangen-Nürnberg, Erlangen, Germany*
ISOBELLE M. EVIE • *Faculty of Life Sciences, The University of Manchester, Manchester, UK*

ILKER Y. EYUPOGLU • *Translational Cell Biology and Neurooncology Laboratory of the Universitätsklinikum Erlangen (UKER), Friedrich-Alexander University of Erlangen–Nürnberg (FAU), and Department of Neurosurgery of the Universitätsklinikum Erlangen, Universitätsklinikum Erlangen (UKER), Friedrich-Alexander University of Erlangen – Nürnberg (FAU), Erlangen, Germany*

CHRISTINA FEY • *Chair Tissue Engineering and Regenerative Medicine, University Hospital Würzburg, Würzburg, Germany*

OLIVER FRIEDRICH • *Friedrich-Alexander University (FAU) Erlangen-Nürnberg, Institute of Medical Biotechnology, Erlangen, Germany*

MARTIN F. FROMM • *Department of Clinical Pharmacology and Clinical Toxicology, Institute of Experimental and Clinical Pharmacology and Toxicology, Friedrich-Alexander-Universität Erlangen-Nürnberg, Erlangen, Germany*

MARTIN C. GARNETT • *School of Pharmacy, University of Nottingham, Nottingham, UK*

DANIEL F. GILBERT • *Friedrich-Alexander University (FAU) Erlangen-Nürnberg, Institute of Medical Biotechnology, Erlangen, Germany*

ANNA M. GRABOWSKA • *Cancer Biology, Division of Cancer and Stem Cells, School of Medicine, Queen's Medical Centre, University of Nottingham, Nottingham, UK*

TEJA W. GROEMER • *Department of Psychiatry and Psychotherapy, Friedrich-Alexander University of Erlangen-Nuremberg, Erlangen, Germany*

STEWART I. HEAD • *School of Medical Sciences (SOMS), University of New South Wales (UNSW), Sydney, NSW, Australia*

HOLGER HÜBNER • *Institute of Bioprocess Engineering, Friedrich-Alexander University Erlangen-Nürnberg, Erlangen, Germany*

NOBORU INOUE • *National Research Center for Protozoan Diseases, Obihiro University of Agriculture and Veterinary Medicine, Hokkaido, Japan*

DELYAN P. IVANOV • *Cancer Biology, Division of Cancer and Stem Cells, School of Medicine, Queen's Medical Centre, University of Nottingham, Nottingham, UK*

BARBARA KAPPES • *Institute of Medical Biotechnology, University of Erlangen-Nürnberg, Erlangen, Germany*

SARAH L. KESSEL • *Department of Technology R&D, Nexcelom Bioscience LLC, Lawrence, MA, USA*

JÖRG KÖNIG • *Department of Clinical Pharmacology and Clinical Toxicology, Institute of Experimental and Clinical Pharmacology and Toxicology, Friedrich-Alexander-Universität Erlangen-Nürnberg, Erlangen, Germany*

ANNA KOSTAREVA • *ITMO University, Saint Petersburg, Russia*

KATHARINA KUENZEL • *Institute of Medical Biotechnology, Friedrich-Alexander-Universität Erlangen-Nürnberg, Erlangen, Germany*

MARIA LEIDENBERGER • *Institute of Medical Biotechnology, University of Erlangen-Nürnberg, Erlangen, Germany*

BRITTANY S. LYNCH • *Integrated Systems Toxicology Division, NHEERL, US EPA, NC, USA*

CHRISTIAN MAERCKER • *Esslingen University of Applied Sciences, Esslingen am Neckar, Germany; German Cancer Research Center (DKFZ), Genomics and Proteomics Core Facilities, Heidelberg, Germany*

SERENA MANGANELLI • *Department of Environmental Health Sciences, Laboratory of Environmental Chemistry and Toxicology, IRCCS—Istituto di Ricerche Farmacologiche "Mario Negri", Milan, Italy*

KELSEY J. MCCULLEY • *Department of Technology R&D, Nexcelom Bioscience LLC, Lawrence, MA, USA*
ANN-KATRIN MENZNER • *Department of Internal Medicine 5, University Medical Center Erlangen, Friedrich-Alexander-Universität Erlangen-Nürnberg, Erlangen, Germany*
MARCO METZGER • *Translational Center Würzburg "Regenerative Therapies for Oncology and Musculosceletal Diseases", Branch of Fraunhofer Institute for Interfacial Engineering and Biotechnology IGB, Würzburg, Germany; Chair Tissue Engineering and Regenerative Medicine, University Hospital Würzburg, Würzburg, Germany*
NTHATISI INNOCENTIA MOLEFE • *National Research Center for Protozoan Diseases, Obihiro University of Agriculture and Veterinary Medicine, Hokkaido, Japan*
SIMIN ÖZ • *German Cancer Research Center (DKFZ), Epigenomics and Cancer Risk Factors, Heidelberg, Germany*
KONSTANTIN PRÄBST • *Institute of Bioprocess Engineering, Friedrich-Alexander University Erlangen-Nürnberg, Erlangen, Germany*
STEFAN RINGGELER • *Institute of Bioprocess Engineering, Friedrich-Alexander University Erlangen-Nürnberg, Erlangen, Germany*
ROBIM M. RODRIGUES • *Department of In Vitro Toxicology and Dermato-Cosmetology, Faculty of Medicine and Pharmacy, Vrije Universiteit Brussel, Brussels, Belgium*
VERA ROGIERS • *Department of In Vitro Toxicology and Dermato-Cosmetology, Faculty of Medicine and Pharmacy, Vrije Universiteit Brussel, Brussels, Belgium*
HEIKO RUNZ • *Institute of Human Genetics, University of Heidelberg, Heidelberg, Germany; Molecular Medicine Partnership Unit (MMPU), University of Heidelberg/EMBL, Heidelberg, Germany; Department of Genetics and Pharmacogenomics, Merck Research Laboratories, Boston, MA, USA*
NICOLAI SAVASKAN • *Translational Cell Biology and Neurooncology Laboratory of the Universitätsklinikum Erlangen (UKER), Friedrich-Alexander University of Erlangen–Nürnberg (FAU), and Department of Neurosurgery of the Universitätsklinikum Erlangen, Universitätsklinikum Erlangen (UKER), Friedrich-Alexander University of Erlangen – Nürnberg (FAU), Erlangen, Germany; BiMECON Ent., www.savaskan.net, Berlin, Germany*
THOMAS SEJERSEN • *Karolinska Institutet, Stockholm, Sweden*
TIMOTHY J. SHAFER • *Integrated Systems Toxicology Division, NHEERL, US EPA, NC, USA*
NINA SIMON • *Institute of Medical Biotechnology, University of Erlangen-Nürnberg, Erlangen, Germany*
NATALIA SMOLINA • *Karolinska Institutet, Stockholm, Sweden; Federal Almazov North-West Medical Research Centre, Russia*
JENNA D. STRICKLAND • *Axion Biosystems, Atlanta, GA, USA; Department of Pharmacology and Toxicology, Michigan State University, MI, USA*
KEISUKE SUGANUMA • *National Research Center for Protozoan Diseases, Obihiro University of Agriculture and Veterinary Medicine, Hokkaido, Japan*
EMIR TAGHIKHANI • *Department of Clinical Pharmacology and Clinical Toxicology, Institute of Experimental and Clinical Pharmacology and Toxicology, Friedrich-Alexander-Universität Erlangen-Nürnberg, Erlangen, Germany*
AENNE S. THORMÄHLEN • *Institute of Human Genetics, University of Heidelberg, Heidelberg, Germany; Molecular Medicine Partnership Unit (MMPU), University of Heidelberg/EMBL, Heidelberg, Germany*

Tamara Vanhaecke • *Department of In Vitro Toxicology and Dermato-Cosmetology, Faculty of Medicine and Pharmacy, Vrije Universiteit Brussel, Brussels, Belgium*
Martin Vielreicher • *Friedrich-Alexander University (FAU) Erlangen-Nürnberg Institute of Medical Biotechnology, Erlangen, Germany*
Cornelia Voigtländer • *Institute of Medical Biotechnology, University of Erlangen-Nürnberg, Erlangen, Germany; Erlangen Graduate School of Advanced Optical Technologies (SAOT), Erlangen, Germany*
Heike Walles • *Translational Center Würzburg "Regenerative Therapies for Oncology and Musculosceletal Diseases", Branch of Fraunhofer Institute for Interfacial Engineering and Biotechnology IGB, Würzburg, Germany; Chair Tissue Engineering and Regenerative Medicine, University Hospital Würzburg, Würzburg, Germany*
Andreas Wierschem • *Institute of Fluid Mechanics, Friedrich-Alexander-Universität Erlangen-Nürnberg, Erlangen, Germany*
Jana K. Wrosch • *Department of Psychiatry and Psychotherapy, Friedrich-Alexander University of Erlangen-Nuremberg, Erlangen, Germany*

Chapter 1

Basic Colorimetric Proliferation Assays: MTT, WST, and Resazurin

Konstantin Präbst, Hannes Engelhardt, Stefan Ringgeler, and Holger Hübner

Abstract

This chapter describes selected assays for the evaluation of cellular viability and proliferation of cell cultures. The underlying principle of these assays is the measurement of a biochemical marker to evaluate the cell's metabolic activity. The formation of the omnipresent reducing agents NADH and NADPH is used as a marker for metabolic activity in the following assays. Using NADH and NADPH as electron sources, specific dyes are biochemically reduced which results in a color change that can be determined with basic photometrical methods. The assays selected for this chapter include MTT, WST, and resazurin. They are applicable for adherent or suspended cell lines, easy to perform, and comparably economical. Detailed protocols and notes for easier handling and avoiding pitfalls are enclosed to each assay.

Key words Viability assay, MTT, WST, Resazurin, Tetrazolium salts, Colorimetric proliferation assay, Metabolic assay

1 Introduction

The development of new drugs is closely related to the cultivation of cells. In high-throughput screening approaches large-molecule libraries, natural extracts, or isolates are investigated in cytotoxicity studies in matters of, for example, antitumoral activity. In order to identify effective substances, it is necessary to differentiate viable, dead, or impeded cells. There is a multitude of methods to determine cell number and viability, including ^{3}H-thymidine incorporation, cell counting with trypan blue, fluorometric DNA assays, or flow cytometry. Most of these methods entail some problems, like producing toxic or radioactive waste, or being time consuming, difficult, or expensive in performance. Therefore these methods are only of limited use for high-throughput screening approaches as well as for small pilot studies [1].

Cellular viability and metabolic activity can also be determined by measuring NADH and NADPH content, as these pyridine

Daniel F. Gilbert and Oliver Friedrich (eds.), *Cell Viability Assays: Methods and Protocols*, Methods in Molecular Biology, vol. 1601, DOI 10.1007/978-1-4939-6960-9_1, © Springer Science+Business Media LLC 2017

nucleotides are formed in the course of metabolic activity. Direct measurement of these reducing agents is possible, but absolute levels are not an optimal indicator for metabolic activity as their turnover rate is more important. The turnover rate can be evaluated by selective reduction of certain compounds, such as different tetrazolium salts (MTT, MTS, XTT, or WST) or resazurin as the enzymatic reduction of these compounds by dehydrogenases uses NADH/NADPH as co-substrate. The reduced form of these compounds results in a colored product which can be measured by basic spectroscopic methods. When cellular metabolic activity is maintained during cultivation, cell density can be set proportional to the concentration of the resulting colored product in a certain range [2]. Here, different assays have been developed with the aim of making them easy to handle and fast to perform. In this chapter, we are focusing on two tetrazolium salt assays forming (a) a water-insoluble formazan (MTT) and (b) a water-soluble formazan (WST) and (c) on the resazurin assay. Each of these assays shows different characteristics, each one with its advantages and disadvantages. Viability assays containing MTT form a solid crystalline product, whose crystal spikes eventually destroy the cell's integrity, which ultimately leads to cell death. As a result formazan formation is stopped and the endpoint of the reaction is used to evaluate cell culture viability. Obvious disadvantages are unavoidable cell death and the additional dissolving step necessary for measuring formazan absorbance. In WST-based assays a soluble formazan product is formed and therefore there is no need for an additional solvation step. However, formazan formation follows a reaction kinetic of the pseudo first order, whose reaction rate is used to evaluate metabolic activity. This makes constant reaction conditions crucial for these assays. Even small changes in incubation time, temperature, or pH value can largely influence measured values. Viability assays containing resazurin also initially show a pseudo first-order reaction kinetic but in these assays a fluorescent product is formed which greatly enhances sensitivity and range of measurement, especially for small cell concentrations. However, resazurin-based assays inherit more pitfalls beyond those of MTT or WST.

1.1 MTT Assay

Tetrazolium salt solutions are colorless or only weakly colored which change to a strong colored solution when forming the formazan product. Over the years different tetrazolium salts have been developed for various applications in histochemistry, cell biology, biochemistry, and biotechnology. Concerning cell culture applications the most important tetrazolium salts are MTT, XTT, MTS, and WST [2].

In cell culture, the first and most commonly used tetrazolium salt is MTT (3-(4,5-dimethylthiazol-2-yl)-2,5-diphenyltetrazolium bromide) that was introduced by Mosmann to measure proliferation and cytotoxicity in high-throughput screening approaches in 96-well plates [3]. Due to its lipophilic side groups and positive net charge MTT is able to pass the cell membrane and is reduced in viable cells by mitochondrial or cell plasma enzymes like oxidoreductases, dehydrogenases, oxidases, and peroxidases using NADH, NADPH, succinate, or pyruvate as electron donor. This results in a conversion of MTT to the water-insoluble formazan (*see* Fig. 1) [2].

Besides enzymatic reactions there are different nonenzymatic reactions with reducing molecules like ascorbic acid, glutathione, or coenzyme A that are able to interact with MTT forming the formazan product and produce a higher absorbance accordingly [4]. The formation of needlelike formazan crystals destroys the cell's integrity and thus leads to cell death. The metabolism breaks down and so the reaction of MTT to formazan is interrupted very quickly. Due to the cell death-associated reaction stop this kind of assay is called an endpoint determination. Because the crystals are formed intracellularly, MTT-based assay protocols usually include a cell lysis step and a formazan-dissolving step before a spectroscopic measurement can be performed. In spite of its advantages of being rapid and simple, the formation of an insoluble product and the necessity to dissolve it exclude this assay for any real-time assays [2]. That is why constitutive work based on the studies of Mosmann proposed some modifications that improve the performance and sensitivity of this assay, but the problem of dissolving solid formazan crystals still exists [5–8].

1.2 WST-8 Assay

To overcome this time-consuming post-reaction processing some tetrazolium derivatives that produce water-soluble products have been developed, such as MTS (3-(4,5-dimethylthiazol-2-yl)-5-(3-carboxymethoxyphenyl)-2-(4-sulfophenyl)-2H-tetrazolium) [9, 10], XTT (2,3-bis-(2-methoxy-4-nitro-5-sulfophenyl)-2H-tetrazolium-5-carboxanilide) [11, 12], or WST (2-(2-methoxy-4-nitrophenyl)-3-(4-nitrophenyl)-5-(2,4-disulfophenyl)-2H-tetrazolium) [13, 14].

This solubility is generally achieved by introducing negative-charged sulfone groups to the phenyl rings in order to compensate the positive charge of the tetrazolium ring. These derivatives have

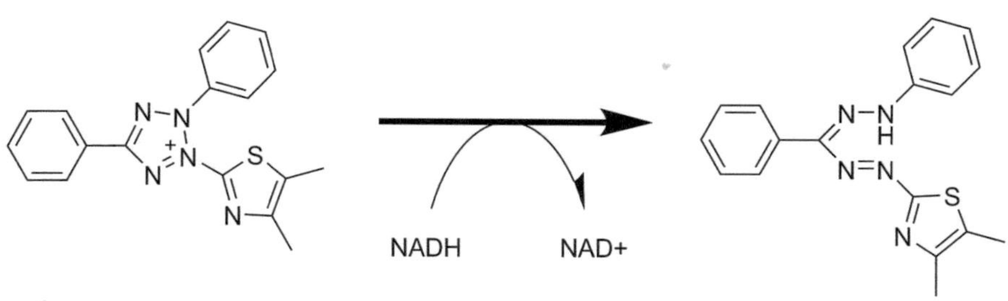

Fig. 1 Enzymatic reduction of MTT to formazan. Formazan forms solid crystals that pierce the cell's membrane after a certain growth and lead to cell death, disrupting further formation of formazan

a neutral or negative net charge which hinders their passage through cell membranes. The reduction of WST is mainly performed extracellularly and the electron transfer necessary for reduction of the tetrazolium needs to be transduced by intermediate electron acceptors like 5-methyl-phenazinium methyl sulfate (PMS) or phenazine ethyl sulfate (PES). These electron carriers facilitate the transmembrane electron transfer to link intracellular metabolism and extracellular reduction of the tetrazolium [10].

WST-8 as a second-generation tetrazolium salt was first synthesized by Tominaga in 1999 [15]. The dye carries a negative net charge and is therefore largely cell impermeable. WST-8 as a viability indicator also requires the use of an intermediate electron acceptor for its extracellular reduction, for example mPMS (*see* Fig. 2).

The amount of reduced WST-tetrazolium can be quantified with an absorption measurement at 450 nm in the culture medium. This allows to perform real-time assays [2]. The dye reduction is proportional to the number of viable cells. This is a good approximation for cells in the exponential growth phase. But this can become problematic when nutrients are depleted or substances that affect the metabolic activity are tested; therefore optimal culture conditions are required and a thorough calibration has to be performed with the desired cell lines and culture approach to evaluate linear range and cell concentration/formazan absorbance relation [2, 16].

1.3 Resazurin Reduction Assay

Resazurin, discovered by Weselsky [17], is an indicator of cellular metabolic ability that has been used since the late 1920s to estimate bacterial infestation of milk [18]. Since then, this redox dye is used as an indicator of active metabolism in cell cultures in

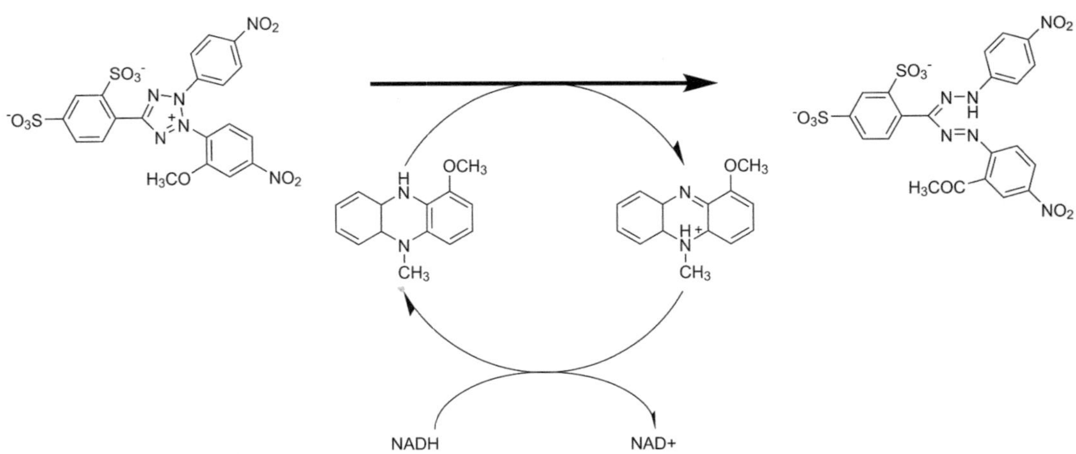

Fig. 2 Reduction of WST-8 to formazan by NADH via the electron mediator mPMS. Reaction takes place extracellularly, while mPMS mediates the electron transfer across the cell's membrane from NADH to WST

various applications. These include cell viability [19, 20], culture proliferation, or cytotoxicity studies [21] and to a certain extent also high-throughput screenings [22, 23]. The resazurin assay is based on the intracellular reduction of resazurin to resorufin by viable, metabolically active cells [20]. Various mechanisms for resazurin reduction by viable cells are described that use NADH and NADPH as electron source. These include reduction by mitochondrial [24] or microsomal enzymes [25], by enzymes in the respiratory chain [26], or by electron transfer agents, preferably *N*-methylphenazinium methosulfate (PMS) [27]. Direct reduction of resazurin with NADH was not observed [27].

Resazurin can be dissolved in physiological buffers, which allows direct use in cell cultures. The resazurin solution is a deep blue-colored solution which shows little to no intrinsic fluorescence. When resazurin diffuses through cell membranes it is metabolically reduced by viable cells to the fluorescent, pink-colored product, resorufin, which is also permeable [4, 20, 22, 28, 29]. The formation of this water-soluble, fluorescent product is the major advantage compared to the tetrazolium salt-based assays. When excited at a wavelength of 579 nm, resorufin emits a fluorescent signal at 584 nm. Resazurin and resorufin also show different spectral properties; the absorbance maximum of resazurin lies at 605 nm and that of resorufin at 573 nm. But only resorufin can be determined fluorimetrically, in opposition to resazurin.

Other advantages of the resazurin assay are comparably low costs and the possibility to multiplex it with other assays, for example with a caspase assay for the determination of apoptosis in cell cultures [30]. Resazurin assays are reported to be more sensitive and reliable than other assays using tetrazolium dyes but there are several factors that have to be considered before using a resazurin assay. The resorufin increase curve has only a limited linear range that is highly dependent on the temperature, pH, and initial resazurin concentration. These parameters have to be kept constant especially during incubation and measurement to avoid creating artifacts. The temperature naturally has an effect on the reaction rate. Furthermore, the equilibrium of the resazurin-resorufin reaction shifts towards resazurin with decreasing pH values. Moreover, the reduction of resazurin to resorufin is not the final step of the reaction in some cases. Resorufin can be further reduced actively to dihydroresorufin by some cells (*see* Fig. 3) [31]. This compound does not show any fluorescence and is highly toxic to cells. Dihydroresorufin can spontaneously be reverted back to resorufin but the reaction rate of this reverse reaction is much slower.

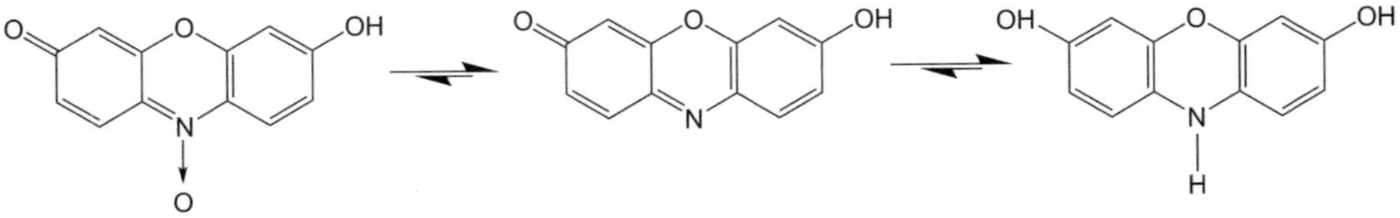

Fig. 3 Reduction of resazurin to resorufin and further to dihydroresorufin by NADH. First reverse reaction back to resazurin is favored by low pH values. Further reduction to dihydroresorufin can be performed by some cell lines, resulting in a cytotoxic colorless molecule

2 Materials

2.1 Calibration Protocol

1. Sodium chloride solution, 0.9% (w/w): Dissolve 9 g of sodium chloride (NaCl) in 1000 ml of deionized water. Afterwards, this solution can be sterilized in an autoclave for 15 min at 121 °C for long term storage.
2. Trypan blue stock solution: Dissolve 4 g of trypan blue in 1000 ml of 0.9% NaCl solution. Filter with 0.2 μm pore size to remove undissolved trypan blue crystals. Aliquots can be stored frozen at −20 °C.
3. Phosphate-buffered saline solution (PBS): Dissolve the following salts in 1000 ml of deionized water: 8 g NaCl, 0.2 g potassium chloride (KCl), 1.44 g disodium phosphate ($Na_2HPO_4*2H_2O$), 0.2 g monopotassium phosphate (KH_2PO_4). Adjust pH to 7.4 using sodium hydroxide (NaOH) or hydrogen chloride (HCl).
4. Accutase: Accutase solution can be purchased as a ready-to-use sterile filtered solution and should be stored at −20 °C. For frequent use aliquots can be stored at 4 °C.
5. Hemocytometer: For determination of cell density a counting chamber is required. The following protocols refer to the Neubauer or Neubauer improved format.

2.2 MTT Assay

MTT can be purchased either as a ready-to-use kit or as a pure tetrazolium salt (i.e., thiazolyl blue tetrazolium bromide). The salt can be dissolved and stored in aliquots. Both MTT stock solution and MTT solution kit should be stored light protected at −20 °C. Avoid refreezing of thawed aliquots to prevent accumulation of formazan by unspecific conversion of MTT [1].

1. MTT-Medium Mastermix solution: Dissolve 0.5 g MTT in 100 ml 0.9% NaCl solution, which results in a final concentration of 5 mg/ml (assay concentration: 1 mg/ml). Filtrate the solution using a filter with a pore size of 0.2 μm in order to sterilize the MTT solution and to remove all solid particles like unspecifically formed formazan crystals. Make a 20% (v/v) MTT-Medium Mastermix solution for the desired amount of wells to be measured (e.g., 20 μl of MTT solution and 80 μl of fresh medium per well in a 96-well plate).

2. Igepal solution: Mix 400 µl of Igepal (Nonidet P40) with 100 ml of deionized water.
3. Dimethyl sulfoxide (DMSO): A purity of 99.5% is sufficient.

2.3 WST-8 Assay

The WST-8 or Cell Counting Kit-8 (CCK-8) is a one-bottle solution and should be stored at −20 °C. For frequent use aliquots can be stored light protected at 4 °C, although quick usage is recommended. Repeated thawing and freezing may cause an increase in unspecific formazan reduction.

1. WST-8 Medium Mastermix solution: Aliquot sterile fresh culture medium and preincubate, e.g., 37 °C, 5% CO_2 (*see* **Notes 1** and **2**: if using a CO_2 atmosphere slightly loosen the screw cap of the medium tube to allow gas exchange for pH adjustment). Prepare a 10% (v/v) WST-8 Medium Mastermix solution for the desired amount of wells to be measured with the incubated culture medium (e.g., 10 µl of WST-8 solution and 90 µl of fresh incubated medium per well in a 96-well plate) and keep at incubated conditions.

2.4 Resazurin Reduction Assay

Resazurin can be purchased in a ready-to-use form but resazurin content and purity can differ depending on supplier and storage time. Therefore it is recommended to use high-purity resazurin salts (i.e., resazurin sodium salt). Long-term storage of resazurin in aqueous solutions should be avoided as well as repeated freezing/thawing cycles (*see* **Note 3**).

1. Medium/Resazurin Mastermix solution: Aliquot sterile fresh culture medium and preincubate at desired culture or measurement conditions (*see* **Note 4**). Prepare a Medium/Resazurin Mastermix solution with a predefined total volume depending on the number of measurements (100 µl per well in a 96-well plate) and a resazurin concentration of 4 mg/ml and keep Mastermix solution at desired conditions (*see* **Notes 5** and **6**). Filter-sterilize the Mastermix solution, if necessary, through a 0.2 µm pore filter into a sterile, light-protected container.

3 Methods

3.1 Calibration Assay

A calibration for each cell line and different culture conditions is crucial for the following viability assays. The conversion of indicators such as MTT, WST, and resazurin is highly dependent on cellular metabolic activity. As a general rule of thumb, cells should show a doubling time smaller than 36 h. Determining the viability of slower growing cell cultures with these methods is limited. The following protocol refers to adherent cells cultivated in cell culture flasks with a growth area of 75 cm^2 and a medium volume of 22.5 ml. In case of suspension cells start with **step 6**. In any case the preculture should be in the exponential growth phase (*see* **Note 7**).

1. Transfer supernatant medium into a sterile 50 ml conical centrifuge tube (Falcon tube) and save it for later use.
2. After washing the cell layer with 10 ml of PBS, remove and discard the PBS.
3. Add 3 ml of Accutase and wait until cells are detached.
4. Resuspend the cells with the medium of **step 1** and transfer the cell suspension to a sterile 50 ml Falcon tube.
5. Pellet cells by centrifugation with 180 × *g* for 8 min, discard the supernatant, and add 10 ml of fresh medium to remove the old culture medium and Accutase.
6. Resuspend cells and take a sample of 200 μl of well mixed cell suspension.
7. Mix 100 μl of the cell culture sample with 100 μl of 0.4% trypan blue solution.
8. After resuspension fill both chambers of a hemocytometer with 10 μl each (*see* **Note 8**).
9. Count the total number of cells (both stained and not stained by trypan blue) in each of the eight corner squares of the hemocytometer (*see* **Note 9**). Calculate the cell density using the following formula:

$$\frac{Cells}{ml} = \frac{Total\ number\ of\ cells\ in\ 8\ squares}{8} \bullet Dilution\ factor \bullet 10^{-4.}$$

When using adherent cells it is suitable to calculate a cell density per cm^2 by using the latter formula:

$$\frac{Cells}{cm^2} = \frac{Cells}{ml} \bullet \frac{Suspension\ volume\ in\ ml}{Growth\ area\ in\ cm^2}$$

10. Calculate viability of your cell culture by counting stained cells exclusively and use this value in the following formula (*see* **Note 10**):

$$Viability\,(\%) = \frac{Total\ number\ of\ cells - number\ of\ stained\ cells}{Total\ number\ of\ cells} \bullet 100\%$$

11. Make an equidistant serial dilution of the cell suspension (e.g., 100, 80, 60, 40, 20, and 0% of original cell density) with culture medium.
12. Pipette cells in 96-well plates and, for adherent cells, allow them to adhere for about 4 h at constant conditions (*see* **Notes 11** and **12**).

13. Proceed with desired viability assay protocol (MTT, WST, or resazurin).
14. Plot absorbance/fluorescence signal over a course of incubation time for each dilution step to determine linear range and possible absorbance maximum of the assay for each specific cell line or different conditions.

3.2 MTT Assay

1. Remove the cell culture medium from the wells that need to be measured and replace with Mastermix solution (100 µl per well) (*see* **Notes 13** and **14**). Always carry a blank control without cells to assess unspecific formazan conversion.
2. Incubate for a period of 2–4 h (*see* **Note 15**) under cell type-specific culture conditions.
3. After incubation centrifuge the well plate for 10 min at 3220 × *g* to concentrate formazan crystals and discard the supernatant medium.
4. For cell lysis add 30 µl of Igepal and incubate the assay for 10 min on a well-plate shaker till crystals are detached from the solid surface of the well (*see* **Note 16**).
5. Add 170 µl of DMSO and repeat the incubation using a well-plate shaker until the formazan crystals are completely dissolved. If necessary use a pipette for complete dissolving of the crystals (*see* **Note 17**).
6. Measure the absorption using a plate reader at 570 nm. Use a wavelength of 650 nm as reference to determine the background noise caused by undissolved particles and cell debris.
7. Plot absorbance signal at 570 nm versus cell number for cell concentration calibration. Calculate the cell density with the absorbance signal from the previously done calibration for creating the growth curve. Calculate the ratio of signal intensity of the sample and the control culture in % to determine cytotoxicity (*see* **Note 18**).

3.3 WST-8 Assay

1. Remove the culture medium from the cells and replace it with WST-Medium Mastermix (*see* **Note 13**). Always carry a blank control without cells to determine unspecific formazan conversion. Avoid bubble formation since it will highly interfere with the absorption measurement.
2. Incubate cells for 1–4 h (*see* **Notes 19** and **20**).
3. Measure absorbance at 450 nm for WST signal. A second measurement at 650 nm is recommended to assess influencing factors like bubbles, light scattering of cells, or condensing water on the lid. Prior to the measurement shake the plate for 10 s to evenly distribute formed formazan throughout the well.
4. Plot absorbance signal at 450 nm versus cell number for cell concentration calibration. Calculate the cell density with the

absorbance signal from previously done calibration for creating the growth curve. Calculate the ratio of signal intensity of sample and control culture in % to determine cytotoxicity (*see* **Notes 18** and **21**).

5. Remove the WST-containing solution and add 100 μl of fresh, culture condition-incubated medium if cells are needed for further experiments (*see* **Notes 13** and **22**).

3.4 Resazurin Assay

Some cells are able to reduce resorufin further to dihydroresorufin (*see* **Note 23**). This has to be ruled out before the resazurin assay can be used for a specific cell line.

1. Remove the cell culture medium from the well and add 100 μl of Mastermix solution to each well. Avoid bubble formation. An optional set of wells can be prepared with medium-only and medium plus Mastermix solution for background subtraction and instrument gain adjustment (*see* **Notes 13** and **24**).
2. Incubate for a desired amount of time at constant conditions, depending on cell line and linear range (*see* **Notes 4** and **25**). Incubation time and cell concentration range have to be determined prior with a calibration for each specific cell line and environmental parameters (*see* **Note 26**).
3. Record fluorescence using a 560 nm excitation wavelength and a 590 nm emission wavelength. Prior to the measurement shake the plate for 10 s to evenly distribute formed resorufin throughout the well.
4. Plot fluorescence signal at 590 nm versus cell number for cell concentration calibration. Calculate the cell density with the fluorescence signal from previously done calibration for creating the growth curve. Calculate the ratio of signal intensity of the samples and control culture in % to determine cytotoxicity (*see* **Note 27**).
5. Remove the resazurin-containing solution and add 100 μl of fresh, culture condition-incubated medium if cells are needed for further experiments (*see* **Notes 13**, **28**, and **29**).

4 Notes

1. Wrong-tempered culture medium affects the absorbance signal. Since all enzymatic reactions in the cell are highly temperature dependent, cold medium results in decreased signal intensity. Also keep temperature fluctuations of your incubator in mind for error analysis. Frequent opening of the incubator door may result in an overall lower mean temperature. Also temperature regulation and distribution inside the incubator typically fluctuate. According to Arrhenius' law a temperature

change of 2 °C leads to a 14% difference in reaction rates and should be considered when calculating cell numbers.

2. When quantifying cellular proliferation in growth curves or toxicity assays it is important to use fresh culture medium in order to guarantee good nutrient supply for the cells. Poor nutrient supply (e.g. glucose, glutamine, or oxygen) may lead to lower signal intensity.
3. The reaction of resazurin and resorufin always tends to reach a state of equilibrium. Therefore resazurin solutions stored for a longer period of time always contain unspecifically formed resorufin that can affect the outcome.
4. A constant temperature is of high importance when using resazurin as viability indicator. Small changes in temperature affect the reaction rate and can generate different signals when measuring after a constant incubation time. So it is necessary to keep the temperature constant even during measurement periods. Furthermore, pH is also important to maintain. Resorufin can react back to resazurin. This reaction is favored at lower pH values. This is also the reason why CO_2-buffered media are not optimal for this viability method, as a defined CO_2 environment mostly cannot be maintained during measurement periods.
5. The reduction of resazurin does not require an intermediate electron acceptor such as PMS, but it can enhance signal generation [4].
6. Increased resazurin concentrations do not change resazurin turnover, but may change the endpoint [24].
7. As the viability assays with MTT, WST-8 and resazurin are highly dependent on the cell metabolism rate, the cell culture should be in the exponential growth phase. If it is desired to measure high cell densities in the assay, the culture should be in the late exponential phase to harvest a sufficient amount of cells.
8. Avoid longer contact times of trypan blue as it has cytotoxic effects, leading to stained cells that were viable before exposure to trypan blue. When determining viability, exposure time to trypan blue should not exceed 30 min.
9. Cell numbers per corner square should be between 60 and 100 cells. Dilute the original sample if necessary with 0.9% sodium chloride solution and return to **step 7** of the protocol. A volume of 200 μl of the sample should provide enough material for another test if necessary. If the sample has to be diluted use the appropriate dilution factor in the formula (e.g., diluting 100 μl of sample with 100 μl of 0.9% NaCl results in a dilution factor of 2; the further mixture of 100 μl diluted sample with 100 μl 0.4% trypan blue solution results also in a dilution factor of 2, which gives an overall dilution factor of 4).

10. For a representative calibration the viability should be near to 100%, in any case over 95%.
11. When adapting the assay to other well dimensions keep height of medium constant to allow good oxygen supply. A medium height of 3 mm is recommended resulting in a 100 μl volume for 96-well plates (growth area 0.3 cm^2) or a 600 μl volume for 24-well plates (growth area 2 cm^2).
12. Some cell lines with poor adhesion may need longer to attach. However, longer adherence times may distort the result since cell growth may take place in the meantime. As a rule of thumb do not exceed 20% of the doubling time for adherence (e.g., 4 h of adherence for cells with a doubling time of 20 h).
13. For suspension cells a centrifugation step (e.g., 180 × *g* for 8 min) is sufficient to separate cells from supernatant medium.
14. Replacing the old medium with fresh medium prevents a lack of nutrients that would affect the metabolism and therefore would have an impact on the performance of the MTT assay. Especially during cultivation conditions like the availability of glucose [1] or a change in pH [32] influence the reliability of the MTT assay.
15. The conversion rate of MTT is closely connected to the cell type used. Depending on the conversion rate the necessary exposure time of the cells to MTT may vary to reach an end-point (*see* Fig. 4). That is why a close look on the reaction kinetics is needed for every single cell type [32], which should be performed during calibration.
16. Igepal is recommended for cell lysis. In other protocols SDS is used as detergent [6]. But we have observed better cell lysis when using Igepal and thus a decreasing background noise in comparison to SDS.
17. If solvents other than DMSO should be used it has to be considered that depending on the type of solvent a shift in the absorbance spectrum and sensitivity can be observed [7]. Thus the wavelength to apply may change. Furthermore, pure organic solvents may precipitate and serum proteins which disturb the spectroscopic measurement of formazan [5]. Under the microscope it can be observed that precipitated proteins on the crystals' surfaces hinder their dissolving and therefore elongate the necessary time for this step.
18. Check linear measuring range of photometer or plate reader. High cell densities can lead to absorption signals >3. This may require diluting the sample with DMSO to ensure a reliable absorption measurement (*see* Fig. 4).
19. Extending the incubation time increases the signal intensity. This may be necessary for small cell densities (*see* Fig. 5). However for high cell densities or fast-proliferating cells this

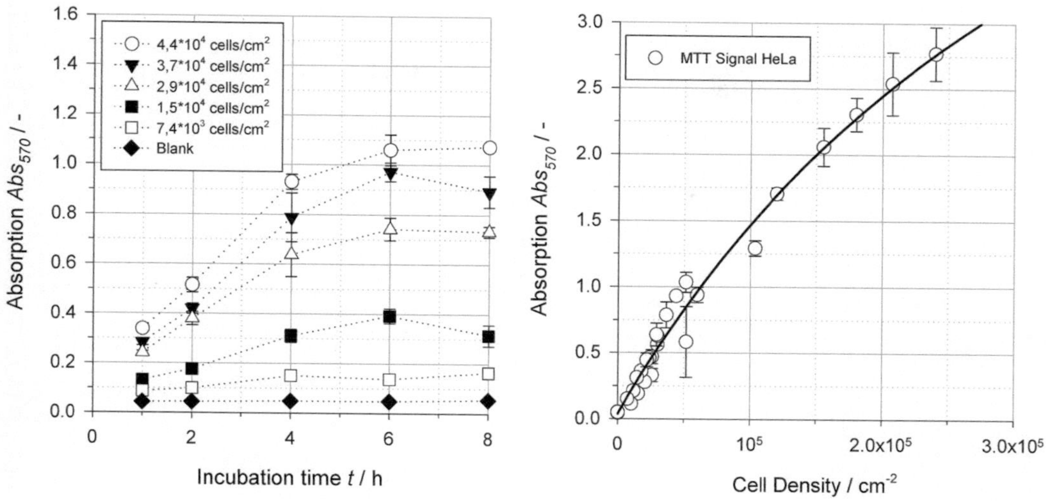

Fig. 4 Reduction of MTT to formazan by HeLa cells in RPMI 1640 supplemented with 10% (v/v) FCS and 4 mM glutamine. *Left*: Kinetic conversion of MTT to formazan. *Right*: Calibration with fit for HeLa after 4 h of incubation. Symbols represent mean of $n = 6$ single measurements with respective standard deviation. Line represents mathematical fit of the form $Abs_{450} = Abs_0 + Abs_{\max} \bullet \frac{Cell\ density}{a + Cell\ density}$

may lead to a loss of signal, resulting in decreased accuracy. To face the latter problem incubation times may be reduced, but should not be below 1 h. With incubation times shorter than 1 h, pipetting and preparation times have a larger influence on the measurement resulting in a poor reproducibility.

20. For best comparison measurements should be performed with the same incubation time. Deviations in time can cause inaccuracies especially for higher cell densities, since the overall conversion rate is much higher.

21. The mathematical fit for the cell calibration is nonlinear and is of the form $Abs_{450} = Abs_0 + Abs_{\max} \bullet \frac{Cell\ density}{a + Cell\ density}$. For slow-proliferating cells or small cell densities a linear fit of the form $Abs_{450} = Abs_0 + a \bullet Cell\ density$ can be a good approximation as a pseudo first-order reaction can be assumed. Keep in mind that due to the nonlinear function, the discrepancy between calculated and real cell densities is increasing for higher absorption signals.

22. Although it is stated that the WST-8 does not show any cytotoxicity on most cell lines, cellular metabolism can be inhibited. The reduction of WST-8 consumes reducing agents such as NADH and NADPH, which are then no longer available for the cell's metabolism. For that reason, it is recommended to remove residuals after the measurements for further cell usage.

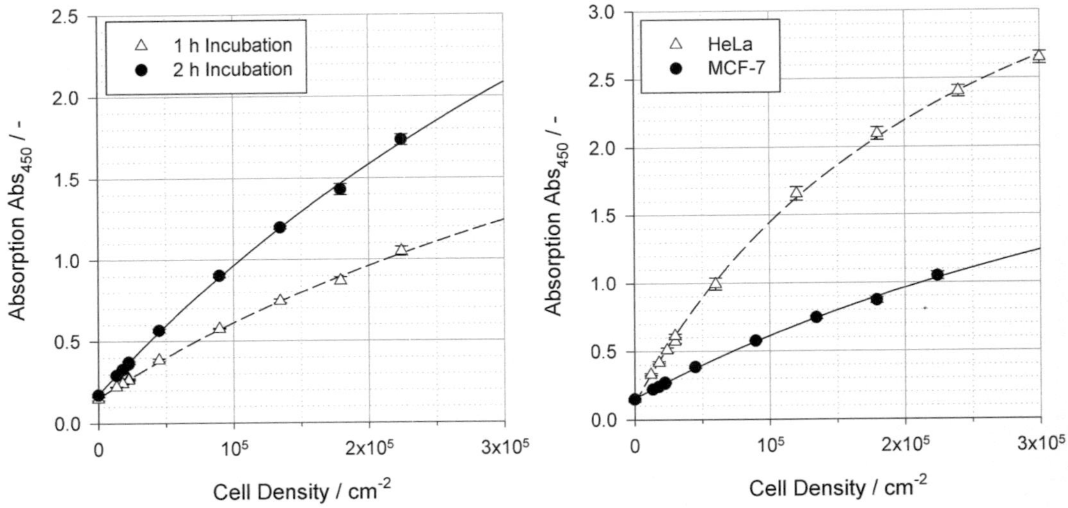

Fig. 5 Calibration curve for WST-8 assay: *Left*: Calibration with fit for MCF-7 cells in RPMI 1640 supplemented with 10% (v/v) FCS and 4 mM glutamine for 1 h (*open triangle*) and 2 h (*filled circle*) Incubation time with WST Mastermix. *Right*: Calibration for HeLa (*open triangle*) and MCF-7 (*filled circle*) cells. Symbols represent mean of $n = 6$ single measurements with respective standard deviation lines. Mathematical fit of the form $Abs_{450} = Abs_0 + Abs_{max} \bullet \frac{Cell\ density}{a + Cell\ density}$

23. Some cells show the ability to reduce resorufin further to the colorless dihydroresorufin (*see* Fig. 6). This compound is highly toxic to cells and drastically affects cell viability. Exposing cells to resazurin for long periods or elevated concentrations may result in cytotoxicity that can mask or interfere with the experimental outcome. Therefore concentration and incubation time must be optimized beforehand. The cytotoxicity of the resazurin assay can be determined by comparing this method with a different method, for example an ATP assay [22].
24. Cells have to be incubated with an adequate amount of substrate for a sufficient amount of time to generate a detectable signal as metabolic activity has to be maintained during resazurin reduction [22].
25. The reaction can be stopped using the addition of 3% SDS and the signal can be measured in between 24 h [29].
26. The number of cells per well and the length of incubation must be determined empirically beforehand. Typical incubation times usually lie between 1 and 4 h and minimal cell numbers can be as low as 40 cells [29], 80 cells [20], or between 200 and 50,000 cells/well in a 96-well plate [28]. The linear range has to be determined during calibration. This is highly dependent on cell concentrations, especially during the late exponential phase and stationary phase of batch cultures as well as on the resazurin concentration.

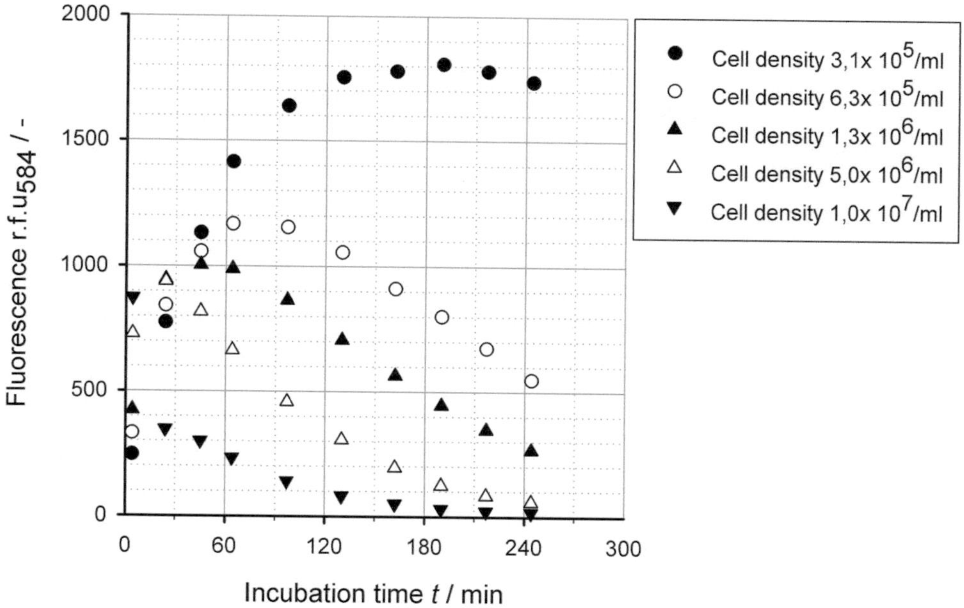

Fig. 6 Relative fluorescent units (r.f.u.) for different cell concentration of Sf21 insect cells at different times of incubation. While at lower cell lines concentrations an increase of r.f.u. can be observed, higher cell concentrations facilitate further reduction of resorufin to dihydroresorufin which leads to a decreasing signal. Sf 21 cells were incubated at 27 °C and a pH of 6.4. This can also lead to decreasing r.f.u. due to the shift of the resazurin/resorufin equilibrium to resazurin

27. Another way to gain even more accurate results is to record resazurin formation over a sufficient period of time and calculate initial reaction rate for $t = 0$ from the slope of the curve. Plot this reaction rate versus cell concentration.

28. Although it is stated that resazurin does not show any cytotoxicity on most cell lines, cellular metabolism can be inhibited by resazurin. The reduction of resazurin consumes reducing equivalents such as NADH and NADPH, which are then no longer available for the cell's metabolism. Because resorufin can react back to resazurin, this constant dissipation of reducing agents can have an impact on the cell's viability. For that reason, it is recommended to remove residual resazurin/resorufin after the measurements.

29. The resazurin assay is one of the few assays that allows to multiplex assays, for example a resazurin with a combined caspase assay. This may, however, require a sequential protocol to avoid color quenching by resazurin [30]. As with all methods that use fluorescence, interference and color quenching from other assays have to be considered [22].

References

1. Sylvester PW (2011) Optimization of the tetrazolium dye (MTT) colorimetric assay for cellular growth and viability. In: Drug design and discovery. Methods and protocols, p 157–168
2. Berridge MV, Herst PM, Tan AS (2005) Tetrazolium dyes as tools in cell biology: New insights into their cellular reduction. In: El-Gewely MR (ed) Biotechnology annual review, vol 11. Elsevier, p 127–152
3. Mosmann T (1983) Rapid colorimetric assay for cellular growth and survival: application to proliferation and cytotoxicity assays. J Immunol Methods 65(1–2):55–63. doi:10.1016/0022-1759(83)90303-4
4. Riss TL, Moravec RA, Niles AL et al (2013) Cell viability assays. In: Sittampalam GS, Coussens NP, Nelson H et al (eds) Assay guidance manual. Eli Lilly & Company and the National Center for Advancing Translational Sciences, Bethesda, MD
5. Denizot F, Lang R (1986) Rapid colorimetric assay for cell growth and survival: modifications to the tetrazolium dye procedure giving improved sensitivity and reliability. J Immunol Methods 89(2):271–277
6. Tada H, Shiho O, Kuroshima K et al (1986) An improved colorimetric assay for interleukin 2. J Immunol Methods 93(2):157–165
7. Carmichael J, DeGraff WG, Gazdar AF et al (1987) Evaluation of a tetrazolium-based semiautomated colorimetric assay: assessment of chemosensitivity testing. Cancer Res 47(4):936–942
8. Hansen MB, Nielsen SE, Berg K (1989) Re-examination and further development of a precise and rapid dye method for measuring cell growth/cell kill. J Immunol Methods 119(2):203–210
9. Barltrop JA, Owen TC, Cory AH et al (1991) 5-(3-carboxymethoxyphenyl)-2-(4,5-dimethyl thiazolyl)-3-(4-sulfophenyl)tetrazolium, inner salt (MTS) and related analogs of 3-(4,5-dimethyl thiazolyl)-2,5-diphenyltetrazolium bromide (MTT) reducing to purple water-soluble formazans As cell-viability indicators. Bioorg Med Chem Lett 1(11):611–614. doi:10.1016/S0960-894X(01)81162-8
10. Cory AH, Owen TC, Barltrop JA et al (1991) Use of an aqueous soluble tetrazolium/formazan assay for cell growth assays in culture. Cancer Commun 3(7):207–212
11. Paull KD, Shoemaker RH, Boyd MR et al (1988) The synthesis of XTT: a new tetrazolium reagent that is bioreducible to a water-soluble formazan. J Heterocycl Chem 25(3):911–914. doi:10.1002/jhet.5570250340
12. Scudiero DA, Shoemaker RH, Paull KD et al (1988) Evaluation of a soluble tetrazolium/formazan assay for cell growth and drug sensitivity in culture using human and other tumor cell lines. Cancer Res 48(17):4827–4833
13. Ishiyama M, Tominaga H, Shiga M et al (1996) A combined assay of cell vability and in vitro cytotoxicity with a highly water-soluble tetrazolium salt, neutral red and crystal violet. Biol Pharm Bull 19(11):1518–1520. doi:10.1248/bpb.19.1518
14. Ishiyama M, Shiga M, Sasamoto K et al (1993) A new sulfonated tetrazolium salt that produces a highly water-soluble formazan dye. Chem Pharm Bull 41(6):1118–1122. doi:10.1248/cpb.41.1118
15. Tominaga H, Ishiyama M, Ohseto F et al (1999) A water-soluble tetrazolium salt useful for colorimetric cell viability assay. Anal Commun 36(2):47–50. doi:10.1039/A809656B
16. Weyermann J, Lochmann D, Zimmer A (2005) A practical note on the use of cytotoxicity assays. Int J Pharm 288(2):369–376. doi:10.1016/j.ijpharm.2004.09.018
17. Weselsky P (1871) Ueber die Azoverbindungen des Resorcins. Ber Dtsch Chem Ges 4(2):613–619. doi:10.1002/cber.18710040230
18. Palmer LS, Weaver M et al (1930) Milchuntersuchung. Zeitschrift für analytische Chemie 82(6):268–271. doi:10.1007/BF01362069
19. Rampersad SN (2012) Multiple applications of Alamar Blue as an indicator of metabolic function and cellular health in cell viability bioassays. Sensors 12(9):12347–12360
20. O'Brien J, Wilson I, Orton T et al (2000) Investigation of the Alamar Blue (resazurin) fluorescent dye for the assessment of mammalian cell cytotoxicity. Eur J Biochem 267(17):5421–5426. doi:10.1046/j.1432-1327.2000.01606.x
21. Riss TL, Moravec RA (2004) Use of multiple assay endpoints to investigate the effects of incubation time, dose of toxin, and plating density in cell-based cytotoxicity assays. Assay and Drug Dev Technol 2(1):51–62
22. Chen T (2009) A practical guide to assay development and high-throughput screening in drug discovery. In: Critical reviews in combinatorial chemistry. CRC Press, Boca Raton, FL
23. Hamid R, Rotshteyn Y, Rabadi L et al (2004) Comparison of Alamar Blue and MTT assays for high through-put screening. Toxicol In Vitro 18(5):703–710
24. Fries R d, Mitsuhashi M (1995) Quantification of mitogen induced human lymphocyte proliferation:

Comparison of alamarBlue™ assay to 3H-thymidine incorporation assay. J Clin Lab Anal 9(2):89–95. doi:10.1002/jcla.1860090203

25. Gonzalez RJ, Tarloff JB (2001) Evaluation of hepatic subcellular fractions for Alamar blue and MTT reductase activity. Toxicol In Vitro 15(3):257–259. doi:10.1016/S0887-2333(01)00014-5
26. McMillian MK, Li L, Parker JB et al (2002) An improved resazurin-based cytotoxicity assay for hepatic cells. Cell Biol Toxicol 18(3):157–173. doi:10.1023/A:1015559603643
27. Candeias L, MacFarlane DS, McWhinnie SW et al (1998) The catalysed NADH reduction of resazurin to resorufin. J Chem Soc Perkin Trans 2(11):2333–2334
28. Celis JE (2006) Cell biology: a laboratory handbook. Elsevier, Ansterdam
29. Wu G (2010) Assay development: fundamentals and practices. Wiley, New York
30. Węsierska-Gądek J, Gueorguieva M, Ranftler C et al (2005) A new multiplex assay allowing simultaneous detection of the inhibition of cell proliferation and induction of cell death. J Cell Biochem 96(1):1–7. doi:10.1002/jcb.20531
31. Twigg RS (1945) Oxidation-reduction aspects of resazurin. Nature 155(3935):401–402. doi:10.1038/155401a0
32. Sieuwerts AM, Klijn JGM, Peters HA et al (1995) The MTT tetrazolium salt assay scrutinized: how to use this assay reliably to measure metabolie activity of cell cultures in vitro for the assessment of growth characteristics, IC50-values and cell survival. Clin Chem Lab Med 33(11):813–824

Chapter 2

Assaying Cellular Viability Using the Neutral Red Uptake Assay

Gamze Ates, Tamara Vanhaecke, Vera Rogiers, and Robim M. Rodrigues

Abstract

The neutral red uptake assay is a cell viability assay that allows in vitro quantification of xenobiotic-induced cytotoxicity. The assay relies on the ability of living cells to incorporate and bind neutral red, a weak cationic dye, in lysosomes. As such, cytotoxicity is expressed as a concentration-dependent reduction of the uptake of neutral red after exposure to the xenobiotic under investigation. The neutral red uptake assay is mainly used for hazard assessment in in vitro toxicology applications. This method has also been introduced in regulatory recommendations as part of 3T3-NRU-phototoxicity-assay, which was regulatory accepted in all EU member states in 2000 and in the OECD member states in 2004 as a test guideline (TG 432). The present protocol describes the neutral red uptake assay using the human hepatoma cell line HepG2, which is often employed as an alternative in vitro model for human hepatocytes. As an example, the cytotoxicity of acetaminophen and acetyl salicylic acid is assessed.

Key words Viability assay, Neutral red uptake, HepG2

1 Introduction

The neutral red uptake (NRU) assay is a viability assay based on the ability of living cells to incorporate and bind neutral red (NR) [1]. This weak cationic eurhodine dye can penetrate cells by nonionic diffusion at physiological pH. Once NR is in the cell, it accumulates intracellularly in lysosomes, where a proton gradient assures a more acidic pH and the dye becomes charged [2]. Xenobiotics can lead to alterations of the cell surface or lysosomal membrane, which results in a decreased uptake and binding of NR. As such, the NRU assay allows to assess membrane permeability and lysosomal activity, making it possible to differentiate viable, damaged, or dead cells. Cytotoxicity is expressed as a concentration-dependent reduction of the uptake of NR after exposure to the xenobiotic, thus providing a sensitive, integrated signal of both cell integrity and cell growth inhibition [1]. The NRU has miscellaneous biological applications and is commonly used to evaluate the cytotoxicity

Daniel F. Gilbert and Oliver Friedrich (eds.), *Cell Viability Assays: Methods and Protocols*, Methods in Molecular Biology, vol. 1601, DOI 10.1007/978-1-4939-6960-9_2, © Springer Science+Business Media LLC 2017

of a variety of chemical substances such as pharmaceuticals and cosmetics [3, 4]. Several validation studies have been set up for the NRU as a test for cytotoxicity [5]. In 2000 a NRU test on Balb/c 3T3 mouse fibroblasts to assess phototoxicity, was regulatory accepted in all EU member states and in 2004 it was adopted as an official Organisation for Economic Co-operation and Development (OECD) test guideline (TG 432) [6]. In 2013, the European Commission Joint Research Centre has published a recommendation on the use of the 3T3 NRU assay in which it stresses the validity of the NRU in a weight-of-evidence approach to predict acute oral toxicity of chemicals in a regulatory setting [7]. The facility of the NRU assay permits automation, which improves throughput and allows fast and reliable screening of a large amount of test compounds in a relatively short time span [4, 8].

For the purpose of this book chapter, the NRU is described on HepG2 cells. This human hepatoma cell line originates from a 15-year-old Caucasian male and is widely employed in hepatotoxicity studies. Under proper culture conditions, HepG2 cells display (limited) hepatocyte-like features and are therefore often utilized as an alternative in vitro model for human hepatocytes [9–11].

2 Materials

2.1 General Equipment

1. Incubator: 37 ± 1 °C, 90 ± 5% humidity, 5.0 ± 1% CO_2/air.
2. Laminar flow clean bench/cabinet (standard: "biological hazard").
3. Water bath: 37 ± 1 °C.
4. Inverse-phase contrast microscope.
5. Laboratory balance.
6. 96-Well plate spectrophotometer (i.e., plate reader) equipped with 540 ± 10 nm filter.
7. Shaker for microtiter plates.
8. Cell counter or hemocytometer.
9. Pipettes, pipettors (multichannel and single channel; multichannel repeater pipette).
10. 96-Well flat-bottom tissue culture microtiter plates.
11. Multichannel reagent reservoir.
12. Vortex mixer.
13. Filters/filtration devices.

2.2 Reagents

1. HepG2 cells.
2. Dulbecco's modification of Eagle's medium (DMEM) with L-glutamine (preferably) and high glucose (4.5 g/l).

3. 200 mM L-Glutamine (in case DMEM does not contain L-glutamine).
4. Fetal bovine serum (FBS) (*see* **Note 1**).
5. TripLE™ Express Enzyme.
6. Phosphate-buffered saline (PBS) without Ca^{2+} and Mg^{2+}.
7. Penicillin/streptomycin solution.
8. Neutral Red (NR) Dye liquid form.
9. Dimethyl sulfoxide (DMSO, cell culture grade).
10. Ethanol (EtOH), U.S.P. analytical grade (100%, non-denatured for test chemical preparation; 95% can be used for the desorption solution).
11. Glacial acetic acid, analytical grade.
12. Distilled water or any purified water suitable for cell culture (sterile).
13. Test compounds (acetaminophen and acetyl salicylic acid were used in this protocol).

3 Methods

All solutions, glassware, pipettes, etc. have to be sterile and all procedures should be carried out under aseptic conditions and in the sterile environment of a laminar flow cabinet (biological hazard standard).

3.1 HepG2 Cell Culture

HepG2 cells are routinely grown as a monolayer in tissue culture-grade flasks (e.g., 75 cm^2) at 37 ± 1 °C, 90 ± 5% humidity, and 5.0 ± 1% CO_2/air (*see* **Note 2**). When cells exceed 50% confluence (but do not reach 80% confluence) they should be passaged by removing them from the flask using TripLE™ Express Enzyme as follows:

1. Prepare Routine Culture Medium by supplementing DMEM with 10% FBS, 100 IU/ml penicillin, 100 μg/ml streptomycin, and 4 mM glutamine (if not already present in media) (*see* **Note 3**).
2. Pre-warm the routine culture medium and PBS in a water bath at 37 °C.
3. Aspirate routine culture medium from the flask, briefly rinse cultures with 10 ml PBS, and add 5 ml TripLE™ Express Enzyme.
4. Incubate flask in the incubator for 2–5 min. Tap the flask on the side, to make sure that all cells are detached (check with microscope).
5. Carefully resuspend the cells and transfer the cell suspension into a Falcon tube (*see* **Note 4**).

6. Rinse the culture flask two times with 10 ml pre-warmed routine culture medium and add to the Falcon tube. Centrifuge the cell suspension at 385 × *g* for 5 min.
7. Remove supernatant. To avoid aspiration of cells, leave some routine culture medium on top of the pellet. Resuspend pellet in 5 ml routine culture medium and count the cells using a cell counter or a hemocytometer.
8. Prepare a cell suspension of 2.0–3.0 × 10^4 cells/ml in routine culture medium. Using a multichannel pipette, dispense 100 μl routine culture medium only into the peripheral wells (blanks) of a 96-well tissue culture microtiter plate. In the remaining wells, seed between 7.5 and 10.0 × 10^3 cells/well (*see* **Note 5**).
9. Incubate cells for 24 ± 2 h (37 ± 1 °C, 90 ± 5% humidity, 5.0 ± 1% CO_2/air). This incubation period assures cell recovery and adherence and progression to the exponential growth phase.
10. Refresh routine culture medium every 2–3 days (*see* **Note 6**).
11. Examine each plate under a phase-contrast microscope to assure that cell growth is relatively even across the microtiter plate. This check is performed to identify experimental and systemic cell seeding errors.

3.2 Preparation of Test Chemicals

1. Prepare test chemical immediately prior to use. Test chemical solutions should not be prepared in bulk for use in subsequent tests. The test chemical should be completely soluble and the solutions must not be cloudy nor have noticeable precipitate. Each stock dilution should have a minimal volume of at least 1–2 ml.
2. For chemicals dissolved in DMSO or EtOH, the final DMSO or EtOH concentration for application to the cells must not exceed 0.5% (v/v). All test concentrations and vehicle controls should contain the same concentration of DMSO or EtOH.
3. The stock solution for each test chemical should be prepared at the highest concentration found to be soluble. The lower concentrations in a range-finding experiment would then be prepared by successive dilutions that decrease by, e.g., one log unit each. Once the toxicity range for a compound is found, smaller concentration intervals should be tested.
4. Prior to exposure of the test chemicals, the stock solutions must be diluted in pre-warmed (37 °C) routine culture medium.
5. Table 1 shows an example of the concentration gradients for acetaminophen and acetylsalicylic acid. Hereby a dilution factor of 2.15 was used to prepare the serial dilutions.

3.3 Cell Culture Treatment

1. Aspirate the routine culture medium from the plates.
2. Add 100 μl of each concentration according to a predefined plate layout. An example of a plate layout is given in Fig. 1 (*see* **Note 7**).

Table 1
Test concentration gradients of acetaminophen and acetylsalicylic acid

Concentration (μg/ml)	C1	C2	C3	C4	C5	C6	C7	C8
Acetaminophen	5000	2325.6	1081.7	503.1	234.0	108.8	50.6	23.5
Acetylsalicylic acid	2000	930.2	432.7	201.2	93.6	43.5	20.2	9.4

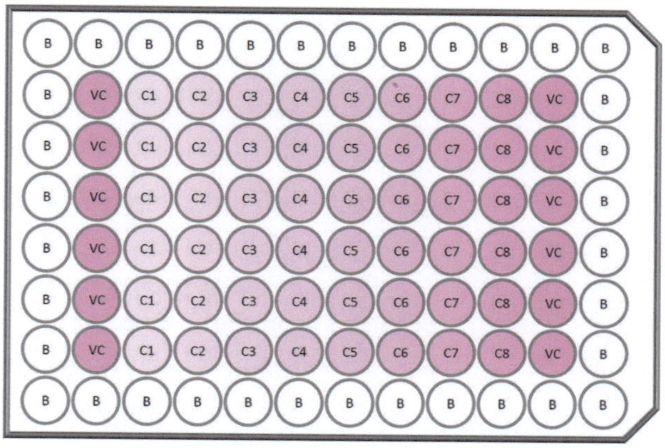

Fig. 1 Plate layout for cell exposure. B blank, VC vehicle control, C1–C8 test concentration in descending order

3. Incubate the plate at proper conditions, for 24 ± 1 h. Longer incubation times, e.g., 48 or 72 h, can also be used.

3.4 Neutral Red Uptake Assay

1. Prepare the NR medium (25 μg/ml) in routine culture medium (pre-warmed to 37 °C) (*see* **Note 8**).
2. Carefully aspirate the routine culture medium with test chemical and rinse the cells very carefully with 250 μl pre-warmed PBS. Aspirate the rinsing solution (*see* **Note 9**).
3. Add 250 μl NR medium (to all wells including the blanks) and incubate (37 ± 1 °C, 90 ± 5% humidity, and 5.0 ± 1% CO_2/air) for 3 ± 0.1 h.
4. After incubation, remove the NR medium, and carefully rinse the cells with 250 μl pre-warmed PBS. Figure 2 illustrates HepG2 cells with intracellularly bound NR.
5. Prepare the desorption solution (1% glacial acetic acid solution, 50% EtOH, 49% H_2O) and add exactly 100 μl desorption solution to all wells, including blanks.
6. Shake microtiter plate on a microtiter plate shaker (e.g., 80 rpm, Stuart mini orbital shaker SSM1) for 20–45 min to extract

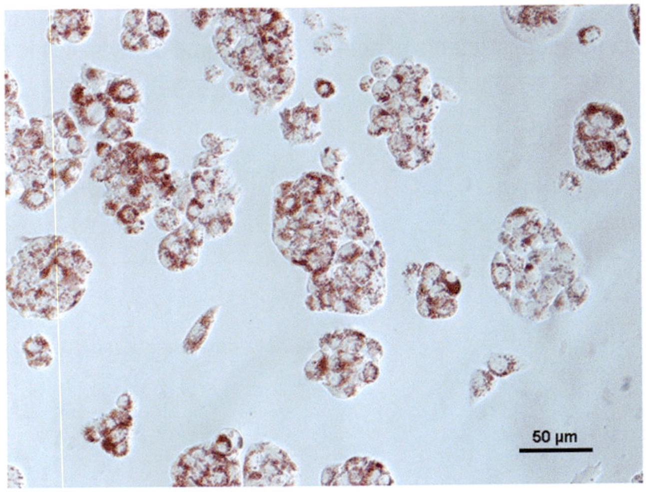

Fig. 2 HepG2 cells with intracellularly bound neutral red

NR from the cells and form a homogeneous solution. Protect plates from light by covering them, e.g., with aluminum foil.

7. Plates should be still for at least 5 min after removal from the plate shaker. Measure the absorption (within 60 min of adding the desorption solution) of the resulting colored solution at 540 ± 10 nm in a microtiter plate reader.

3.5 Data Analysis

The obtained spectrophotometric data is mostly presented as a concentration-response curve (often referred to as dose-response curve) in which the effect caused by the xenobiotic can be visualized following a concentration gradient. Analysis of this curve provides information on the cytotoxic effect caused in the cells exposed for a determined period of time. This analysis is based on four-parameter logistic nonlinear regression that can be conducted using several mathematical software packages. Some of those, e.g., MasterPlex® (MiraiBio Group of Hitachi Solutions America, Ltd.), have been specifically designed for analysis of life sciences assays. Figure 3 illustrates the concentration-response of HepG2 cells exposed during 24 h to a concentration gradient of acetaminophen and acetylsalicylic acid.

The main endpoint readout of concentration-response curves is the determination of the concentration at which a particular percentage of the cells show a decrease in viability. Hereby, the IC_{50} (inhibitory concentration 50) or the concentration of test substance at which 50% of cell death is observed is conventionally used as a parameter of in vitro cytotoxicity. The IC_{10}, representing the concentration that induces 10% cell death, or lower concentrations are on the other hand often referred to as subcytotoxic concentrations.

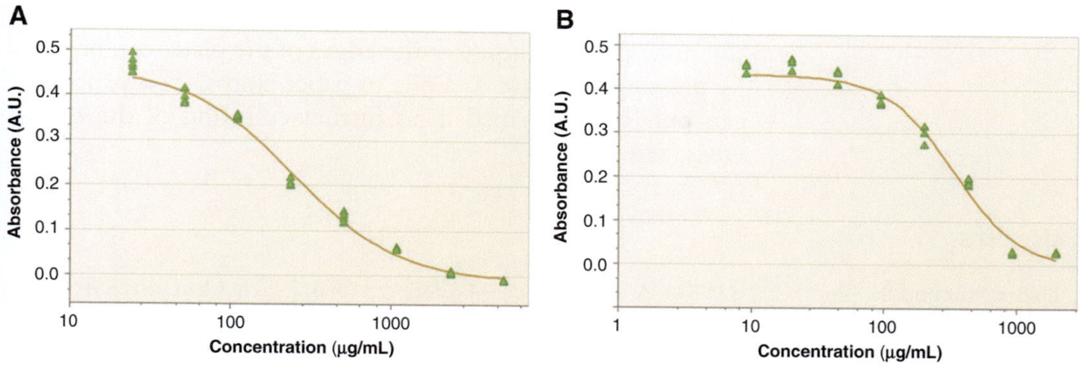

Fig. 3 Concentration-response curves of HepG2 cells exposed for 24 h to (**a**) acetaminophen and (**b**) acetylsalicylic acid

4 Notes

1. Due to the inconsistency of different sera, the cytotoxicity of different batches of FBS should be investigated. A sufficient amount of the same batch FBS should be reserved and used within the same experiments.
2. Completed media formulations should be kept at approximately 2–8 °C and stored for no longer than 1 month.
3. For the purpose of this book chapter, the NRU assay is documented using the hepatic human cell line HepG2. This assay can, however, be applied using other animal or human cell systems.
4. Trypsinization is inhibited by the presence of serum in cell culture media. Therefore, the dissociation of the cells should be complete before adding routine culture medium.
5. Other plate formats than 96-well plates might also be used for the NRU assay. The incubation volumes should be adapted according to each recipient.
6. At higher cell densities, slight acidification of the routine culture medium may occur (observed by an orange color shift of phenol red indicator). In this case the frequency of routine culture medium refreshment should be increased and/or the cells should be passaged.
7. In the presented plate layout each concentration is tested in sixfold. These technical repeats can be decreased to three, which makes it possible to test two different compounds in the same plate.
8. The NR medium can be filtered to reduce potentially formed NR crystals (e.g., Millipore filtering, 0.2–0.45 μm pore size).
9. To reduce the time of this step, the aspiration of NR medium and PBS can be replaced by “dumping” the content of the

plate by a flip movement into a recipient with large opening or the sink. Eventual liquid at the edges of the plates can be dried by pressing the plate to a pile of paper cloths. This procedure can only be performed if no further culturing of the cells is envisaged.

References

1. Borenfreund E, Puerner J (1984) A simple quantitative procedure using monolayer cultures for cytotoxicity assays (HTD/NR90). J Tissue Cult Methods 9(1):7–9
2. Repetto G, del Peso A, Zurita JL (2008) Neutral red uptake assay for the estimation of cell viability/cytotoxicity. Nat Protoc 3(7): 1125–1131
3. Zuang V (2001) The neutral red release assay: a review. Altern Lab Anim 29(5):575–599
4. Rodrigues RM, Bouhifd M, Bories G, Sacco M, Gribaldo L, Fabbri M, Coecke S, Whelan MP (2013) Assessment of an automated in vitro basal cytotoxicity test system based on metabolically-competent cells. Toxicol In Vitro 27(2):760–767
5. The National Toxicology Program Interagency Center for the Evaluation of Alternative Toxicological Methods (2003) Test method protocol for the BALB/c 3T3 neutral red uptake cytotoxicity test. A test for basal cytotoxicity for an in vitro validation study phase III. https://ntp.niehs.nih.gov/iccvam/methods/acutetox/invidocs/phiiiprot/3t3phiii.pdf. Accessed Jun 2016
6. Organisation for Economic Co-operation and Development (2004) Guideline 432: in vitro 3T3 NRU phototoxicity test. OECD guidelines for the testing of chemicals. http://www.oecd-ilibrary.org/environment/test-no-432-in-vitro-3t3-nru-phototoxicity-test_9789264071162-en. Accessed Jun 2016
7. European Commission Joint Research Center (2013) EURL ECVAM recommendation on the 3T3 neutral red uptake (3T3 NRU) cytotoxicity assay for the identification of substances not requiring classification for acute oral toxicity. https://eurl-ecvam.jrc.ec.europa.eu/eurl-ecvam-recommendations/3t3-nru-recommendation. Accessed Jun 2016
8. Bouhifd M, Bories G, Casado J, Coecke S, Norlén H, Parissis N, Rodrigues RM, Whelan MP (2012) Automation of an in vitro cytotoxicity assay used to estimate starting doses in acute oral systemic toxicity tests. Food Chem Toxicol 50(6):2084–2096
9. Knowles BB, Aden DP (1983) Human hepatoma derived cell line, process for preparation thereof, and uses therefor. US Patent 4,393,133
10. Schoonen WG, Westerink WM, de Roos JA, Débiton E (2005) Cytotoxic effects of 100 reference compounds on HepG2 and HeLa cells and of 60 compounds on ECC-1 and CHO cells. I mechanistic assays on ROS, glutathione depletion and calcein uptake. Toxicol In Vitro 19(4):505–516
11. Chiu JH, Hu CP, Lui WY, Lo SJ, Chang CM (1990) The formation of bile canaliculi in human hepatoma-cell lines. Hepatology 11:834–842

Chapter 3

Assessment of Cell Viability with Single-, Dual-, and Multi-Staining Methods Using Image Cytometry

Leo Li-Ying Chan, Kelsey J. McCulley, and Sarah L. Kessel

Abstract

The ability to accurately measure cell viability is important for any cell-based assay. Traditionally, viability measurements have been performed using the trypan blue exclusion method on a hemacytometer, which allows researchers to visually distinguish viable from nonviable cells. While the trypan blue method can work for cell lines or primary cells that have been rigorously purified, in more complex samples such as PBMCs, bone marrow, whole blood, or any sample with low viability, this method can lead to errors. In recent years, advances in optics and fluorescent dyes have led to the development of automated benchtop image-based cell counters for rapid cell concentration and viability measurement. In this work, we demonstrate the use of image-based cytometry for cell viability detection using single-, dual-, or multi-stain techniques. Single-staining methods using nucleic acid stains such as EB, PI, 7-AAD, DAPI, SYTOX Green, and SYTOX Red, and enzymatic stains such as CFDA and Calcein AM, were performed. Dual-staining methods using AO/PI, CFDA/PI, Calcein AM/PI, Hoechst/PI, Hoechst/DRAQ7, and DRAQ5/DAPI that enumerate viable and nonviable cells were also performed. Finally, Hoechst/Calcein AM/PI was used for a multi-staining method. Fluorescent viability staining allows exclusion of cellular debris and nonnucleated cells from analysis, which can eliminate the need to perform purification steps during sample preparation and improve efficiency. Image cytometers increase speed and throughput, capture images for visual confirmation of results, and can greatly simplify cell count and viability measurements.

Keywords Image cytometry, Viability, Enzymatic stain, Nucleic acid stain, Multi-stain method, Fluorescent stain, Trypan blue, Cellometer, Celigo

1 Introduction

It is important to accurately measure cell viability for any cell-based assay performed in immuno-oncology, stem cell, and toxicology research, or for traditional cell culture and plating for downstream assays [1, 2]. In the past decade, a new generation of affordable chip-based image cytometry systems, such as the Cellometer [3–5] (Nexcelom Bioscience), Countess II [6] (Life Technologies), and NucleoCounter [7] (Chemometec), have been introduced to address the known issues of traditional cell

Daniel F. Gilbert and Oliver Friedrich (eds.), *Cell Viability Assays: Methods and Protocols*, Methods in Molecular Biology, vol. 1601, DOI 10.1007/978-1-4939-6960-9_3, © Springer Science+Business Media LLC 2017

viability detection methods such as manual counting and flow cytometry [8–10]. Manual counting using a hemacytometer is time consuming and has high operator-dependent variations [11]. Flow cytometry systems require a considerable amount of maintenance and highly skilled operators. In addition, the lack of imaging capability may generate uncertainties in the results [12, 13]. In contrast, automated image cytometers can quickly and easily capture and analyze bright-field and fluorescent images of trypan blue (TB) or fluorescently stained target cells to measure the number of live and dead cells and determine viability [14–17]. High-throughput plate-based image cytometry systems such as the Celigo [18] (Nexcelom Bioscience), Opera [19, 20] (Perkin Elmer), and IN Cell Analyzer 2200 [20, 21] (GE) have also been developed to measure cell viability in standard multi-well microplates. These high-throughput image cytometers can be used to screen potential cancer drug candidates for drug discovery research.

Depending on the cell sample, image cytometers are used to measure cell viability by staining cells with one, two, or three dyes for optimal measurements. For cell lines with high viability or primary cells that have been rigorously purified, a single-dye staining method such as trypan blue or propidium iodide can be used, where the image cytometer measures total and dead cell counts in bright-field and fluorescent images [22–24]. In contrast, primary cell samples that contain a high level of red blood cells (RBC) and platelets require staining with multiple dyes. The total, live, and dead cell counts are measured in fluorescent images, which eliminates the potential of counting nonnucleated cells or nonspecific particles in the samples.

In this work, we describe the Cellometer (chip-based) and Celigo (plate-based) image cytometry protocols to rapidly assess cell viability using trypan blue, fluorescent nucleic acid, and enzymatic dyes, as well as utilizing dual- and multi-fluorescent staining methods. Jurkat cells were stained with ethidium bromide (EB), propidium iodide (PI), 7-aminoactinomycin D (7-AAD), 4′,6-diamidino-2-phenylindole (DAPI), SYTOX Green, or SYTOX Red nucleic acid dyes to measure cell viability [4, 25–29]. Similarly, Jurkat cells were stained with carboxyfluorescein diacetate (CFDA) or Calcein AM enzymatic dyes to measure cell viability [4, 27, 30]. Next, Jurkat cells were dual-stained with acridine orange, Hoechst 33342 (Hoechst), CFDA, or Calcein AM in combination with PI to enumerate live and dead cells [31–34]. In addition, DRAQ5™ [35] and DRAQ7™ [36] were used in combination with DAPI and Hoechst, respectively, to stain MCF7 GFP cells. Finally, Hoechst, Calcein AM, and PI were used to stain HeLa cells to measure total, live, and dead cells, respectively [37].

2 Materials

2.1 Target Cells

1. Jurkat cells (TIB-152, American Type Culture Collection (ATCC)).
2. MCF7 GFP cells.
3. HeLa cells.

2.2 Cell Culture Media

1. RPMI medium supplemented with 10% fetal bovine serum (FBS) and 1% pen/strep antibiotics.
2. EMEM medium supplemented with 10% FBS, 1× NEAA, and 1× GlutaMAX.
3. DMEM medium supplemented with 10% FBS.
4. 0.25% Trypsin.

2.3 Imaging Vessels

1. Nexcelom disposable counting chamber (CHT4-SD100) (*see* **Note 1**).
2. 96-Well TC-treated black wall, clear-bottomed microplate (e.g., Greiner 655090) (*see* **Notes 2** and **3**).

2.4 Viability Stains

1. Viability stains commonly used with image cytometry for assessment of viability using single, dual, or multiple stains are shown in Table 1.

Table 1
List of single-, dual-, and multi-stain combinations

Single stain	*Component*	*Company*	*Catalog #*	*Excitation (nm)*	*Emission (nm)*
Trypan blue		Sigma-Aldrich	T8154	N/A	N/A
EB		Sigma-Aldrich	E1510	522	606
7-AAD		Thermo Fisher Scientific	A1310	543	647
SYTOX Green		Thermo Fisher Scientific	S7020	504	523
SYTOX Red		Thermo Fisher Scientific	S34859	640	658
CFDA		Thermo Fisher Scientific	C195	492	517
PI		Nexcelom Bioscience	CS1-0109	538	617
Calcein AM		Nexcelom Bioscience	CS1-0119	496	516
DAPI		Nexcelom Bioscience	CS1-0127	358	461

(continued)

Table 1
(continued)

Dual stain	*Component*	*Company*	*Catalog #*	*Excitation (nm)*	*Emission (nm)*
AO/PI	AO	Nexcelom Bioscience	CS2-0106	501	527
	PI			538	617
CFDA/PI	CFDA	Thermo Fisher Scientific	C195	492	517
	PI	Nexcelom Bioscience	CS1-0109	538	617
Calcein AM/PI	Calcein AM	Nexcelom Bioscience	CSK-0118	496	516
	PI			538	617
Hoechst/PI	Hoechst	Nexcelom Bioscience	CSK-V0001-1	352	455
	PI			538	617
DRAQ5/DAPI	DRAQ5	Biostatus	DR50050	646	697
	DAPI	Nexcelom Bioscience	CS1-0127	358	461
DRAQ7/Hoechst	Hoechst	Nexcelom Bioscience	CS1-0128	352	455
	DRAQ7	Biostatus	DR70250	644	697
Multi-stain	*Component*	*Company*	*Catalog #*	*Excitation (nm)*	*Emission (nm)*
Hoechst/Calcein AM/PI	Hoechst	Nexcelom Bioscience	CSK-V0001-1	352	455
	Calcein AM			496	516
	PI			538	617

The company, catalog #, excitation, and emission wavelengths are shown for each stain

2.5 Image Cytometers

1. Cellometer image cytometer.
2. Fluorescence Optics Module (FOM) for Cellometer:
   (a) VB-450-302, VB-535-402, VB-595-502, VB-660-502, VB-695-602.
3. Celigo image cytometer.

3 Methods

3.1 Cellometer Image Cytometer

1. Select and install the correct fluorescent filters into the Cellometer depending on the stains used (*see* **Note 4**):
   (a) VB-450-302: DAPI and Hoechst.
   (b) VB-535-402: AO, CFDA, Calcein AM, SYTOX Green, and TB.

   (c) VB-595-502: EB and PI (used alone without AO).
   (d) VB-660-502: 7-AAD and PI (used in combination with AO).
   (e) VB-695-602: SYTOX Red, DRAQ5, and DRAQ7.
2. Turn on the Cellometer image cytometer.
3. Open the Cellometer software.
4. Allow the system to connect and register.

3.2 Celigo Image Cytometer

1. Turn on the Celigo image cytometer.
2. Turn on the computer.
3. Open the Celigo software.
4. Allow the system to connect and register.

3.3 Viability Stain Preparation

1. After acquiring the viability stains, use the diluents indicated in Table 2 to prepare the stains to the working concentrations for Cellometer and Celigo image cytometers (*see* **Note 5**).

Table 2
List of diluents and working concentrations for single-, dual-, and multi-stain combinations for cellometer

Single stain	*Component*	*Diluent*	*Working concentration*
Trypan blue		PBS	0.2%
EB		PBS	200 μg/mL
7-AAD		PBS	100 μg/mL
SYTOX Green		PBS	1 μM
SYTOX Red		PBS	0.5 μM
CFDA		PBS	10 μM
PI		N/A	Use directly
Calcein AM		H_2O	10 μM
DAPI		H_2O	20 μg/mL
Dual stain	*Component*	*Diluent*	*Working concentration*
AO/PI	AO	N/A	Use directly
	PI		
CFDA/PI	CFDA	PBS	20 μM
	PI		100 μg/mL
Calcein AM/PI	Calcein AM	H_2O	20 μM
	PI	PBS	100 μg/mL
Hoechst/PI	Hoechst	PBS	20 μM
	PI		100 μg/mL

Table 3
List of diluents and working concentrations for single-, dual-, and multi-stain combinations for celigo

Single stain	*Component*	*Diluent*	*Working concentration*
PI		N/A	8 μg/mL
Calcein AM		PBS	2 μM
DAPI		PBS	4 μM
Dual stain	*Component*	*Diluent*	*Working concentration*
AO/PI	AO	PBS	5× and 10× diluted
	PI		
Calcein AM/PI	Calcein AM	PBS	1 μM
	PI	PBS	2 μg/mL
Hoechst/PI	Hoechst	PBS	8 μg/mL
	PI		2 μg/mL
DRAQ 5/DAPI	DRAQ 5	Media	10 μM
	DAPI		4 μM
Hoechst/DRAQ 7	Hoechst	Media	8 μg/mL
	DRAQ 7		3 μM
Multi-stain	*Component*	*Diluent*	*Working concentration*
Hoechst/Calcein AM/PI	Hoechst	PBS	8 μg/mL
	Calcein AM		2 μM
	PI		2 μg/mL

2. Stain preparation for Cellometer image cytometry is shown in Table 2.
3. Stain preparation for Celigo image cytometry is shown in Table 3.

3.4 Jurkat Cell Preparation

1. Culture the Jurkat cells (TIB-152, American Type Culture Collection (ATCC)) in RPMI medium supplemented with 10% fetal bovine serum (FBS) and 1% pen/strep antibiotics.
2. Maintain the cell culture in an incubator at 37 °C and 5% CO_2.
3. Passage cells at 1:5 dilution every 2–3 days by replenishing with fresh medium.

3.5 MCF7 GFP Cell Preparation

1. Culture adherent cells in EMEM medium supplemented with 10% FBS, 1× NEAA, and 1× GlutaMAX, in T25 Tissue Culture (TC)-treated flasks until 90% confluent.
2. Maintain the cell cultures in an incubator at 37 °C and 5% CO_2.
3. Passage cells at 1:10 and 1:20 dilutions every 3–4 days with 0.25% trypsin.

3.6 HeLa Cell Preparation

1. Culture the adherent HeLa cells in DMEM medium supplemented with 10% FBS, in T25 TC-treated flasks until 90% confluent.
2. Maintain the cell cultures in an incubator at 37 °C and 5% CO_2.
3. Passage cells at 1:10 and 1:20 dilutions every 3–4 days with 0.25% trypsin.

3.7 Viability Analysis and Calculation Methods

1. Each stain is used to determine the number of total, live, or dead cell counts.
2. The stain combinations for viability analysis are shown in Table 4.
3. The total, live, or dead cell counts are used to directly calculate viability:
   (a) $\text{Viability\%} = \dfrac{\text{Live \#}}{\text{Total \#}}$
   (b) $\text{Viability\%} = \dfrac{\text{Total} - \text{Dead \#}}{\text{Total \#}}$

Table 4
Viability analysis methods for each stain combination

Single stain	*Total*	*Live*	*Dead*
Trypan blue		Bright field	Trypan blue
EB	Bright field		EB
7-AAD	Bright field		7-AAD
SYTOX Green	Bright field		SYTOX Green
SYTOX Red	Bright field		SYTOX Red
CFDA	Bright field	CFDA	
PI	Bright field		PI
Calcein AM	Bright field	Calcein AM	
DAPI	Bright field		DAPI
Dual stain	*Total*	*Live*	*Dead*
AO/PI		AO	PI
CFDA/PI		CFDA	PI
Calcein AM/PI		Calcein AM	PI
Hoechst/PI	Hoechst		PI
DRAQ5/DAPI	DRAQ5		DAPI
Hoechst/DRAQ7	Hoechst		DRAQ7
Multi-stain	*Total*	*Live*	*Dead*
Hoechst/Calcein AM/PI	Hoechst	Calcein AM	PI

(c) $\text{Viability}\% = \dfrac{\text{Live}\,\#}{\text{Live} + \text{Dead}\,\#}$

3.8 Suspension and Trypsinized Adherent Cell Staining Preparation for Chip-Based Image Cytometer

1. To measure the viability of cells on Cellometer, mix suspension cells or trypsinized adherent cells 1:1 with the viability stains directly in centrifuge tubes (*see* **Note 6**).
2. Follow Table 5 for mixing volumes and incubation time. If staining requires incubation, incubate the cells in 37 °C and 5% CO_2.
3. Pipette 20 μL of stained cell sample into one of the two openings on the counting chamber (Fig. 1a).
4. Wait for 15 s until the cells settle to the bottom of the chamber (*see* **Note 7**).

3.9 Suspension Cell Staining Preparation for Plate-Based Image Cytometer

1. To measure the viability of suspension cells on Celigo, stain the cells directly in the 96-well microplate at 1:1 with the working concentration of the fluorescent dyes.
2. Pipette 100 μL of the stain to each well and then add 100 μL of the cells (*see* **Notes 8** and **9**).
3. For AO/PI only, pipette 50 μL of 10× diluted AO/PI to each well and then add 10 μL of cells.

Table 5
Cell and stain mixing volume and incubation time for each stain for Cellometer

Single stain	*Cell volume*	*Stain volume*	*Incubation time*
Trypan blue	20 μL	20 μL	None, image immediately
EB			None, image immediately
7AAD			5 min
SYTOX Green			None, image immediately
SYTOX Red			15 min
CFDA			15 min
PI			None, image immediately
Calcein AM			15 min
DAPI			5 min
Dual stain	*Cell volume*	*Stain volume*	*Incubation time*
AO/PI	20 μL	20 μL	None, image immediately
CFDA/PI			15 min
Calcein AM/PI			15 min
Hoechst/PI			15 min

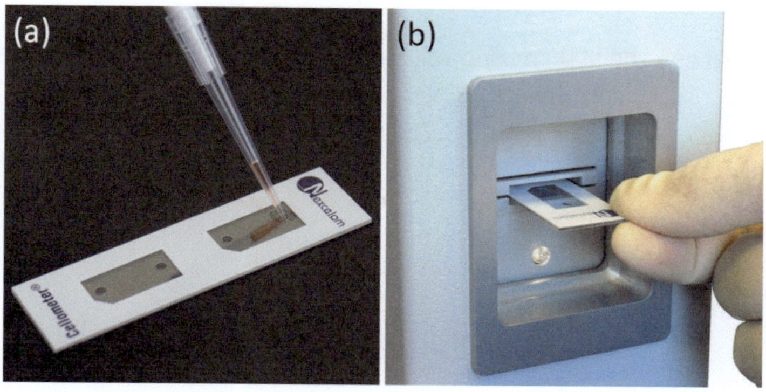

Fig. 1 Sample preparation steps for counting cells using Cellometer image cytometer. (**a**) Pipette 20 μL of stained cells into a counting chamber. (**b**) Insert the counting chamber into the image cytometer for image acquisition and data analysis

Table 6
Suspension cell and stain mixing volumes and incubation time for each stain for celigo

Single stain	*Cell volume*	*Stain volume*	*Incubation time*
PI	100 μL	100 μL	None, image immediately
Calcein AM			30 min
DAPI			15 min
Dual stain	*Cell volume*	*Stain volume*	*Incubation time*
AO/PI	10 μL	50 μL	None, image immediately
Calcein AM/PI	100 μL	100 μL	30 min
Hoechst/PI			30 min
DRAQ5/DAPI			30–45 min
Hoechst/DRAQ7			30–45 min
Multi-stain	*Cell volume*	*Stain volume*	*Incubation time*
Hoechst/Calcein AM/PI	100 μL	100 μL	30 min

4. Centrifuge the microplate to make the suspension cells settle to the bottom (*see* **Note 10**).
5. Follow Table 6 for mixing volumes and incubation time. If staining requires incubation, incubate the cells in 37 °C and 5% CO_2.

3.10 Adherent Cell Staining Preparation for Plate-Based Image Cytometer

1. To measure the viability of adherent cells on Celigo, stain cells directly in the 96-well microplate by adding 100 μL of the fluorescent stains at the working concentration to wells containing 100 μL of cells.

Table 7
Adherent cell and stain mixing volume and incubation time for each stain for Celigo

	Cell volume	Stain volume	Incubation time
Single stain			
PI	100 μL	100 μL	None, image immediately
Calcein AM			30 min
DAPI			15 min
Dual stain			
AO/PI	50 μL	50 μL	None, image immediately
Calcein AM/PI	100 μL	100 μL	30 min
Hoechst/PI			30 min
DRAQ5/DAPI			30–45 min
Hoechst/DRAQ7			30–45 min
Multi-stain			
Hoechst/Calcein AM/PI	100 μL	100 μL	30 min

2. Adherent cells are initially plated in 96-well microplate and then allowed to adhere overnight in 37 °C and 5% CO_2.
3. For AO/PI, pipette 50 μL of 5× diluted AO/PI to each well on top of the existing 50 μL of PBS.
4. The final volume in each well should be 200 μL prior to imaging.
5. Follow Table 7 for mixing volumes and incubation time. If staining requires incubation, incubate the cells in 37 °C and 5% CO_2.

3.11 Assessing Cell Viability Using Chip-Based Image Cytometry Method

1. Select the correct "Assay Type" depending on the viability stain used (Table 4):
   (a) Bright-field assay using trypan blue.
   (b) Fluorescent assay using single-fluorescent staining.
   (c) Fluorescent assay using dual-fluorescent staining.
2. Insert the counting chamber filled with stained cells into the Cellometer (Fig. 1b).
3. Click on "Preview Brightfield Image" to view target cells in the chamber (*see* **Note 11**).
4. Use the focus knob located on the right side of the instrument and focus until the center of the target cell becomes bright and the membrane is dark and thin.

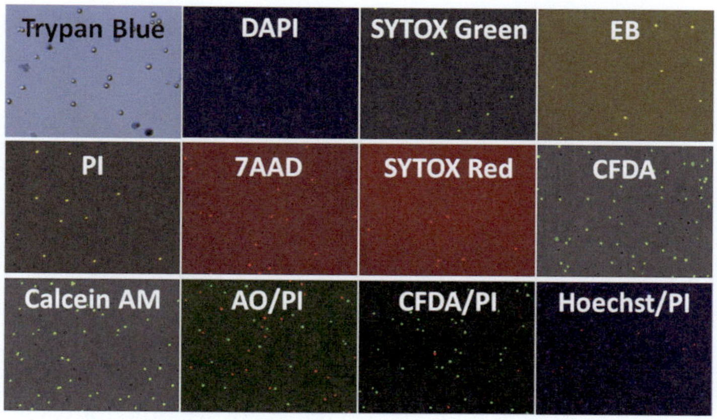

Fig. 2 Examples of bright-field (trypan blue), bright-field/fluorescent overlay (DAPI, SYTOX Green, EB, PI, 7-AAD, SYTOX Red, CFDA, Calcein AM), and fluorescent images (AO/PI, CFDA/PI, Hoechst/PI) using Cellometer image cytometer

5. If fluorescent stains are used, stop the preview and click on "Preview Fl Image" or "Preview F2 image" to set up the exposure time depending on the fluorescent stains.
6. Adjust the exposure time for either Fl or F2 channels until the cell intensity reaches 100% and the background is low visually.
7. After the cells are focused in the bright-field image and the fluorescent exposure times are adjusted, click on "Count."
8. Wait until the bright-field and fluorescent images are captured and analyzed for the results to be displayed (*see* **Note 12**).
9. Save the images and results (Fig. 2).

3.12 Assessing Cell Viability Using Plate-Based Image Cytometry Method

1. Click on "Create a New Scan" under the START tab.
2. Select the plate type and vendor.
3. Type in a unique plate identification name.
4. Insert the 96-well microplate containing stained cells.
5. In the SCAN tab, select the Celigo Application under "Cell Viability" depending on the viability stain used:
   (a) Live + Total, Dead + Total, or Live + Dead for single- or dual-fluorescent staining.
   (b) Live + Dead + Total for multi-fluorescent staining.
6. Set up the Total channel for bright-field or fluorescent illumination and set up the exposure time.
7. For bright-field images, adjust exposure for the background to be between 125 and 150.
8. For fluorescent images, adjust exposure for cell object pixel intensity to be between 150 and 175 and the background to be low between 1 and 40 (*see* **Note 13**).

9. Perform focus registration procedure either using "Hardware-based" or "Image-based" autofocus on the Total channel in a well near the center of the group of wells used (*see* **Note 14**). Avoid using the outer wells as a focus registration well.
10. Select the Live/Dead/Total channel to its corresponding fluorescence depending on the viability stain used (Table 4):
    (a) Blue: DAPI and Hoechst.
    (b) Green: AO and Calcein AM.
    (c) Red: PI and 7-AAD.
    (d) Far red: DRAQ5 and DRAQ7.
11. Set up the exposure time for each fluorescent channel depending on the fluorescent stains.
12. Adjust the exposure time until the cell object pixel intensity reaches between 150 and 175 and the background is low between 1 and 40.
13. Adjust and set Focus Offset for each additional fluorescent channel, in reference to Total channel or the focus registration channel.
14. Select the wells for image acquisition and analysis.
15. Click on "Start Scanning."
16. In the ANALYZE tab, set up analysis parameters for each channel.
17. Select a well that has been scanned and click on the Live/Dead/Total Image and Graphic Overlays to observe the counting of the target cells of each channel.
18. If the cells are not counted properly, adjust the parameters on the left from the "General," "Identification," and "Pre-filtering" sections to properly count the cells.
19. After the analysis parameters are set up, click on "Start Analysis." Scanning and analyzing microplates can be performed simultaneously.
20. In the RESULT tab, double-click the thumbnail images to observe the whole well images and counted cells.
21. Export the images and results (Fig. 3).

4 Notes

1. Placing the counting chamber on a clean surface will prevent debris from blocking the imaging area.
2. It is important to use high-quality microplates, such as Greiner 655090, because of consistent flatness between different wells and plates.

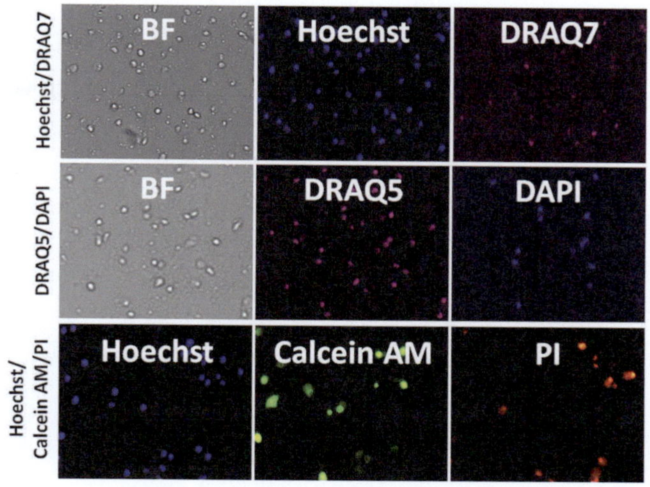

Fig. 3 Examples of bright-field and fluorescent images of Hoechst/DRAQ7, DRAQ5/DAPI, and Hoechst/Calcein AM/PI using Celigo image cytometer

3. Preparing the microplate in the BSL2 tissue culture hood will allow cells to be seeded and stained in a sterile environment.
4. Ensure that the fluorescent filters are selected that correspond to the viability stains used in chip-based image cytometers.
5. Ensure that the correct diluent is used for stain preparation. Reconstitute any powder with DMSO prior to diluting with diluent.
6. For chip-based image cytometers, suspension cells can be directly pipetted into the counting chambers. In contrast, adherent cells require trypsinization prior to pipetting into the counting chambers. Cell density should be between 1×10^5 and 1×10^7 cells/mL to fall in the linear range for counting with Cellometer.
7. Ensure that enough time is allowed for the cells to settle to the bottom to prevent moving of cells during image acquisition.
8. Final volume in a 96-well should be 200 μL to prevent meniscus from interfering with bright-field image analysis.
9. Cell densities for 96-well microplate should be around 15,000–30,000 for suspension, and 10,000–20,000 for adherent. If experiment is performed over multiple days, reduce the number of cells to accommodate cell proliferation, where cells are not overconfluent at the experiment endpoint.
10. To settle cells to the bottom of the plate, use a swinging bucket centrifuge that is balanced with a similar weighted plate. Bring speed up to $160 \times g$, then turn off, and let come to stop gradually. Sometimes, the cells may not completely settle; thus longer centrifugation time is required, which may push more cells towards the edge of the well.

11. Check the bright-field focus prior to image acquisition to make sure that the cells are in focus.
12. Check if the cells are counted (circled) correctly in the images.
13. The exposure time is dependent on the fluorescent stains used.
14. Hardware-based autofocus is used for plate types that are flat bottomed and highly consistent, while Image-based autofocus is used for plate types that may not be flat and wherein thickness is not consistent from well to well.

Declaration of Conflicting Interests

The authors, LLC, KM, and SK, declare competing financial interests. The work performed in this chapter is for reporting on product performance of Nexcelom Bioscience, LLC. The performed experiments were to demonstrate cell viability detection methods using image cytometry.

References

1. Cook JA, Mitchell JB (1989) Viability measurements in mammalian cell systems. Anal Biochem 179:1–7
2. Oh H, Livingston R, Smith K et al (2004) Comparative study of the time dependency of cell death assays. MURJ 11:53–62
3. Chan LL, Zhong XM, Qiu J et al (2011) Cellometer vision as an alternative to flow cytometry for cell cycle analysis, mitochondrial potential, and immunophenotyping. Cytometry Part A 79A:507–517
4. Chan LL, Wilkinson AR, Paradis BD et al (2012) Rapid image-based cytometry for comparison of fluorescent viability staining methods. J Fluoresc 22:1301–1311
5. Saldi S, Driscoll D, Kuksin D et al (2014) Image-based cytometric analysis of fluorescent viability and vitality staining methods for ale and lager fermentation yeast. J Am Soc Brew Chem 72:253–260
6. Han X, Liu Z, Mc J et al (2015) CRISPR-Cas9 delivery to hard-to-transfect cells via membrane deformation. Sci Adv 1:1–8
7. Shah D, Naciri M, Clee P et al (2006) NucleoCounter—an efficient technique for the determination of cell number and viability in animal cell culture processes. Cytotechnology 51:39–44
8. Al-Rubeai M, Welzenbach K, Lloyd DR et al (1997) A rapid method for evaluation of cell number and viability by flow cytometry. Cytotechnology 24:161–168
9. Strober W (2001) Monitoring cell growth. In: Current protocols in immunology, vol APPENDIX 3A
10. Shapiro HM (2004) "Cellular Astronomy" - a foreseeable future in cytometry. Cytometry Part A 60A:115–124
11. Stoddart M (2011) Cell viability assays: introduction. Methods Mol Biol 740:1–6
12. Davey HM, Kell DB (1996) Flow cytometry and cell sorting of heterogeneous microbial populations: the importance of single-cell analyses. Microbiol Rev 60:641–696
13. Michelson AD (1996) Flow cytometry: a clinical test of platelet function. Blood 87:4925–4936
14. Tibbe AGJ, de Grooth BG, Greve J et al (2002) Imaging technique implemented in CellTracks system. Cytometry Part A 47:248–255
15. Shapiro HM, Perlmutter NG (2006) Personal cytometers: slow flow or no flow? Cytometry Part A 69A:620–630
16. Gerstner AOH, Mittag A, Laffers W et al (2006) Comparison of immunophenotyping by slide-based cytometry and by flow cytometry. J Immunol Methods 311:130–138
17. Mital J, Schwarz J, Taatjes DJ et al (2005) Laser scanning cytometer-based assays for measuring host cell attachment and invasion by the human pathogen *Toxplasma gondii*. Cytometry Part A 69A:13–19
18. Hall A, Wu L-P, Parhamifar L et al (2015) Differential modulation of cellular bioenerget-

ics by poly(l-lysine)s of different molecular weights. Biomacromolecules 16:2119–2126

19. Siqueira-Neto JL, Moon S, Jang J et al (2012) An image-based high-content screening assay for compounds targeting intracellular *Leishmania donovani* amastigotes in human macrophages. PLoS Negl Trop Dis 6:e1671
20. Zanella F, Lorens JB, Link W (2010) High content screening: seeing is believing. Trends Biotechnol 28:237–245
21. Schepers K, Pietras EM, Reynaud D et al (2013) Myeloproliferative neoplasia remodels the endosteal bone marrow niche into a self-reinforcing leukemic niche. Cell Stem Cell 13:285–299
22. Szabo SE, Monroe SL, Fiorino S et al (2004) Evaluation of an automated instrument for viability and concentration measurements of cryopreserved hematopoietic cells. Lab Hematol 10:109–111
23. Macfarlane RG, Payne AM-M, Poole JCF et al (1959) An automatic apparatus for counting red blood cells. Br J Haemacytol 5:1–15
24. Verso ML (1971) Some nineteenth-century pioneers of haematology. Med Hist 15:55–67
25. Falzone N, Huyser C, Franken D (2010) Comparison between propidium iodide and 7-amino-actinomycin-D for viability assessment during flow cytometric analyses of the human sperm acrosome. Andrologia 42:20–26
26. Gordon KM, Duckett L, Daul B et al (2003) A simple method for detecting up to five immunofluorescent parameters together with DNA staining for cell cycle or viability on a benchtop flow cytometer. J Immunol Methods 275:113–121
27. Jarnagin JL, Luchsinger DW (1980) The use of fluorescein diacetate and ethidium bromide as a stain for evaluating viability of mycobacteria. Biotech Histochem 55:253–258
28. Roth B, Poot M, Yue S et al (1997) Bacterial viability and antibiotic susceptibility testing with SYTOX green nucleic acid stain. Appl Environ Microbiol 63:2421–2431
29. Wlodkowic D, Skommer J, Faley S et al (2009) Dynamic analysis of apoptosis using cyanine SYTO probes: from classical to microfluidic cytometry. Exp Cell Res 315:1706–1714
30. Bratosin D, Mitrofan L, Palii C et al (2005) Novel fluorescence assay using calcein-AM for the determination of human erythrocyte viability and aging. Cytometry Part A 66A:78–84
31. Jones KH, Senft JA (1985) An improved method to determine cell viability by simultaneous staining with fluorescein diacetate-propidium iodide. J Histochem Cytochem 33:77–79
32. Donoghue AM, Garner DL, Donoghue DJ et al (1995) Viability assessment of Turkey sperm using fluorescent staining and flow cytometry. Poult Sci 74:1191–1200
33. Mascotti K, McCullough J, Burger SR (2000) HPC viability measurement: trypan blue versus acridine orange and propidium iodide. Transfusion 40:693–696
34. Cai K, Yang J, Guan M et al (2005) Single UV excitation of Hoechst 33342 and propidium iodide for viability assessment of rhesus monkey spermatozoa using flow cytometry. Arch Androl 51:371–383
35. Smith PJ, Wiltshire M, Davies S et al (1999) A novel cell permeant and far red-fluorescing DNA probe, DRAQ5, for blood cell discrimination by flow cytometry. J Immunol Methods 229:131–139
36. Akagi J, Kordon M, Zhao H et al (2013) Real-time cell viability assays using a new anthracycline derivative DRAQ7®. Cytometry Part A 83A:227–234
37. Sutkeviciene N, Andersson MA, Zilinskas H et al (2005) Assessment of boar semen quality in relation to fertility with special reference to methanol stress. Theriogenology 63:739–747

Chapter 4

High-Throughput Spheroid Screens Using Volume, Resazurin Reduction, and Acid Phosphatase Activity

Delyan P. Ivanov, Anna M. Grabowska, and Martin C. Garnett

Abstract

Mainstream adoption of physiologically relevant three-dimensional models has been slow in the last 50 years due to long, manual protocols with poor reproducibility, high price, and closed commercial platforms. This chapter describes high-throughput, low-cost, open methods for spheroid viability assessment which use readily available reagents and open-source software to analyze spheroid volume, metabolism, and enzymatic activity. We provide two ImageJ macros for automated spheroid size determination—for both single images and images in stacks. We also share an Excel template spreadsheet allowing users to rapidly process spheroid size data, analyze plate uniformity (such as edge effects and systematic seeding errors), detect outliers, and calculate dose-response. The methods would be useful to researchers in preclinical and translational research planning to move away from simplistic monolayer studies and explore 3D spheroid screens for drug safety and efficacy without substantial investment in money or time.

Key words Alamar blue, Viability assays, Overlay culture, Hanging drop, FiJi ImageJ, Image analysis, Three-dimensional cell culture, In vitro model, Preclinical screening, Drug sensitivity

1 Introduction

The purpose of this chapter is to describe the detailed practical procedures behind three complementary techniques (volume, resazurin reduction, and acid phosphatase activity) for spheroid viability assessment in high-throughput 96-well format [1]. Spheroids are self-organized three-dimensional (3D) aggregates of cells displaying physiologically relevant gradients of oxygen, nutrients, and cell–cell and cell matrix interactions [2, 3]. Aggregate cultures were first described in the 1950s by Moscona [4], and the advantages of using spheroids in cancer research were recognized in the 1970s by Sutherland [5]. However, poor reproducibility, large variation in spheroid size, lengthy hand-operated manipulations, and low throughput precluded the introduction of spheroids in mainstream drug screens. The introduction of plate-based platforms for spheroid culture in hanging drop [6, 7] or liquid overlay

Daniel F. Gilbert and Oliver Friedrich (eds.), *Cell Viability Assays: Methods and Protocols*, Methods in Molecular Biology, vol. 1601, DOI 10.1007/978-1-4939-6960-9_4, © Springer Science+Business Media LLC 2017

[8, 9] has enabled researchers to produce a single spheroid per well and control spheroid size in a high-throughput format. This has stimulated the adoption of spheroid screens for drugs targeting dormant tumor cells [10], spheroid assay modeling chemo- and radio-resistance [11, 12] strategies for sensitizing hypoxic cells in the tumor core [13, 14], and drug safety assays [15, 16].

We present a suite of three multiplexable methods to assess spheroid health in overlay spheroid cultures. The plate format used is compatible with standard plate readers, and the methods rely on generic reagents available at lower costs compared to ready-made kits. The techniques use open-source software for image analysis (Fiji distribution of ImageJ [17, 18]) and do not require investments in new equipment or software. The practical application of these methods and the characterization of linearity and sensitivity have been discussed in our recent publication [1]. These methods are suitable for monocultures (one cell type) and cocultures (mixed cell types). In cocultures, the volume or metabolic activity of the coculture spheroid can be used as a proxy measure for the total number of cells; subsequently the proportion of each cell type can be quantified with microscopy or flow cytometry [19].

Manual spheroid size measurements and morphological characterization of 3D aggregates have been extensively used from the earliest days of spheroid research [4, 5, 20]. With the development of overlay and hanging-drop plates, many proprietary platforms (such as Celigo, Zeiss, Perkin-Elmer) have been employed in measuring spheroid size and estimating volume in high throughput [8, 21]. These platforms require substantial investment in equipment and image analysis software, which may not be available to all research labs and may slow down the adoption of spheroid screens. We have written two macros which automate spheroid size analysis on the bioimaging Fiji [17] distribution of the open-source image analysis platform ImageJ [18]. The first macro works on manually acquired images from simple setups of camera-equipped bright-field microscopes, where spheroid images are recorded as separate files in a folder. It is compatible with computers with less than 2GB of RAM which may struggle to load all images in a single image sequence. The second macro is substantially faster and works with image sequences (stacks), for example images taken with automated-stage microscopes, which often produce stacks of multiple images in one file. It can also be used on separate images imported as an image sequence in ImageJ (*see* **Note 10**). Estimating spheroid viability solely based on spheroid size can be misleading because:

1. Cell densities can be different in the different regions of the spheroid (e.g., hypoxic or necrotic core).
2. Cells can shrink without dying in response to drug treatment.
3. Dead cells can form a large proportion of the spheroid at drug concentrations above IC50.

Methods to alleviate these shortcomings use, for example, dyes which detect dead cells (propidium iodide, 7-aminoactinomycin-D, ethidium homodimer) or employ immunohistochemistry assays on fixed spheroids with markers of apoptosis (e.g., caspase-3). However, there are currently no high-throughput methods to probe spheroid histology and protein expression. Another approach would be to use metabolism-based assays for spheroid health assessment.

The resazurin (Alamar Blue) assay measures spheroid metabolism of live cells, which reduce resazurin to the fluorescent product resorufin [22]. If purchased as resazurin powder and prepared in-house, the assay is thousands of times cheaper than the ATP-based kits. The resazurin assay does not kill or lyse the spheroids and the latter can be harvested for dissociation and cell counts, flow cytometry, or histology 4h after resazurin addition [1, 19]. Its main limitation compared to the ATP assay is that both resazurin and the fluorescent resorufin need to distribute throughout the spheroid and back into the medium to be quantified. For spheroids where the cells form tight junctions and impede dye diffusion, treatment with chelators (ethylene glycol-bis(2-aminoethylether)-*N*,*N*,*N*′,*N*′-tetraacetic acid—EGTA) has been suggested to improve the reliability of the assay [23]. Regardless of dye diffusion, all metabolism-based assays (ATP included) would detect less metabolically active cells (e.g., in the hypoxic core) as "less alive" and can potentially underestimate dormant-cell viability. Furthermore, Chan et al. have demonstrated that certain anticancer compounds elicit a spike in metabolic activity at cytotoxic concentrations resulting in overestimation of viability [24].

The third assay described is the acid-phosphatase assay advocated by Friedrich [21, 25]. It measures acid phosphatase activity as a proxy for cell viability and involves lysing the cells, thus eliminating the problems with diffusion and substrate penetration. While excellent results have been reported with tumor cell lines with this assay [21, 25], it overestimated spheroid viability in etoposide-treated primary cultures of human fetal neurospheres [1].

We have used the three viability assays in a variety of human medulloblastoma cells lines [26], patient-derived early-passage medulloblastoma cultures, and human fetal brain-derived neurospheres [1, 19]. The linearity of assay signal in the physiologically relevant size range for each cell line has been determined by plotting assay readout against the number of cells per spheroid. The protocols below give an in-depth view of the practical requirements for optimal assay performance.

2 Materials

2.1 General Cell Culture Reagents

1. DMEM (Dulbecco's modified Eagle's medium): 4.5 g/L High glucose with 1 mM sodium pyruvate.
2. Ham's F12 medium with 25 mM 4-(2-hydroxyethyl)-1-piperazineethanesulfonic acid (HEPES).

3. 200 mM l-Glutamine solution.
4. Accutase cell-dissociation solution: Aliquot and store at −20 °C. Thaw in the fridge before use.
5. Fetal bovine serum (FBS).
6. Full cell culture medium: For 100 mL: 44.5 mL DMEM, 44.5 mL Ham's F12, 1 mL l-glutamine solution, 10 mL fetal bovine serum.
7. Phosphate-buffered saline (Ca^{2+} and Mg^{2+} free), sterile solution (PBS).
8. Cell culture-treated flasks.
9. Ultra-low-attachment round-bottom 96-well plates (Corning, 7007).
10. Plastic paraffin film.

2.2 Fetal Brain Culture

1. B27 supplement without insulin (50×).
2. N2 supplement (100×).
3. Heparin sodium salt from porcine intestinal mucosa (Sigma Aldrich, Grade I-A, cell culture tested): Prepare a 5 mg/mL stock solution in sterile-filtered deionized water and store at 2–8 °C for up to 6 months.
4. Human serum albumin: Lyophilized powder.
5. Recombinant human epidermal growth factor (EGF): Prepare a 20 μg/mL stock solution in sterile-filtered 0.3% wt/vol (human serum albumin in PBS), aliquot, and store at −20 °C for up to 1 year.
6. Recombinant human basic-fibroblast growth factor (FGF): Prepare a 10 μg/mL stock solution in sterile-filtered 0.3% wt/vol (human serum albumin in PBS), aliquot, and store at −20 °C for up to 1 year.
7. Neural stem cell medium: For 100 mL: Filter 47.5 mL DMEM and 47.5 mL F12 through a 0.22 μm filter and then add 2 mL B27, 1 mL N2, 100 μL heparin, 100 μL EGF stock (20 μg/mL), 100 μL FGF stock (10 μg/mL) (*see* **Note 1**).
8. Non-treated cell culture suspension flasks with filter cap, 25 cm^2.
9. Polystyrene conical tubes, 15 mL (*see* **Note 2**).

2.3 Resazurin

1. Hanks' Balanced Salt Solution (HBSS) with Ca^{2+} and Mg^{2+}.
2. Resazurin (high-purity biological stain quality). Resazurin stock solution: Dissolve 12 mg resazurin in 100 mL Hanks' Balanced Salt Solution with Ca^{2+} and Mg^{2+} (HBSS). Filter through 0.22 μm filter to sterilize. Aliquot and store protected

from light at −20 °C for up to a year. Thaw just before use and store any leftover stock solution at 2–8 °C for no more than 2 weeks. Avoid repeated freeze-thawing.

2.4 Acid Phosphatase

1. Triton-X 100 solution.
2. Para-nitrophenyl phosphate (PNPP, N4645, Bioreagent, Sigma-Aldrich). Store at −20 °C.
3. Sodium acetate 3 M solution: Dissolve 20.41 g sodium acetate trihydrate in 40 mL deionized water. Adjust pH to 5.2 with glacial acetic acid in the fume hood. Adjust volume to 50 mL and filter-sterilize (0.22 μm filter).
4. Sodium hydroxide 1 M: Make 4 g NaOH up to 100 mL with deionized water.
5. Acid phosphatase buffer (APH buffer): 0.1 M Sodium acetate and 0.1% Triton X (45 mL). Prepare 40 mL and store at 2–8 °C for up to 4 weeks. Add 1.5 mL of 3 M sodium acetate buffer to 45 μL Triton X. Make up to 45 mL with sterile deionized water.
6. APH assay solution: Prepare immediately before the assay. Dissolve 18 mg of 4-nitrophenyl phosphate in 9 mL APH buffer. Use the solution immediately. Do not store and protect from light.

3 Methods

3.1 Propagation of Fetal Brain Tissue as Neurospheres

1. Seed 1–2 × 10^6 cells in 5 mL NSC medium in a non-treated 25 cm^2 flask (*see* **Note 3**).
2. Feed the neurospheres by adding 2 mL of fresh medium every 2–3 days and change the medium and the flask if the neurospheres start attaching to the surface of the flask (*see* **Note 4**).
3. Dissociate the neurospheres every 7 days or when they reach a diameter of around 100–300 μm (*see* **Note 5**).

3.2 Dissociation of Neurospheres

1. Remove neurosphere cell suspension from culture flask, transfer to a 15 mL polystyrene conical tube, and centrifuge (50 × *g*, 30 s).
2. Aspirate medium leaving a maximum of 100 μL of medium remaining.
3. Resuspend neurospheres in 2–5 mL PBS and centrifuge (50 × *g*, 30 s).
4. Aspirate the PBS leaving a maximum of 100 μL of PBS remaining.

5. Resuspend the neurospheres in 1 mL Accutase and shake in a water bath (37 °C, 5 min).
6. Pre-wet a 1 mL pipette tip and pipette the cloudy neurosphere suspension up and down to get a single-cell suspension.
7. Add 2 mL medium to stop the dissociation reaction and centrifuge (300 × *g*, 5 min)
8. Gently aspirate the supernatant; resuspend cells in 200 μL PBS, using the 200 μL pipette; then add 800 μL fresh medium; and perform a cell count.

3.3 Propagation of Tumor Cell Lines

Culture as required by the cell line. For UW 228-3 medulloblastoma cells:

1. Seed 250,000–500,000 cells in a 75 cm^2 cell culture-treated flask, add fresh medium every other day, and completely change the medium every 4 days. Subculture to 75–85% confluency.
2. Dissociate by removing the medium, washing with PBS, incubating with 4 mL trypsin (5 min), stopping the reaction with 6 mL fresh medium, centrifuging (300 × *g*, 5 min), and resuspending in 1 mL PBS. Add 1–4 mL fresh medium and perform a cell count.

3.4 Spheroid Formation in Ultra-low-Attachment Plates (Day 0)

1. Prepare a single-cell suspension from the cell line/primary cells to be used and perform a cell count.
2. Pipette 200 μL/well PBS in all wells of lines A and H (top and bottom) of the 96-well plate (*see* **Note 6**).
3. Add 100 μL/well medium in column 1 (wells B to G).
4. Seed cells in 100 μL/well medium in columns 2–11 (wells B to G) (*see* **Note 7**).
5. Wrap each plate in plastic paraffin film along its perimeter to securely hold the lid and bottom together.
6. Centrifuge in a plate centrifuge (100 × *g*, 3 min), remove plastic paraffin film, and place in the incubator.

3.5 Optimizing Spheroid Seeding Density

1. Prepare serial dilutions of the cell suspensions with cell counts ranging from 5000 to 200,000 cells per mL.
2. Seed 100 μL from the different dilutions as six technical replicates per plate column.
3. Centrifuge plate (100 × *g*, 3 min) and place in incubator.
4. Monitor spheroid volume and perform resazurin and APH assay at the appropriate time point (see the acceptance criteria section for picking spheroid size) (*see* **Note 8**).

3.6 Spheroid Drug Sensitivity Screens

1. Prepare a single-cell suspension of the primary cells or cell line of interest and dilute to the predetermined seeding concentration.
2. Seed 100 μL/well medium without cells in column 1 and 100 μL/well of the cell suspension in columns 2–12.
3. Centrifuge plate (100 × *g*, 3 min) and place in incubator for the preset spheroid growth period (3–4 days).
4. Image plates on day 1 and every other day if needed. Image just before drug addition.
5. On day 3 or 4, prepare a 2× serial dilution series of the drug to be analyzed in medium. Use nine half-logarithmically spaced concentrations (0.3, 1, 3 μM, etc.) centered along the estimated IC50 or the potency values reported in the literature (*see* **Note 9**).
6. Add 100 μL to the appropriate wells as shown in Table 1.
7. Incubate for 48–72 h, and exchange medium (refresh drug solutions every 48–72 h, depending on drug stability in medium).
8. Proceed to volume, resazurin, and APH analysis.

3.7 Spheroid Volume Analysis

1. Image spheroids from day 1 onwards (24 h after seeding) when required.
2. Collect phase-contrast bright-field images either manually or with a microscope equipped with an automated programmable stage (preferred).
3. Download the Fiji distribution of ImageJ (http://fiji.sc/) and start the program.

3.7.1 Image Analysis Steps for Macro Ver1 (Slower)

Use for manually collected images (collection of spheroid images in a folder)

1. Load macro by dragging and dropping the **Macro1.ijm** file onto the main ImageJ window. Alternatively open the Macro editor (*Plugins/New/Macro* paste the **macro1** text, and select "*Language/ImageJ1 macro.*"
2. Measure the scale of your image (number of pixels equal to 100 μm) and enter the value after "*distance=*" in line 6 in macro1.
3. Start the macro by pressing the *Run* button in the Macro editor window. There should be no images open when starting the macro.
4. Select an input folder containing the original spheroid images. The input folder should only contain images and no other files.

Table 1
Example plate template including medium-only (contamination) control, untreated (negative or solvent) control, and complete cell death (positive) control

PBS wells											
Medium only	Untreated control (drug solvent)	$0.01 \times \mathbf{C}$	$0.03 \times \mathbf{C}$	$0.1 \times \mathbf{C}$	$0.3 \times \mathbf{C}$	Conc **C** = Reported IC50	$3 \times \mathbf{C}$	$10 \times \mathbf{C}$	$30 \times \mathbf{C}$	$100 \times \mathbf{C}$	25% DMSO (complete cell kill control)
PBS wells											

The middle concentration (**C**) can be chosen based on literature-reported values or approximate potency within an order of magnitude (e.g., micromolar, nanomolar)

5. Select an output folder which is going to contain the quality control images with spheroid outlines.
6. Wait for the macro to finish; progress is displayed in the status bar of the Main window.
7. Analyze the results as described below.

3.7.2 Image Analysis Steps for Macro Ver2 (Faster)

Use for automatically acquired images in an image sequence (collection of images in a single image file—image stack or hyperstack)

1. Load macro by dragging and dropping the *Macro2.ijm* file onto the main ImageJ window. Alternatively open the Macro editor (Plugins/New/Macro), paste the *macro2* text, and select "*Language/ImageJ1 macro.*"
2. Load image stack (sequence) containing spheroid images. Drag the image file onto ImageJ's main window or open using "*File/Import/BioFormats.*" In the BioFormats Import options use the "*Hyperstack*" option and tick "*Open all series,*" "*Concatenate series…,*" and "*Autoscale.*" A spheroid image sequence appears with a slider at the bottom to view all spheroid images (*see* **Notes 10** and **11**).
3. If the image displays size in pixels, rather than metric units, apply the correct predetermined scale ("*Analyze/Set Scale*").
4. Start the macro by pressing the *Run* button in the Macro editor window (*see* **Note 12**).
5. A window will appear showing an overlay of all spheroids. Use the rectangular selection tool to select the smallest area of the image containing all spheroids and press "*OK*" in the "*Action Required*" window (*see* **Note 13**).
6. A second window of the spheroid image sequence will appear and another "Action required" pop-up window will prompt you to select a threshold. Go to *Image/Adjust/Threshold* (shortcut: Select the image window and press Shift + T). (For manual thresholding continue to **step 7**; for the code to automate the next two steps *see* **Note 14**.)
7. Use the sliders to select a threshold which identifies all spheroids (red) and selects little background (*see* **Note 15**).
8. Press Apply in the "*Select Threshold window,*" and then answer "*Yes*" in the window asking "*Process all images?*" Some Fiji versions (e.g., 64-bit) display a "Convert stack to binary" window, where you just need to click "OK." Afterwards press "*OK*" in the "*Action required*" window (*see* **Note 16**).
9. Use the ROI manager to cycle through all outlined spheroids in the stack, check for failed image recognition, and remove false-positive spheroid outlines (*see* **Note 17**). The macro will produce two new files in the same folder as the original image sequence: a "Filename Quality control (QC)" file containing

the spheroids along with their recognized outlines and a "Filename Results" file containing all measurements.

3.7.3 Analyzing the Spheroid Measurements from ImageJ in Excel (See Note 18)

To streamline data analysis we have developed an Excel template which calculates the radius and volume of an equivalent sphere from spheroid area. The template is optimized for plate designs incorporating six consecutive technical replicates, first condition untreated control, and up to ten treatment conditions.

Measurement values from the ImageJ analysis need to be pasted in cells B1 to F67. The panels in blue are calculated values based on the measured spheroid area (*see* **Note 19**).

1. Open the "Filename Results.csv" file; select columns "FeretX," "FeretY," and "FeretAngle"; and delete them leaving only the column with image numbers, and the ones labeled "Label," "Area," "Feret," and "MinFeret" in columns "A" to "E" in Excel.
2. Click on any cell containing data, select all cells with data (Ctrl + A), and copy (Ctrl + C).
3. Open the "Volume analysis.xlsx" template file.
4. Paste the copied values from the Results file into the template by selecting cell B1 and pasting from the clipboard (Ctrl + V) (*see* **Note 20**).
5. Insert empty rows for missing spheroids in columns C to F (Ctrl Shift +), remove outliers and paste data in GraphPad Prism.

3.8 Spheroid Metabolism Analysis (Resazurin Assay)

1. Thaw the resazurin stock solution.
2. Prepare a working resazurin solution (12 mL per plate) by adding 1.6 mL resazurin stock to 10.4 mL of fresh medium. Wrap in aluminum foil to protect from light.
3. Remove spent medium from the plate wells to leave 50 μL in each well.
4. Add 150 μL from the working resazurin solution to all wells.
5. Leave the plate in the incubator (37 °C, 4 h).
6. Read fluorescence with a plate reader at 530–570 nm excitation and 580–600 nm emission (*see* **Note 21**).

3.9 Acid Phosphatase Assay

1. Remove old medium from the plate wells leaving a maximum of 50 μL and replace with 150 μL PBS.
2. Repeat **step 1**.
3. Remove 100 μL of PBS from each well, leaving 100 μL of PBS.
4. Add 100 μL fresh APH assay solution.
5. Leave in the incubator (37 °C, 90 min).
6. Add 10 μL 1 M NaOH solution to each well and record absorbance (405 nm) within 10 min.

3.10 Dissociating Spheroids to Determine the Number of Cells Per Spheroid.

1. Collect 6–12 technical replicate spheroids per condition in 1.5 mL centrifuge tubes.
2. Leave the spheroids to settle to the bottom of the tube (1–2 min) or centrifuge (100 × *g*, 30 s).
3. Replace culture medium with PBS and centrifuge (100 × *g*, 30 s).
4. Add 200 μL Accutase and incubate at 37 °C (5–20 min) with shaking.
5. Using a 200 μL pipette take up and dispense the cloudy suspension up and down to get a single-cell suspension.
6. Stop the dissociation reaction by adding 800 μL fresh medium and centrifuge (200 × *g*, 5 min).
7. Remove the supernatant and resuspend in 50–100 μL of suitable Ca- and Mg^{2+}-free buffer (PBS) to perform a cell count.

3.11 Data Analysis in GraphPad Prism (Versions 6 and 7)

1. Open GraphPad prism and select an X–Y Graph with six replicates for Y.
2. Write "[Drug name]" for "X-title" and paste the decimal logarithms of the concentrations from the Excel template file cells M16 to M26 in "X."
3. Name Group A with the experiment or cell line name and paste the individual well readings for volume, resazurin, or acid phosphatase in "Y1 to Y6."
4. Normalize data by clicking "*Analyze/Normalize*" and define 100% as untreated control and 0% as positive control (25% DMSO-treated wells, 0 for volume measurements).
5. Fit a sigmoidal dose-response curve: "*Analyze/Nonlinear Regression/Sigmoidal, 4P, X is log (concentration).*" Fitting method: *least squares*, constrain top to 100, constrain bottom >0, adjust range to fit points only up to the highest concentration of the drug used.
6. Inspect curve and reject if inappropriate or goodness-of-fit value "R square" is lower than 0.7.

3.12 Acceptance Criteria

1. The readouts from the APH and resazurin assay would only be useful if there is good separation between the signal in wells containing spheroids from the background of medium-only wells or wells with dead spheroids (positive cell death control, 25% DMSO). Use the readout values for the untreated (medium-only or drug carrier) and the positive cell death controls (25% DMSO) to calculate the *Z*-factor [27]:

$$Z = 1 - \frac{3 * \left(SD_{untreated\_control} + SD_{positive\_control}\right)}{Mean_{untreated\_control} - Mean_{positive\_control}}.$$

Acceptable separation between the two readouts is considered when $Z > 0.4$. For spheroid size optimization select spheroid sizes which consistently give signals with Z-factors above 0.4 compared to cell-free medium controls.

2. For volume, resazurin and APH-accept readouts with coefficient of variation <20% after outlier elimination (empty wells).

4 Notes

1. Both EGF and FGF are unstable in cell medium and the latter should be used within a week or 2 after making.
2. Fetal brain cells can adhere to polypropylene plastics. Use only polystyrene tubes when handling these cells.
3. Always pre-wet tips and pipettes before aspirating the spheres to prevent them from sticking to the plastic.
4. Healthy cultures without extensive cell death do not require complete medium exchange. If there are many single cells 72 h post-seeding or extensive amounts of debris it is recommended to change the medium completely by gently spinning the small neurospheres in a 15 mL polystyrene conical tube (100 × g, 3–4 min). Often some of the spheres will try to attach to the surface of the flask but they can be dislodged by gently tapping the flask every 1–2 days or through adding fresh medium and gently washing the adherent spheres. In cases of extensive attachment change both the medium and the flask.
5. If there is a large discrepancy in sphere sizes you can separate the small ($d < 200$ μm) spheres from the large ($d > 500$ μm) by placing the spheroid suspension in a 15 mL polystyrene conical tube leaving the spheres to settle for 1 min. The supernatant will contain small spheres and the large ones will settle to the bottom. The large spheres can be dissociated and the small ones can be cultured until they reach $d \approx 300$ μm.
6. The outer wells of the plate can suffer from increased evaporation, affecting medium osmolarity and cell viability. That is why it is recommended to fill them with PBS (lines A and H), and use them as cell-free medium controls (column 1, wells B to G) or complete cell kill controls (column 12, wells B to G).
7. We use six technical replicates per plate as often spheroids can be lost in medium exchanges.
8. Most often spheroid size of 300–500 μm is chosen based on the diffusion distance of oxygen and the formation of nutrient gradients. However the formation of hypoxic and nutrient-deprived regions would depend on the cell type, extracellular matrix produced, presence of tight junctions, and nature of the

cell–cell interactions. These differ for each cell line and hypoxia can be ascertained by staining for markers (hypoxia-inducible factor-1α, pimonidazole, carbonic anhydrase-IX).

9. Aim for at least two drug concentrations with little effect on cell viability (top for cytotoxic drugs) and at least two with the maximum effect (bottom) to get a good curve fit in the non-linear regression analysis. Make sure that the concentration of drug solvent is the same in every well and kept below 0.5% for DMSO. Also note that DMSO can deactivate some drugs such as the platinum complexes [28].
10. This macro can also work on separate images in a folder by importing them as an image sequence by selecting "*File/Import/Image Sequence.*"
11. The macro tries to analyze the most recently opened file and that is why users should open the macro first and then the image.
12. The only open image should be the sequence you want to analyze.
13. The macro will produce a minimum-intensity projection of all images in the stack to help you identify the smallest area of the image which contains all spheroids. A pop-up window will appear prompting you to select an area to crop.
14. Thresholding has been left to manual selection in this macro but it can be automated if the same pixel intensity values recognize the spheroids in a series of image sequences. This further speeds up the process and allows the user to skip **steps** 7 and **8**. To use automatic thresholding replace lines 28–31 of the macro with the following code:

    //This is an example of automatic thresholding using the "Yen" algorithm and selecting all pixels with intensity values from 234 to 1222 as spheroid on a light background:

    ```
    run("Threshold...");
    setAutoThreshold("Yen");
    setThreshold(234, 1222);
    setOption("BlackBackground", false);
    run("Convert to Mask", "stack");
    ```
    //end of automated thresholding code
15. Untick "*Dark background*," and try different thresholding algorithms to see which one does the best job in recognizing spheroids from background in your particular imaging set. The macro can be tweaked to recognize spheroids within a certain size limit in line 39:

    ```
    run("Analyze Particles...", "size=50000-Infinity show=[Overlay Outlines] display summarize add in_situ stack");
    ```

The default setting would recognize all spheroids with area larger than 50,000 μm² ($d \approx 250$ μm). The user can change those values to suit their spheroid size and eliminate small specks of false-positive dark background or to include spheroids smaller than 250 μm.

16. The "*Process all images?*" window can appear hidden behind the "*Action required*" window. Always press "*Yes*" in the former before clicking "*OK*" in the latter. Do not close the Threshold window manually; the macro will close it automatically after you press OK in the "*Action required*" window.
17. You can remove incorrectly recognized spheroids by selecting and deleting their outlines in the ROI manager. Missed spheroids can be outlined with the selection tools and added to the ROI manager (keyboard shortcut T). Update spheroid outlines on the image by selecting "Image/Overlay/Remove overlay" and "Image/Overlay/From ROI manager." If you make any deletions or additions to the outlines in the ROI manager, remeasure the spheroids by selecting all ROIs. In the ROI manager window click on one of the serial numbers, press Ctrl + A, and then press the Measure button. Save the results file; the first number in the label corresponds to slide number.
18. The macros produce image files in the output folder (macro1) or a single quality control "Filename QC.tiff" image in the parent directory (macro2). They also save a "Filename Results" file, which can be opened in Excel. The most useful measurement outputs are spheroid area, Feret (maximum) diameter, and MinFeret (minimum) diameter. Spheroid area is a more robust (less variable) measure of spheroid size compared to the maximum and minimum diameters and we have used it throughout our experiments to calculate spheroid diameter and volume of an equivalent sphere.
19. The top-right panel (N2 to X7) reconstructs the plate arrangement and color-codes spheroid volume values compared to the median spheroid volume for the whole plate. White cells encode empty wells, green cells are used for spheroids with 30% larger volume than the median plate volume, and red cells are for spheroids smaller than 70% of the median spheroid volume for the whole plate. This panel can be used to determine plate uniformity and the reproducibility of seeding the same number of cells in each well. The panel is useful before drug additions (days in vitro 1–4) and can identify patterns of uneven cell seeding (drift, edge effects, or other systematic sources of variability).

    The bottom-right section of the spreadsheet (L14 to AF38) is used when multiple conditions are compared, for example in a dose-response assay. The template uses the first six replicates as the untreated control (used to normalize as 100%) and 0 for no spheroids (0% when normalized). It calculates the average and the median volumes for each condition as well as the coefficient of variation, normalized volumes, and area under the dose-response curve (AUC). The AUC is given as a percentage fraction of the whole dose-response area (the AUC if all conditions had 100% viability).

Panels 016 to T26 are color-coded red (larger) or blue (smaller) for spheroids deviating in volume with more than two standard deviations from the average volume for each condition.

Panels N29 to T39 show the relative deviation percentage from the mean for each condition, with values above 20% appearing as amber and above 30% as red. Both panels can be used to identify outliers and eliminate them from the analysis. The decision to eliminate outliers has to be based on documented data of the normal variation in size for the particular cell line and seeding procedure.

The drug concentrations in L16 to L26 can be changed to match the experiment in question. Untreated control wells are given a "virtual concentration value" 2 decimal logs (100 times) lower than the lowest drug concentration and complete kill "20% DMSO" wells are given a "virtual concentration value" of 2 decimal logs (100 times) higher than the highest drug concentration used. These virtual concentrations will not affect the results of the nonlinear regression analysis as long as the curve fit is capped at the highest real drug concentration used.

20. In case of wells with missing spheroid or wells with multiple spheroids the number in "column A" would be different from the number in "(Series…)" in column B. For the correct operation of the template both numbers need to match.

21. If the plate reader has a temperature-controlled chamber let it warm up to 37 °C before the measurements. Alternatively, wrap the plates in aluminum foil and leave them to equilibrate to room temperature for 20 min before measuring fluorescence.

Acknowledgments

Delyan Ivanov was supported by an EPSRC Doctoral Prize award hosted by the University of Nottingham (DP2014/DI). The authors would like to thank Pamela Collier and Alan McIntyre for their help with manuscript editing. Special thanks to Neli Garbuzanova, Janhavi Apte, Arundhati Dongre, Parminder Dhesi, and Amarnath Pal for testing the macros and providing user feedback.

Supplementary Files

The macro files and Excel spreadsheet are available through the Figshare database:

Macro 1 link: https://figshare.com/s/32f81784ee28e3fde015 (DOI: 10.6084/m9.figshare.3487919).

Macro 2 link: https://figshare.com/s/9952d072c3238a60e134 (DOI: 10.6084/m9.figshare.3487943).

Volume analysis template: https://figshare.com/s/6c57cede1d940f6fd952 (DOI: 10.6084/m9.figshare.3487940).

References

1. Ivanov DP, Parker TL, Walker DA et al (2014) Multiplexing spheroid volume, resazurin and acid phosphatase viability assays for high-throughput screening of tumour spheroids and stem cell neurospheres. PLoS One 9:e103817
2. Astashkina A, Mann B, Grainger DW (2012) A critical evaluation of in vitro cell culture models for high-throughput drug screening and toxicity. Pharmacol Ther 134:82–106
3. Hickman JA, Graeser R, de Hoogt R et al (2014) Three-dimensional models of cancer for pharmacology and cancer cell biology: capturing tumor complexity in vitro/ex vivo. Biotechnol J 9:1115–1128
4. Moscona A, Moscona H (1952) The dissociation and aggregation of cells from organ rudiments of the early chick embryo. J Anat 86:287–301
5. Sutherland RM, McCredie JA, Inch WR (1971) Growth of multicell spheroids in tissue culture as a model of nodular carcinomas. J Natl Cancer Inst 46:113–120
6. Tung Y-C, Hsiao AY, Allen SG et al (2011) High-throughput 3D spheroid culture and drug testing using a 384 hanging drop array. Analyst 136:473–478
7. Kelm JM, Timmins NE, Brown CJ et al (2003) Method for generation of homogeneous multicellular tumor spheroids applicable to a wide variety of cell types. Biotechnol Bioeng 83:173–180
8. Vinci M, Gowan S, Boxall F et al (2012) Advances in establishment and analysis of three-dimensional tumor spheroid-based functional assays for target validation and drug evaluation. BMC Biol 10:1–21
9. Ivascu A, Kubbies M (2006) Rapid generation of single-tumor spheroids for high-throughput cell function and toxicity analysis. J Biomol Screen 11:922–932
10. Wenzel C, Riefke B, Gründemann S et al (2014) 3D high-content screening for the identification of compounds that target cells in dormant tumor spheroid regions. Exp Cell Res 323:131–143
11. Falkenberg N, Höfig I, Rosemann M et al (2016) Three-dimensional microtissues essentially contribute to preclinical validations of therapeutic targets in breast cancer. Cancer Med 5:703–710
12. Anastasov N, Höfig I, Radulović V et al (2015) A 3D-microtissue-based phenotypic screening of radiation resistant tumor cells with synchronized chemotherapeutic treatment. BMC Cancer 15:466
13. da Motta LL, Ledaki I, Purshouse K et al (2016) The BET inhibitor JQ1 selectively impairs tumour response to hypoxia and downregulates CA9 and angiogenesis in triple negative breast cancer. Oncogene 36(1):122–132
14. McIntyre A, Hulikova A, Ledaki I et al (2016) Disrupting hypoxia-induced bicarbonate transport acidifies tumor cells and suppresses tumor growth. Cancer Res 76:3744–3755
15. Bell CC, Hendriks DFG, Moro SML et al (2016) Characterization of primary human hepatocyte spheroids as a model system for drug-induced liver injury, liver function and disease. Sci Rep 6:25187
16. Ivanov DP, Al-Rubai A, Grabowska AM et al (2016) Separating chemotherapy-related developmental neurotoxicity from cytotoxicity in monolayer and neurosphere cultures of human fetal brain cells. Toxicol In Vitro 37:88–96
17. Schindelin J, Arganda-Carreras I, Frise E et al (2012) Fiji: an open-source platform for biological-image analysis. Nat Methods 9:676–682

18. Schindelin J, Rueden CT, Hiner MC et al (2015) The ImageJ ecosystem: an open platform for biomedical image analysis. Mol Reprod Dev 82:518–529
19. Ivanov DP, Parker TL, Walker DA et al (2015) In vitro co-culture model of medulloblastoma and human neural stem cells for drug delivery assessment. J Biotechnol 205:3–13
20. Sutherland RM, Eddy HA, Bareham B et al (1979) Resistance to adriamycin in multicellular spheroids. Int J Radiat Oncol Biol Phys 5:1225–1230
21. Friedrich J, Seidel C, Ebner R et al (2009) Spheroid-based drug screen: considerations and practical approach. Nat Protoc 4:309–324
22. O'Brien J, Wilson I, Orton T et al (2000) Investigation of the Alamar Blue (resazurin) fluorescent dye for the assessment of mammalian cell cytotoxicity. Eur J Biochem 267: 5421–5426
23. Walzl A, Unger C, Kramer N et al (2014) The resazurin reduction assay can distinguish cytotoxic from cytostatic compounds in spheroid screening assays. J Biomol Screen 19: 1047–1059
24. Chan GKY, Kleinheinz TL, Peterson D et al (2013) A simple high-content cell cycle assay reveals frequent discrepancies between cell number and ATP and MTS proliferation assays. PLoS One 8:e63583
25. Friedrich J, Eder W, Castaneda J et al (2007) A reliable tool to determine cell viability in complex 3-d culture: the acid phosphatase assay. J Biomol Screen 12:925–937
26. Ivanov DP, Coyle B, Walker DA et al (2016) In vitro models of medulloblastoma: choosing the right tool for the job. J Biotechnol 236: 10–25
27. Zhang J-H (1999) A simple statistical parameter for use in evaluation and validation of high throughput screening assays. J Biomol Screen 4:67–73
28. Hall MD, Telma KA, Chang K-E et al (2014) Say no to DMSO: dimethylsulfoxide inactivates cisplatin, carboplatin, and other platinum complexes. Cancer Res 74:3913–3922

Chapter 5

A Protocol for In Vitro High-Throughput Chemical Susceptibility Screening in Differentiating NT2 Stem Cells

Ann-Katrin Menzner and Daniel F. Gilbert

Abstract

The incidence of neurological diseases including learning and developmental disorders has increased in recent years. Concurrently, the number and volume of worldwide registered and traded chemicals have also increased. There is a broad consensus that the developing brain is particularly sensitive to damage by chemicals and that evaluation of chemicals for developmental toxicity or neurotoxicity is critical to human health. Human pluripotent embryonal carcinoma (NTERA-2 or NT2) cells are increasingly considered as a suitable model for in vitro developmental toxicity and neurotoxicity (DT/DNT) studies as they undergo neuronal differentiation upon stimulation with retinoic acid (RA) and allow toxicity assessment at different stages of maturation. Here we describe a protocol for cell fitness screening in differentiating NT2 cells based on the analysis of intracellular ATP levels allowing for the identification of chemicals which are potentially harmful to the developing brain. The described method is suitable to be adapted to low-, medium-, and high-throughput screening and allows multiplexing with other cell fitness indicators. While the presented protocol focuses on cell fitness screening in human pluripotent stem cells it may also be applied to other in vitro models.

Key words High-throughput screening, Cell-based assays, Cell viability, Developmental toxicity and neurotoxicity, DNT, Chemical susceptibility, Differentiation, Neurotoxicity, Cell fitness, Drug screening, Drug discovery, Target validation, In vitro toxicity screening, CellTiter-Glo Luminescent Cell Viability Assay

1 Introduction

The incidence of learning and developmental disorders including attention-deficit or hyperactivity disorder, autism, mental retardation, or other neurological diseases has increased in recent years [1–4]. Concurrently, the number and volume of worldwide registered and traded chemicals have also risen. There is a broad consensus that the developing central nervous system is particularly sensitive to damage by chemical substances [5] and that evaluation of chemicals for developmental toxicity (DT) or neurotoxicity (DNT) is critical to human health [6, 7]. Nevertheless, only a minority of chemicals have yet been tested for DNT [8, 9],

Daniel F. Gilbert and Oliver Friedrich (eds.), *Cell Viability Assays: Methods and Protocols*, Methods in Molecular Biology, vol. 1601, DOI 10.1007/978-1-4939-6960-9_5,

presumably because first there is no legal obligation for DNT testing and second the present guidelines for DNT testing involve animal experimentation [10] (US EPA 712-C-98-239; US EPA, 1998) that is of poor reproducibility and predictive quality, low in throughput, and cost intensive as well as highly limited regarding mechanistic insights into the toxicant's mode of action [11]. For the aforementioned reasons, in vitro-based alternatives to existing animal-dependent methods allowing for systematic DNT studies in high-throughput and large-scale format are urgently needed [12].

Human pluripotent teratocarcinoma (NTERA-2 or NT2) cells are increasingly considered as a suitable model for in vitro DT and DNT studies [13–18]. Upon treatment with retinoic acid, NT2 cells mimic the process of neuronal differentiation in the maturing brain and thus may allow testing for toxic effects at various developmental stages ranging from non-differentiated stem cells and committed neural progenitors to differentiated neuronal cells [19–24].

NT2 and other stem cells have been reported to show specific changes in chemical susceptibility during differentiation potentially leading to consequences with significant impact on the cell's fate, including cellular viability (e.g., metabolic activity), cell death (e.g., necrosis or apoptosis), or ability of the cells to undergo neuronal differentiation [25]. Thus, studying chemical susceptibility in differentiating stem cells in general and in human pluripotent NT2 cells in particular can be of great value in the context of developmental toxicity or neurotoxicity testing and might contribute to a test battery of experimental strategies for stage-specific DNT prediction.

Here, we describe a protocol for elucidating chemical susceptibility in differentiating NT2 cells based on the analysis of cellular viability. The assay quantifies the viability of chemical-exposed differentiating NT2 cells using intracellular ATP as readout.

Intracellular ATP is an energy carrier driving many cell functions. Cell death is typically indicated by low ATP levels and persistent ATP depletion causes a cell to die. Due to its simple applicability, e.g., by an ATP-dependent luciferase-luciferin reaction (e.g., CellTiter-Glo Luminescent Cell Viability Assay Kit from Promega), the intracellular ATP level has been a long-serving, convenient, and robust indicator of cellular viability. Although the approach is based on analysis of metabolic activity, assessed by a specific luminescence-based reporter system, it could also be replaced by as well as combined with any other luminescence- or fluorescence-based and biochemical cell viability indicator. Examples are cellular integrity (membrane integrity: propidium iodide), cell proliferation (nucleus stain: Hoechst 33342), or metabolic activity (reductase-enzyme product: CellTiter 96 Aqueous One Solution Cell Proliferation Assay; ATP content: CellTiter-Glo Luminescent Cell Viability Assay Kit; esterase activity: Calcein-AM).

It is important to note that the fitness indicator employed in the present assay does not clearly and exclusively indicate cellular viability comprehensively, as it yields information on a very specific physiological condition that only indirectly and partially reflects cellular fitness. Although this limitation can be overcome by multiplexing, i.e., the combination of multiple individual cell viability assaying methods in a single experiment, adding a level of efficiency and confidence for hit selection [26, 27], the presented methodology on its own is not suitable to reliably predict DT and/or DNT.

2 Materials

Prepare all solutions using ultrapure water (prepared by purifying deionized water, to attain a sensitivity of 18 MΩ-cm at 25 °C) and analytical grade reagents. Prepare and store all reagents at room temperature (unless indicated otherwise). Diligently follow all waste disposal regulations when disposing waste materials.

2.1 Molecule Library

1. Chemical stock solutions: 10 mM Small molecules (e.g., NIH CC Library or other small-molecule library) in DMSO, prealiquoted into 96-well plates (*see* **Note 1**).
2. Dimethyl sulfoxide.
3. RPMI-1640.
4. 96-Well plates.
5. Aluminum sealing tapes.
6. MilliQ Millipore water.
7. Liquidator 96 liquid-handling system.
8. 200 μl Racked pipet tips.

2.2 Cell Culture

1. NTERA-2 cl. D1, CRL-1973.
2. T75 Flask with filter cap.
3. Cell culture medium: Dulbecco's modified Eagle's medium with phenol-red, 10% fetal bovine serum (FBS), 100 U/ml penicillin, 100 mg/ml streptomycin.
4. Phosphate-buffered saline (PBS): Sterile Ca^{2+}- and Mg^{2+}-free phosphate-buffered saline.
5. Harvesting solution: 0.05% Trypsin containing 0.025% EDTA.
6. Retinoic acid stock solution (RA): 10 mM in ultrapure water.
7. Screening control solution: 1% DMSO in RPMI.
8. 384-Well plates.
9. Hemocytometer.

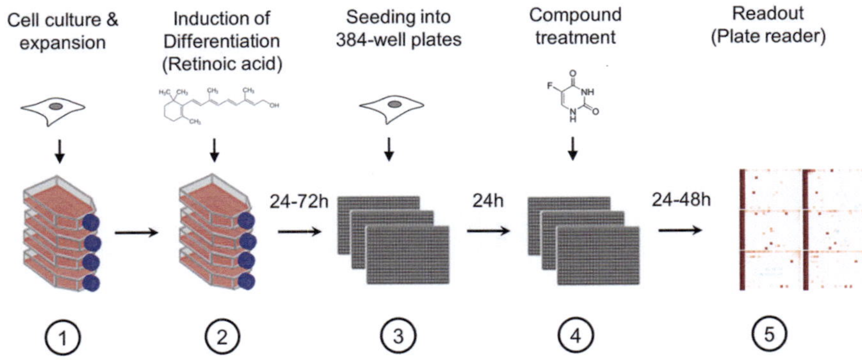

Fig. 1 Work flow for chemical susceptibility screening in early-differentiating NT2 cells. For high-throughput chemical susceptibility screening, NT2 stem cells are expanded in tissue culture flasks until 80–90% confluent (*1*) and are treated with 10 μM retinoic acid (RA) to induce differentiation (*2*). After 24–72 h of RA exposure, cells are seeded into 384-well plates at defined density and are incubated for 24 h (*3*). Chemicals are added to the cells at a final concentration of 10 μM and are incubated for 24–36 h (*4*). Chemical susceptibility in early-differentiating NT2 cells is evaluated via intracellular ATP level using the CellTiter-Glo Luminescent Cell Viability Assay Kit and a plate luminometer (*5*). Reproduced with permission from [25]

2.3 Chemical Susceptibility Screening

Protect the luminescence indicator from light

1. DMEM without phenol red.
2. 24-Channel wand.
3. Multidrop Combi dispensing system with standard tube dispensing cassette.
4. CellTiter-Glo Luminescent Cell Viability Assay Kit (CTG, Promega).
5. CTG solution: 1:1.3 mixture of CTG and DMEM *without* phenol red.

3 Methods

All experimental steps are performed under sterile conditions if not stated otherwise. All equipment used under the sterile bench are sterilized with 70% ethanol before usage. All experimental steps are performed at room temperature and all incubation steps are performed at 37 °C, 5% CO_2 in a humidified incubator, unless indicated otherwise. *See* Fig. 1 for screening workflow. Conduct screening experiments in duplicate or in more replicates.

3.1 Preparation of Chemicals

1. Dilute chemical stock solutions as provided by supplier, e.g., 10 mM in DMSO, for chemical susceptibility screening to a concentration of 100 μM in RPMI (*see* **Note 1**) and distribute diluted chemicals into a new 96-well so-called molecule plate, using the Liquidator liquid-handling system (*see* **Note 2**).

Fill columns 2–11 with diluted chemicals and add screening control solution to columns 1 and 12 for positive and negative controls.

2. Cover plates with aluminum sealing tapes and store at −80 °C until use.

3.2 Cell Culture, Harvesting, and Counting

1. Expansion: Seed 10^6 cells, suspended in 10 ml cell culture medium in T75 (75 cm^2 growth area) flask, and culture cells at 37 °C, 5% CO_2, in a humidified incubator according to standard procedures. Replace medium every 2–3 days. Use cells for screening experiments when approximately 80% confluent (*see* **Note 3**).
2. Harvesting: Remove the medium and wash cells gently with 5 ml PBS. Remove PBS and replace with 2 ml harvesting solution. Incubate cells for 2–4 min. Add 8 ml culture-screening medium and dislodge cells from the flask by pipetting approx. ten times up and down using a 10 ml serological pipet (*see* **Note 4**).
3. Counting: Dilute cell suspension 1:10 in culture-screening medium. Add 10 μl diluted cell suspension to both sides of a hemocytometer and count the number of cells in each of the eight (2 × 4) grids (*see* **Note 5**). Calculate concentration of cells per ml using the following formula: cells per ml = (cell count in eight squares/8) × dilution factor × 10^4, where dilution factor = 10; 10^4 = factor given by the manufacturer.
4. Adjust the cell concentration to either 1.11×10^5 (no RA pretreatment/non-differentiating) or 3.33×10^5 (RA pretreatment/differentiating) cells per ml with cell culture medium (*see* **Note 6**).

3.3 Differentiation and Preparation of Cells for Experiments

1. For induction of differentiation, replace the cell culture medium in the T75 flask by 10 ml cell culture medium supplement with 10 μM retinoic acid (RA) final concentration (*see* **Note 7**).
2. Incubate cells for 24–72 h at 37 °C, 5% CO_2, in a humidified incubator (*see* **Note 8**).
3. Harvest and count cells as described in Subheading 3.2.
4. Seed 5000 non-treated/non-differentiating and 15,000 RA-treated/differentiating cells per well (*see* **Note 9**) in 45 μl cell culture medium into a 384-well plate (*see* **Note 10**) using a Multidrop Combi pipetting system (*see* **Note 11**). Leave the first two columns empty (no cells) as a positive control.
5. Incubate cells for 24 h at 37 °C, 5% CO_2, in a humidified incubator.

3.4 Addition of Small Molecules

1. Thaw frozen diluted chemicals previously distributed into the wells of the molecule plate.
2. Add 5 μl of chemicals (100 μM) to each well containing 45 μl cell culture medium using the liquidator dispensing system (final chemical concentration 10 μM) (*see* **Note 2**).
3. Repeat **step 2** three times to combine four 96-well plates in one 384-well plate.
4. Incubate plates for 24–48 h at 37 °C, 5% CO_2, in a humidified incubator (*see* **Note 12**).

3.5 Cell Fitness Screening

1. Prepare CTG solution using CellTiter-Glo Luminescent Cell Viability Assay Kit according to the manufacturer's instructions and dilute in DMEM *without* phenol red in a ratio of 1:1.3 (*see* **Note 13**). Keep CTG solution on ice and protected from light (*see* **Note 14**).
2. Empty wells of the 384-well plate using a 24-channel wand (*see* **Note 15**).
3. Add 10 μl CTG solution (*see* **Note 16**) into each well of the 384-well plate using a Multidrop Combi dispensing system and incubate for 10 min at room temperature.
4. Measure luminescence intensity with a plate reader, with no filter and an exposure time of 0.1 s (*see* **Notes 17** and **18**).

3.6 Data Analysis and Normalization Using cellHTS2 or web cellHTS2

1. Analyze your data using *cellHTS2* (off-line, *see* **Note 19**) or *web cellHTS2* (online, *see* **Note 20**).
2. When using the online tool, upload the readout data individually to http://web-cellhts2.dkfz.de/.
3. Download analysis results from the online analysis and unpack the archives. Click on the file "index.html" to open an off-line browser-based report of your experiment, including information on plate configuration and sample annotation, quality control data (e.g., replicate correlation, dynamic range of the assay), and a color-coded visualization of scored values providing an intuitive summary of your screening results.
4. Navigate to the folder named "in." The file "toptable.txt" lists all raw and analysis data including scored and normalized values. Use data listed in the columns named "median_ch1" and "median_ch2" for subsequent analysis and assessment of phenotype intersection from non-treated/non-differentiating and RA-treated/differentiating cells.
5. Normalize plate reader data generated from differently treated cells and replicates on the plate median.

3.7 Analysis of Phenotype Intersection

1. Calculate the arithmetic mean of the normalized plate replicates.
2. Rank cell fitness phenotypes in increasing order using, e.g., Microsoft Excel. Top-ranking mean values indicate lowest cell fitness and strongest phenotypes.
3. Calculate the intersection of a defined number of top-ranking phenotypes. You may decide to consider selecting hits from your list based on the degree of intersecting phenotypes.

4 Notes

1. We used a small-molecule library (National Institute of Health Clinical Collections 1 and 2, Evotec) containing 727 (622 accurately annotated) clinical trials I–III compounds, but any other library would also be appropriate. Final drug concentration should be around 10 μM. 96-Well plates can be replaced by any other multititer plate type.
2. Instead of the Liquidator, a multichannel pipet can be used.
3. The initial cell number for expansion of cells as well as the frequency of medium exchange depend on the used cell line, cycling time, and passage number of the cell culture. Refer to the supplier's instructions and recommendations regarding the culture method and cell line's characteristics.
4. Some cell lines require a different detachment procedure and involve other methods such as other reagents or scratching. Refer to the supplier's instructions and recommendations for the appropriate method.
5. For a newly established in vitro model and when there is no experience regarding the properties of the cells, it is recommended to quantify the proportion of viable cells prior to re-seeding into culture flasks or multititer plates, e.g., with trypan blue or another method. Trypan blue is a stain used to selectively color dead cells. When using trypan blue for counting viable cells, follow the manufacturer's instructions with regard to dye concentration and incubation time and count unstained cells only.
6. The cell number depends on the cell line, size of the cells, culture plasticware, cycling time, culture conditions, and duration of the assay and requires specific experimental optimization. Low cell numbers may result in a small dynamic range of the assay as well as strong intra- and inter-multiwell plate variations. In contrast, high cell numbers may cover small effects or result in overgrown monolayer cultures with altered characteristics.

7. Due to instability of retinoic acid at room temperature, freshly thaw stock solutions for each experiment.
8. The incubation time depends on the proliferation rate of the cells. Cells should be maintained as a monolayer culture and used in experiments when approx. 80% confluent.
9. The optimal cell number should be determined prior to chemical susceptibility screening. RA-treated NT2 cells proliferate at a slower rate and thus require to be seeded at lower number compared to non-treated cells in order to achieve comparable cell density on the day of molecule treatment.
10. For readout of CellTiter-Glo, white plates with flat bottom are recommended to maximize light output signal. For multiplexing approaches, e.g., with fluorescent dyes, the use of black plates with transparent flat bottom is recommended to reduce signal cross talk and background fluorescence.
11. Instead of a MultidropCombi dispensing system, a multichannel pipet can be used.
12. Incubation time depends on the drug concentration, metabolic activity of cells, cycling time, and many other parameters and should be optimized prior to chemical susceptibility screening.
13. The ratio may be varied towards smaller CTG amounts.
14. Store the remaining CTG solution at −20 °C. Reconstituted CTG reagent can be stored at RT for up to 8 h with 10% loss of activity, at 4 °C for 48 h with 5% loss of activity, and at −20 °C for 21 weeks with 3% loss of activity (*see* the manufacturer's instructions). The reagent is stable for up to ten freeze-thaw circles with less than 10% loss of activity.
15. If a 24-channel wand is not available, turn the plate upside-down onto a stack of tissue and leave for approx. 30 s until the solution is entirely removed from the wells.
16. Addition of CTG solution will lyse the cells. When used in combination with other staining reagents, CTG should always be added last to avoid interference with other indicators.
17. Half-life of the luminescent signal resulting from this reaction is greater than 5 h, thus eliminating the need for reagent injectors and providing flexibility for continuous or batch-mode processing of many plates, e.g., in high-throughput screening.
18. Readout settings are specific for a Perkin Elmer Victor X4 plate reader and may differ for other instruments.
19. *cellHTS2* provides an end-to-end analysis of cell-based high-throughput screening experiments [28]. Download and documentation are available at http://www.bioconductor.org/packages/release/bioc/html/cellHTS2.html.

20. *web cellHTS2* [29] is a front end for cellHTS2 and provides all features of cellHTS2. Web cellHTS2 guides the user through all analysis options and outputs an HTML including a full quality control report and a ranked hit list. A detailed tutorial and a manual are available at http://web-cellhts2.dkfz.de/.

Acknowledgements

The authors gratefully acknowledge funding of the Staedtler Stiftung. The funders had no role in study design, data collection and analysis, decision to publish, or preparation of the manuscript.

References

1. May M (2000) Disturbing behavior: neurotoxic effects in children. Environ Health Perspect 108:A262–A267
2. Colborn T (2004) Neurodevelopment and endocrine disruption. Environ Health Perspect 112:944–949
3. Rauh VA, Garfinkel R, Perera FP et al (2006) Impact of prenatal chlorpyrifos exposure on neurodevelopment in the first 3 years of life among inner-city children. Pediatrics 118: e1845–e1859
4. Herbert MR (2010) Contributions of the environment and environmentally vulnerable physiology to autism spectrum disorders. Curr Opin Neurol 23:103–110
5. Rice D, Barone S Jr (2000) Critical periods of vulnerability for the developing nervous system: evidence from humans and animal models. Environ Health Perspect 108(Suppl 3): 511–533
6. Grandjean P, Landrigan PJ (2006) Developmental neurotoxicity of industrial chemicals. Lancet 368:2167–2178
7. Grandjean P, Landrigan PJ (2014) Neurobehavioural effects of developmental toxicity. Lancet Neurol 13:330–338
8. Middaugh LD, Dow-Edwards D, Li AA et al (2003) Neurobehavioral assessment: a survey of use and value in safety assessment studies. Toxicol Sci 76:250–261
9. Makris SL, Raffaele K, Allen S et al (2009) A retrospective performance assessment of the developmental neurotoxicity study in support of OECD test guideline 426. Environ Health Perspect 117:17–25
10. OECD (2007) Test Guideline 426. OECD Guideline for testing of Chemicals. In: Developmental neurotoxicity study. Organisation of Economic Co-operation and development, Paris, France
11. Smirnova L, Hogberg HT, Leist M et al (2014) Developmental neurotoxicity - challenges in the 21st century and in vitro opportunities. Altex 31:129–156
12. NRC (2007) Toxicity testing in the 21st century: a vision and a strategy, 1st edn. National Academy Press, Washington, DC
13. Hill EJ, Woehrling EK, Prince M et al (2008) Differentiating human NT2/D1 neurospheres as a versatile in vitro 3D model system for developmental neurotoxicity testing. Toxicology 249:243–250
14. Couillard-Despres S, Quehl E, Altendorfer K et al (2008) Human in vitro reporter model of neuronal development and early differentiation processes. BMC Neurosci 9:31
15. Pallocca G, Fabbri M, Sacco MG et al (2013) miRNA expression profiling in a human stem cell-based model as a tool for developmental neurotoxicity testing. Cell Biol Toxicol 29: 239–257
16. Laurenza I, Pallocca G, Mennecozzi M et al (2013) A human pluripotent carcinoma stem cell-based model for in vitro developmental neurotoxicity testing: effects of methylmercury, lead and aluminum evaluated by gene expression studies. Int J Dev Neurosci 31: 679–691
17. Stern M, Gierse A, Tan S et al (2014) Human Ntera2 cells as a predictive in vitro test system for developmental neurotoxicity. Arch Toxicol 88:127–136
18. Kuenzel K, Friedrich O, Gilbert DF (2016) A recombinant human pluripotent stem cell line stably expressing halide-sensitive YFP-I152L for GABAAR and GlyR-targeted high-throughput drug screening and toxicity testing. Front Mol Neurosci 9:51
19. Lee VM, Andrews PW (1986) Differentiation of NTERA-2 clonal human embryonal

carcinoma cells into neurons involves the induction of all three neurofilament proteins. J Neurosci 6:514–521
20. Pleasure SJ, Page C, Lee VM (1992) Pure, postmitotic, polarized human neurons derived from NTera 2 cells provide a system for expressing exogenous proteins in terminally differentiated neurons. J Neurosci 12:1802–1815
21. Sandhu JK, Sikorska M, Walker PR (2002) Characterization of astrocytes derived from human NTera-2/D1 embryonal carcinoma cells. J Neurosci Res 68:604–614
22. Stewart R, Christie VB, Przyborski SA (2003) Manipulation of human pluripotent embryonal carcinoma stem cells and the development of neural subtypes. Stem Cells 21:248–256
23. Ozdener H (2007) Inducible functional expression of Bcl-2 in human astrocytes derived from NTera-2 cells. J Neurosci Methods 159:8–18
24. Coyne L, Shan M, Przyborski SA et al (2011) Neuropharmacological properties of neurons derived from human stem cells. Neurochem Int 59:404–412
25. Menzner AK, Abolpour Mofrad S, Friedrich O et al (2015) Towards in vitro DT/DNT testing: assaying chemical susceptibility in early differentiating NT2 cells. Toxicology 338:69–76
26. Gilbert DF, Erdmann G, Zhang X et al (2011) A novel multiplex cell viability assay for high-throughput RNAi screening. PLoS One 6: e28338
27. Gilbert DF, Boutros M (2016) A protocol for a high-throughput multiplex cell viability assay. Methods Mol Biol 1470:75–84
28. Boutros M, Bras LP, Huber W (2006) Analysis of cell-based RNAi screens. Genome Biol 7:R66
29. Pelz O, Gilsdorf M, Boutros M (2010) web cellHTS2: a web-application for the analysis of high-throughput screening data. BMC Bioinf 11:185

Chapter 6

Ferroptosis and Cell Death Analysis by Flow Cytometry

Daishi Chen, Ilker Y. Eyupoglu, and Nicolai Savaskan

Abstract

Cell death and its recently discovered regulated form ferroptosis are characterized by distinct morphological, electrophysiological, and pharmacological features. In particular ferroptosis can be induced by experimental compounds and clinical drugs (i.e., erastin, sulfasalazine, sorafenib, and artesunate) in various cell types and cancer cells. Pharmacologically, this cell death process can be inhibited by iron chelators and lipid peroxidation inhibitors. Relevance of this specific cell death form has been found in different pathological conditions such as cancer, neurotoxicity, neurodegeneration, and ischemia. Distinguishing cell viability and cell death is essential for experimental and clinical applications and a key component in flow cytometry experiments. Dead cells can compromise the integrity of the data by nonspecific binding of antibodies and dyes. Therefore it is essential that dead cells are robustly and reproducibly identified and characterized by means of cytometry application. Here we describe a procedure to detect and quantify cell death and its specific form ferroptosis based on standard flow cytometry techniques.

Key words Neurodegeneration, Necrosis Apoptosis, Fluorescence-activated cell sorting, Tumors, Neurooncology, xCT, Ferroptosis, Erastin, Sorafenib

1 Introduction

Flow cytometry performed on fixed and living cells has made a tremendous impact on progress in the biomedical life science field and in drug discoveries [1, 2]. Half a century ago the first cell sorter devices were invented which allowed separation of cells in suspension based on cell volume differences [3–6] or differences in fluorescence signals [7].

These game-changing inventions established the application of flow cytometry into virtually any biomedical field as well as in clinical settings [2]. Recent advances in cytometry-based techniques allow now the detection of heavy-metal isotopes instead of antibody-coupled fluorophores. Signals can be detected in a time-of-flight detector equally as in atomic mass spectrometers, now termed mass cytometry or cytometry by time of flight with the advantage of detecting more than 100 parameters per cell [8–10]. Basically, all common cytometers today are based on the principle

Daniel F. Gilbert and Oliver Friedrich (eds.), *Cell Viability Assays: Methods and Protocols*, Methods in Molecular Biology, vol. 1601, DOI 10.1007/978-1-4939-6960-9_6, © Springer Science+Business Media LLC 2017

of laminar flow (fluid flow in parallel layers) making cells surrounded by the sheath fluid within the cytometer. Len Herzenberg coined the term FACS as fluorescence-activated cell sorter and Becton Dickinson (BD) uses FACS as a trade name.

Cell death, programmed cell death (apoptosis), and ferroptosis can be induced by external signals and are morphologically and biochemically distinct from each other. The characteristic morphological signs of apoptosis (cellular shrinkage, membrane blebbing, nuclear condensation, and fragmentation) are different to those of ferroptosis (small mitochondria with condensed mitochondrial membrane densities, reduction of mitochondria crista, outer mitochondrial membrane rupture, lack of rupture or blebbing of the plasma membrane, rounding up of the cell, normal nuclear size, and lack of chromatin condensation).

To quantify and characterize cell death in its various forms, flow cytometry remains a powerful methodology of choice. Flow cytometry represents a solid technique and allows robust data reproducibility. The dye propidium iodide (PI) is suitable for staining living cells since PI cannot pass through intact cell membranes, but may freely enter cells with compromised ATP/NADPH delivery or damaged cell membranes [11, 12]. 7-Aminoactinomycin (7-AAD) is another avital marker suitable for application in living cells. 7-AAD intercalates only with double-stranded DNA in the case of compromised plasma membranes [13]. Intact cells remain unlabeled. Annexin V is a phospholipid phosphatidylserine-binding protein. Phospholipid phosphatidylserine exposed to the external cellular environment is an early event in cell death and thus is easily accessible with Annexin V serving as a sensitive probe for flow cytometric analysis.

Thus, we describe a straightforward experimental approach to determine cell viability and cell death in human and rodent cells.

2 Materials

All solutions are prepared with Millipore-filtered water (dH_2O) and stored at 4 °C if not otherwise stated. Fluorescent stock solutions are kept in dark-colored tubes at −20 °C. Solved fluorescent working solutions are kept in the dark at 4 °C. Chemicals are analytical grade reagents.

2.1 Buffers and Solutions

1. Propidium iodide (PI, Life Technologies, Darmstadt, Germany) can be applied in living cells to stain dead cells so that PI-positive cells (PI^+) may be excluded from analysis in standard live cell surface staining protocols. PI cannot pass through intact cell membranes, but may freely enter cells with compromised ATP/NADPH delivery or damaged cell

membranes. Upon entering dead cells, PI intercalates into double-stranded DNA or double-stranded RNA in a stoichiometric manner. This intercalation is mediated by non-covalent forces. The excitation maximum of PI is at 535 nm and the emission maximum is 617 nm. Stock solution contains 1 mg PI in 1 ml dH_2O. Store at −20 °C. Prepare a work solution with 5 μg/ml PI.

2. 7-Aminoactinomycin is a peptide and another independent cell death marker (or avital marker) suitable for application in living cells. 7-AAD does not readily pass through intact cell membranes and intercalates only with double-stranded DNA in the case of compromised plasma membranes. Intact cells remain unlabeled. 7-AAD has an absorption maximum at 546 nm and can be efficiently excited at 543 nm. Prepare a stock solution with 1 mg 7-AAD in 1 ml DMSO. The work concentration has a final 7-AAD level of 1 μg/ml.

3. Annexin A5 or V coupled to FITC staining precedes the loss of plasma membrane integrity which accompanies the latest stages of cell death resulting in either apoptosis or necrosis. Thus, FITC-conjugated Annexin V is typically facilitated to monitor the process of apoptotic cell death at an early stage (PI^-/FITC Annexin V^+). The underlying staining mechanism is based on the external exposure and detection of phosphatidylserine. A feature during early apoptotic processes is that the plasma membrane phospholipid phosphatidylserine becomes translocated from the inner to the outer leaflet of the plasma membrane. Annexin V is an approx. 35 kDa Ca^{2+}-dependent phospholipid-binding protein with high affinity for phospholipid phosphatidylserine. Phospholipid phosphatidylserine exposed to the external cellular environment is easily accessible for Annexin V and thus serves as a sensitive probe for flow cytometric analysis of cells that are undergoing apoptosis. Thus, Annexin V^+ cells indicate an earlier apoptosis stage than 7-AAD^+ or PI^+ cells. Note that most fluorescence conjugations to Annexin V retain its high affinity for phospholipid phosphatidylserine. The recommended final concentration for Annexin V is 0.25–1 μg/ml.

4. PBS buffer (pH 7.4): Dissolve 8 g NaCl, 0.2 g KCl, 2.3 g Na_2HPO_4, and 0.22 g KH_2PO_4 in 800 ml dH_2O and adjust the pH to 7.4 (NaOH). Add distilled water (dH_2O) to a total volume of 1 l. Sterilize prior to usage.

5. Annexin V binding buffer should be prepared as 10× buffer with 0.1 M HEPES, pH 7.4; 1.4 M NaCl; and 25 mM $CaCl_2$. Dilute the buffer to 1× with dH_2O prior usage.

6. Flow cytometry staining buffer (FACS buffer) contains 1× PBS with 0.5% BSA or 2–5% FBS, and 0.1% sodium azide.

2.2 Fluorescence-Activated Cell-Sorting Device

The protocol described here has been run on Flow Cytometer BD FACSCanto™ II (BD Bioscience, Heidelberg, Germany). However, any flow cytometer with similar parameter settings can be used. Quantification and analysis of the results were done with Flow Jo 7.6 (Tree Star, Inc., Ashland, USA).

2.3 Cells and Cell Culture Conditions

The rat F98- and human-derived U87 (U87MG) cell lines are established models for malignant glioblastoma multiforme cells and were cultured in Dulbecco's modified Eagle medium (DMEM) (Biochrom AG, Berlin, Germany/Sigma) containing 10% fetal bovine serum (FBS) with 100 U/ml penicillin, 100 μg/ml streptomycin, and 1% l-glutamine. After treatment, cells were trypsinized with 2 ml trypsin +/− EDTA for 5 min. Afterwards cells were washed with 1× PBS, neutralized with DMEM + 10% FBS, and resuspended by mild centrifugation at 900 rpm for 5 min.

3 Methods

When staining cells, use always ice buckets for constant cooling at 4 °C. Protect all tubes from light. Carry out all procedures on ice if not otherwise stated.

3.1 Experimental Procedure

1. Following cell staining for surface antigens, wash cells one to two times with flow cytometry staining buffer.
2. Resuspend cells in an appropriate volume of flow cytometry staining buffer, i.e., 500 μl.
3. For every 100 μl of cell suspensions, add 5 μl of PI working solution (final concentration of PI is 5 μg/ml) or alternatively 7-AAD staining solution (final concentration 1 μg/ml) (*see* **Note 1**).
4. Incubate cells for 5–15 min in light-protected boxes on ice before analyzing cells on a flow cytometer. Alternatively, incubate cells light protected for 5 min at room temperature and proceed immediately with flow cytometry (*see* **Note 2**).

3.2 Staining Procedure for FITC Annexin V and 7-AAD

1. Wash resuspended cells with flow cytometry staining buffer, centrifuge for 5 min twice, and then resuspend cells in Annexin V binding buffer at a concentration of $0.25–1.0 \times 10^7$ cells/ml.
2. Transfer 100 μl of cell suspension in a 5 ml test tube.
3. Add 5 μl of FITC Annexin V.
4. Add 5 μl of 7-AAD (*see* **Note 1**).
5. Gently vortex the cells and incubate for 15 min at room temperature (25 °C) in the dark.
6. Add 400 μl of Annexin V binding buffer to each tube (*see* **Note 2**).

7. Run the cell suspension by flow cytometry (*see* **Note 3**).
8. The following controls are used in the same manner as the experimental groups in order to set up compensation and quadrants (*see* **Note 4**):
   (a) Unstained cells.
   (b) Cells stained with FITC Annexin V (no 7-AAD).
   (c) Cells stained with 7-AAD (no FITC Annexin V).

3.3 Ferroptosis and Cell Death Induction Procedure

1. Treat cells with cell death and ferroptosis-inducing drugs (1–10 μM sorafenib, LC Laboratories, Woburn, USA) or 1–10 μM erastin (Hycultec GmbH, Beutelsbach, Germany) for 72 h (Fig. 1).
2. For mechanistic studies and combined treatment procedures treat cells with 5 μM sorafenib or 5 μM erastin and ±50 μM deferoxamine (DFO, Sigma-Aldrich, Taufkirchen, Germany). Another ferroptosis inhibitor is ferrostatin (Ferr-1, Sigma-Aldrich) which can be used in a concentration of 0.5 μM (Fig. 2).

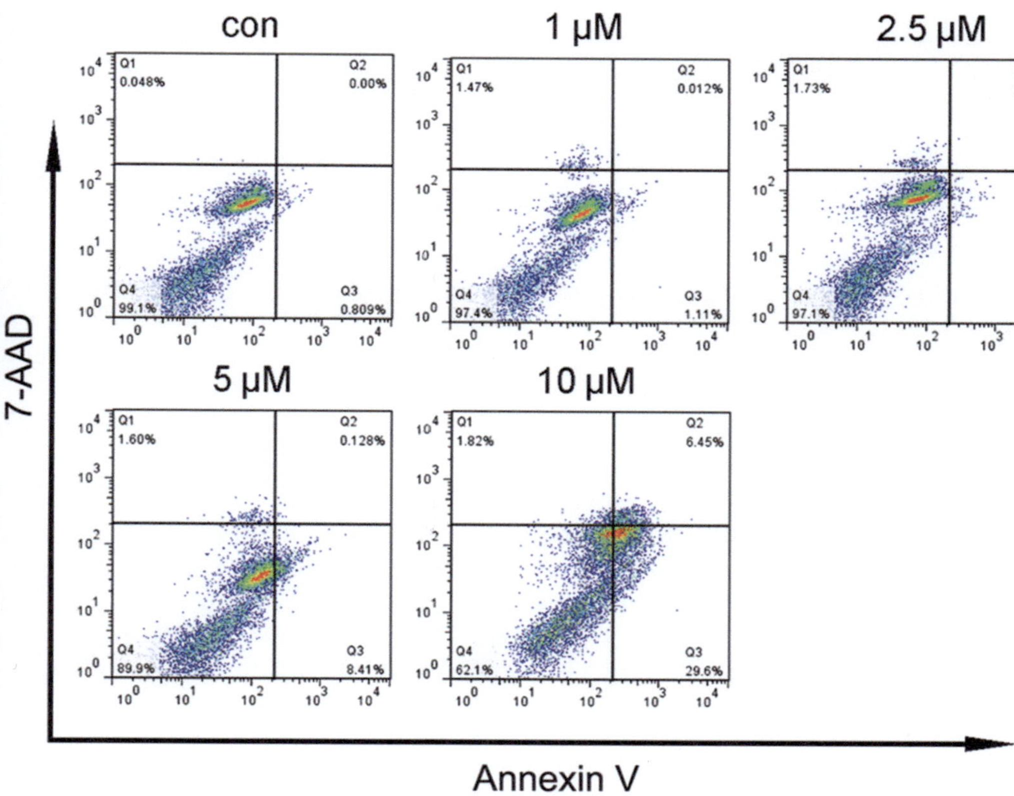

Fig. 1 Ferroptosis induction by sorafenib in glioma cells. U87MG were treated with sorafenib for 72 h. Cell death was detected by flow cytometry using Annexin V and 7-AAD. Sorafenib induced dose-dependent apoptotic cell death in U87MG cells. *Q1*: non-apoptotic or ferroptotic cells (Annexin V negative/7-AAD negative), *Q2*: late-apoptotic and early necrotic cells (Annexin V positive/7-AAD positive), *Q3*: early-apoptotic cells (Annexin V positive/7-AAD negative), *Q4*: viable cells (Annexin V negative/7-AAD negative)

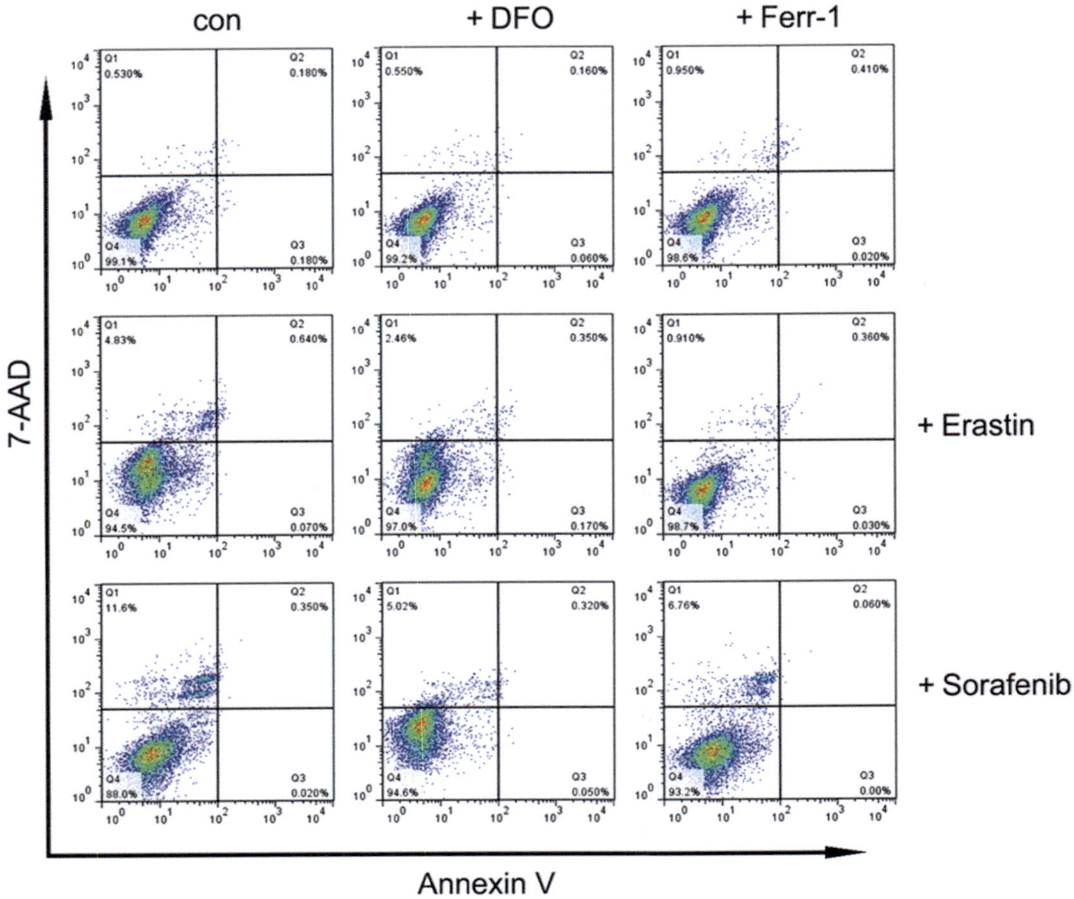

Fig. 2 Ferroptosis inhibitors rescue sorafenib-induced cell death in F98 cells. F98 cells were treated with sorafenib or erastin ± deferoxamine (DFO, 50 μM) or ± ferrostatin-1 (Ferr-1, 0.5 μM) for 16 h. Cell death was detected by flow cytometry using Annexin V and 7-AAD. Sorafenib or erastin in combination with ferroptosis inhibitors decreased non-apoptotic cell death (*Q1*)

3. After treatment, Annexin V/7AAD double staining can subsequently be performed to assess further mechanisms of cell death (Figs. 1 and 2). After incubation with Annexin V/7-AAD, cells were transferred to FACS tubes and analyzed by Flow Cytometer BD FACSCanto II (BD Bioscience, Heidelberg, Germany). Analyses were carried out with Flow Jo 7.6 (Tree Star, Inc., Ashland, USA).

4 Notes

1. Propidium iodide and 7-AAD must remain in the buffer during acquisition. Do not wash cells after the addition of propidium iodide or 7-AAD.

2. Cells should be analyzed as soon as possible after the initial incubation period due to adverse effects on the viability of cells left in the presence of propidium iodide or 7-AAD for too long.
3. We recommend to use flow cytometry tubes from Sarstedt (Nümbrecht, Germany).
4. A minimum of 10,000 cells were counted and analyzed per condition.

References

1. Edwards BS, Sklar LA (2015) Flow cytometry: impact on early drug discovery. J Biomol Screen 20(6):689–707
2. Jaye DL, Bray RA, Gebel HM et al (2012) Translational applications of flow cytometry in clinical practice. J Immunol 188(10):4715–4719
3. Crosland-Taylor PJ (1953) A device for counting small particles suspended in a fluid through a tube. Nature 171(4340):37–38
4. Hallermann L, Thom R, Gerhartz H (1964) Electronic differential counting of granulocytes and lymphocytes after intravital fluorochrome staining with acridine organe. Verh Dtsch Ges Inn Med 70:217–219
5. Fulwyler MJ (1965) Electronic separation of biological cells by volume. Science 150: 910–911
6. Kamentsky LA, Melamed MR, Derman H (1965) Spectrophotometer: new instrument for ultrarapid cell analysis. Science 150:630–631
7. Wilder ME, Cram LS (1977) Differential fluorochromasia of human lymphocytes as measured by flow cytometry. J Histochem Cytochem 25:888–891
8. Bendall SC, Nolan GP, Roederer M, Chattopadhyay PK (2012) A deep profiler's guide to cytometry. Trends Immunol 33:323–332
9. Bandura DR, Baranov VI, Ornatsky OI, Antonov A, Kinach R, Lou X, Pavlov S et al (2009) Mass cytometry: technique for real time single cell multitarget immunoassay based on inductively coupled plasma time-of-flight mass spectrometry. Anal Chem 81: 6813–6822
10. Perfetto SP, Chattopadhyay PK, Roederer M (2004) Seventeen-colour flow cytometry: unravelling the immune system. Nat Rev Immunol 4:648–655
11. Sehm T, Fan Z, Ghoochani A, Rauh M et al (2016) Sulfasalazine impacts on ferroptotic cell death and alleviates the tumor microenvironment and glioma-induced brain edema. Oncotarget. doi:10.18632/oncotarget.8651
12. Sehm T, Rauh M, Wiendieck K, Buchfelder M, EyÃ¼poglu IY, Savaskan NE (2016) Temozolomide toxicity operates in a xCT/SLC7a11 dependent manner and is fostered by ferroptosis. Oncotarget 7(46):74630–74647. doi: 10.18632/oncotarget.11858.
13. Chen D, Fan Z, Rauh M, Buchfeller M, Eyupoglu IY, Savaskan N. ATF4 promotes angiogenesis and neuronal cell death and confers ferroptosis in a xCT-dependent manner. Oncotarget 2017 (in press).

Chapter 7

Assaying Mitochondrial Respiration as an Indicator of Cellular Metabolism and Fitness

Natalia Smolina, Joseph Bruton, Anna Kostareva, and Thomas Sejersen

Abstract

Mitochondrial respiration is the most important generator of cellular energy under most circumstances. It is a process of energy conversion of substrates into ATP. The Seahorse equipment allows measuring oxygen consumption rate (OCR) in living cells and estimates key parameters of mitochondrial respiration in real-time mode. Through use of mitochondrial inhibitors, four key mitochondrial respiration parameters can be measured: basal, ATP production-linked, maximal, and proton leak-linked OCR. This approach requires application of mitochondrial inhibitors—oligomycin to block ATP synthase, FCCP—to make the inner mitochondrial membrane permeable for protons and allow maximum electron flux through the electron transport chain, and rotenone and antimycin A—to inhibit complexes I and III, respectively. This chapter describes the protocol of OCR assessment in the culture of primary myotubes obtained upon satellite cell fusion.

Key words Cellular respiration, Mitochondrial function, Satellite cells, Myotubes, Mitochondrial disease, Cell viability

1 Introduction

Mitochondria act as power plants within almost every eukaryotic cell. Oxidative phosphorylation processes taking part on the mitochondrial membranes provide cell with energy by consumption of oxygen molecules and production of ATP molecules. Therefore, mitochondrial functions delineate the bioenergetics cellular state. It is well known that mitochondria reveal functional plasticity [1] and are capable of adapting to a variety of acute and chronic conditions, thus contributing to vulnerability or resilience of muscle cells [2]. Restriction of ATP production and reserve respiratory capacity reflecting overall mitochondrial dysfunction are often associated with heart failure [3], neurodegenerative disorders [4], diabetes [5], and cell death [6]. Thus, the ability of the mitochondria to make ATP and to consume oxygen in response to energy demands serves as a reliable hallmark of its functional state, reflecting cell

Daniel F. Gilbert and Oliver Friedrich (eds.), *Cell Viability Assays: Methods and Protocols*, Methods in Molecular Biology, vol. 1601, DOI 10.1007/978-1-4939-6960-9_7, © Springer Science+Business Media LLC 2017

viability. Oxidative phosphorylation depends on the reduction of oxygen to water; therefore one of the methods to assess mitochondrial function is by measuring its oxygen consumption rate. For intact cells, the best assay to measure basal cell respiratory rate, ATP production, proton leak, maximum respiration, and spare respiratory capacity is a high-throughput real-time mode approach—Agilent Seahorse XF Analyzers. This chapter gives an overview of a protocol that we have developed to estimate mitochondrial oxygen consumption rate in mouse primary skeletal myotubes using the Seahorse XF96 extracellular flux analyzer.

2 Materials

2.1 Cell Isolation, Plating, and Cultivation Reagents

1. Collagenase type I.
2. DMEM.
3. Horse serum (HS).
4. Fetal bovine serum (FBS).
5. Chicken embryo extract (CEE).
6. 10,000 U/mL Penicillin-streptomycin (Pen-Strep).
7. 200 mM l-Glutamine (l-glu).
8. PBS.
9. Geltrex LDEV-free reduced growth factor basement membrane matrix.
10. FX96 cell culture microplate (Seahorse, Agilent Technologies).
11. 15 and 50 mL tubes.

2.2 Media and Solution Preparation

1. Resuspension medium: DMEM, 10% HS, 1% Pen-Strep.
2. Proliferation medium: DMEM, 20% FBS, 10% HS, 1% CEE, 1% l-Glu, 1% Pen-Strep.
3. Differentiation medium: DMEM, 2% HS, 1% l-Glu, 1% Pen-Strep.
4. Plating surface media: Dilute Geltrex in cold DMEM to give 1% solution without any supplements.

2.3 XF96 Analyzer Assay

1. XF96 sensor cartridge XF Calibrant solution.
2. Assay medium: XF Base Medium, 10 mM sodium pyruvate, 10 mM glucose, 2 mM l-glutamine, adjust a 7.3 pH using 1 M NaOH solution.
3. 1 M NaOH.
4. Seahorse XF Cell Mito Stress Test Kit: Oligomycin, FCCP (carbonyl cyanide 4-(trifluoromethoxy)phenylhydrazone), rotenone, antimycin A: Prepare 8 μM oligomycin (8×) by adding 240 μL of 100 μM oligomycin stock solution to 2760 μL

assay medium; 9 μM FCCP (9×) by adding 270 μL of 100 μM FCCP stock solution to 2730 μL assay medium; and 5 μM rotenone + 5 μM antimycin A (10×) by adding 300 μL of both 50 μM rotenone and antimycin A stock solutions to 2700 μL assay medium (Table 1).

2.4 Equipment Used

1. CO_2 incubator (set to 37 °C, 5% CO_2, 95% humidity).
2. Non-CO_2 incubator (set to 37 °C).
3. Water bath (set to 37 °C).
4. pH meter.
5. Seahorse XFe96 Analyzer.

3 Methods

3.1 Cell Culture Preparation (14 Days Before the Day of Assay)

All procedures are performed under sterile conditions and at room temperature, unless specifically stated. Surgery instruments should be autoclaved. All volumes and quantities are given for carrying out one 96-well plate of Seahorse analysis.

1. Sacrifice three mice (6–8 weeks old) C57Bl/6 by cervical dislocation (*see* **Note 1**).
2. Isolate six soleus muscles from two hind legs and place it in 5 mL of DMEM with 1% Pen-Strep. Amount of media must be sufficient to cover all muscles in order not to let it to dry.
3. Replace muscles for digestion to filtered 0.1% collagenase I solution (1 mL per two muscles) in 15 mL tube and incubate for 90 min at 37 °C. All muscles should be digested together in order to minimize inter-animal variability (*see* **Note 2**).
4. To remove enzyme solution, centrifuge tube with muscles for 5 min at 400 × *g*, and discard supernatant with 1 mL tip.
5. Gently resuspend the pellet in 4 mL of resuspension medium by pipetting approximately 20 times (*see* Subheading 2.2, **step 1**). Use 1 mL cut tip for resuspension. Leave for 3 min to let debris settle. Remove the supernatant to a fresh 15 mL tube.
6. Repeat **step 5** to increase yields of isolated cells.
7. To collect satellite cells, centrifuge the supernatant for 10 min at 1000 × *g*. Discard supernatant and redissolve the pellet in 3 mL of proliferation medium with 1 mL tip (*see* Subheading 2.2, **step 2**).
8. Count the cell number.

3.2 Cell Plating and Cultivation

1. At least 1 h ahead of plating, prepare cultivation surface (*see* Subheading 2.2, **step 4**).
2. Coat all wells on FX96 microplate with approximately 15 μL of 1% Geltrex solution (*see* **Note 3**). Leave for 1 h in incubator (set to 37 °C).

Table 1
Compound preparation for loading sensor cartridge ports

		When applying XF Cell Mito Stress Test Kit			When applying other reagents			
	Final well concentration (μM)	Stock solution concentration (μM)	Stock solution volume (μL)	Medium volume (μL)	Stock solution concentration (μM)	Stock solution volume (μL)	Medium volume (μL)	Add to port (μL)
Oligomycin (port A)	1.0	100 in assay media	240	2760	1000 in DMSO	24	2976	25
FCCP (port B)	1.0	100 in assay media	270	2730	2000 in DMSO	13.5	2986.5	25
Rotenone/antimycin (port C)	0.5	50 in assay media	300	2700	10,000 in DMSO/ 10,000 in ethanol	1.5/1.5	2997	25

Constant loading volume/variable compound concentration approach. Starting well volume: 175 μL assay medium

3. Before plating cells, pipette away residual Geltrex and wash once with 50 μL PBS.
4. After counting the cell number, adjust the volume of cell suspension in order to get concentration 5000 cells per 150 μL proliferation medium. Use prewarmed proliferation medium (*see* **Note 4**).
5. Seed 5000 cells (150 μL) per well of Geltrex-coated FX96 microplate. Do not seed cells in background correction wells (A1, A12, H1, H12); put only cultivation medium in there (*see* **Note 5**).
6. Cultivate cells in an incubator at 37 °C under a 5% CO_2 atmosphere and 95% humidity.
7. Monitor cell adherence and proliferation under microscope. When cells reach 80% of confluence, induce differentiation. Usually it takes 5–7 days.
8. To induce differentiation, remove proliferation medium, wash cells once with 50 μL prewarmed PBS, and add 150 μL of prewarmed differentiation medium (*see* Subheading 2.2, **step 3**).
9. Differentiate until all cells fuse to myotubes. Usually it takes 5–7 days.
10. Proliferation and differentiation media were renewed every other day by replacement of half of medium (*see* **Note 6**).

3.3 Measurement of Cellular O_2 Consumption Rates Prior to the Day of Assay

1. Prepare assay medium (*see* Subheading 2.3, **step 2**). Prepare this medium freshly for each assay (*see* **Note 7**).
2. Hydrate XF96 sensor cartridge with XF Calibrant solution at least 12 h ahead of the assay in non-CO_2 incubator (set to 37 °C).
3. Leave Seahorse XFe96 Analyzer switched on overnight with software running to ensure that temperature is 37 °C on the day of the assay.

3.4 Measurement of Cellular O_2 Consumption Rates on the Day of Assay

1. Warm assay medium to 37 °C.
2. Inspect cells under microscope to ensure that myotubes are evenly distributed over the plate.
3. Gently remove differentiation medium from all wells, wash once with 50 μL assay medium, and add assay medium up to final volume of 175 μL to all wells.
4. Incubate cells for 1 h in non-CO_2 incubator (set to 37 °C).
5. During the incubation, prepare stock solutions of mitochondrial inhibitor compounds (*see* Subheading 2.3, **step 4**).
6. Load XF96 sensor cartridge placed in hydrated utility plate with stock solutions of compounds. Load 25 μL of oligomycin, FCCP, and mixture of rotenone and antimycin A to ports A, B, and C correspondingly. Incubate the loaded cartridge in non-CO_2 incubator (set to 37 °C) up to the starting of the assay (at least 15 min) (*see* **Notes 8–10**).

Table 2
Typical mix and measurement cycle times

Command	Time (min)	Port	Number of repeats
Calibrate	At least 15 min		
Equilibrate	12		
Mix Wait Measure	2 2 3		3
Inject		A	
Mix Wait Measure	2 2 3		3
Inject		B	
Mix Wait Measure	2 2 3		3
Inject		C	
Mix Wait Measure	2 2 3		3

7. Set up a control protocol in the XFe96 Analyzer software. Protocol represents measurement loops consisted of three times for 3-min measurements each separated by 2-min wait and 2-min mix for every parameter (Table 2).
8. To calibrate XFe96 Analyzer, load XF96 sensor cartridge and utility plate into the right side of the instrument tray.
9. After calibration is complete, replace utility plate with the pre-incubated cell culture microplate.
10. Start the assay (Fig. 1).

3.5 Data Analysis and Normalization

1. For comparison, data can be expressed as percent change from baseline values to account potential differences in cell number across wells (*see* **Notes 5** and **11**).

4 Notes

1. We have successfully used C2C12 instead of primary satellite cells.
2. Collagenase I solution could be prepared ex tempore or stored at −20 °C as 10× stock solution. Collagenase I solution must be filtered through a 0.22 μm filter.

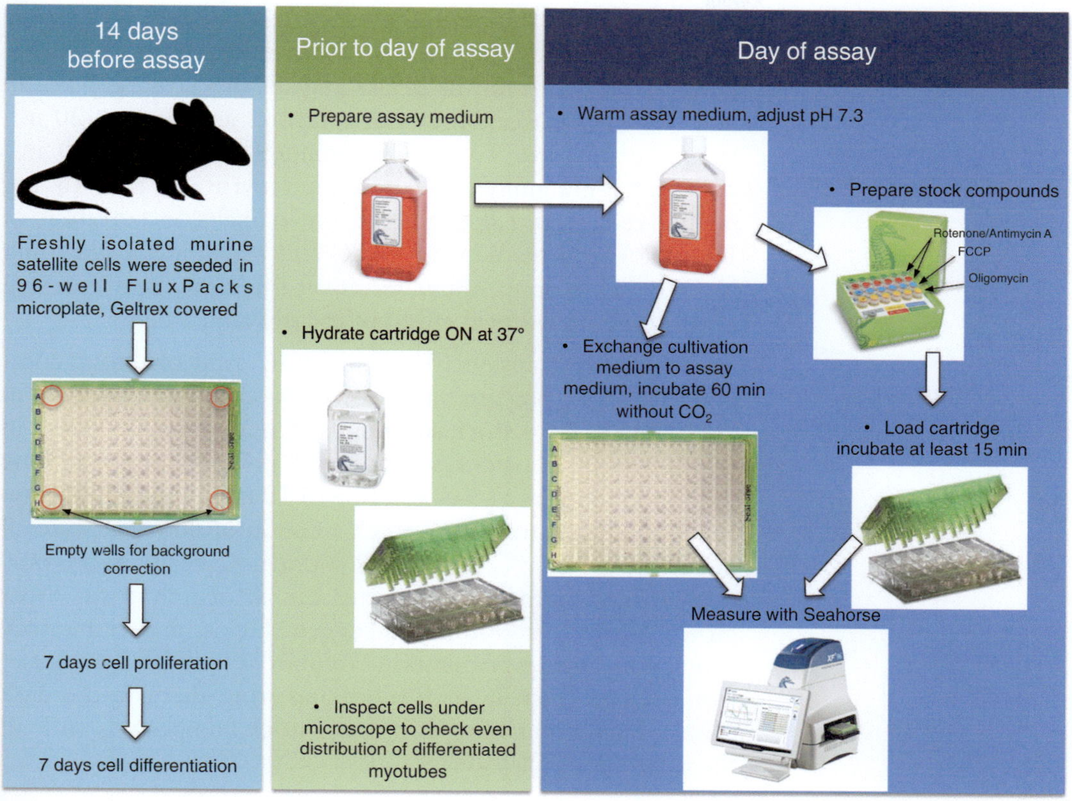

Fig. 1 Flowchart of experiment

3. Geltrex must be diluted in cold DMEM. Instead of Geltrex, surface can be coated with Matrigel, collagen, or laminin. However, Geltrex in our hand revealed the best results.
4. Survival and growth of satellite cells are temperature sensitive. Therefore, to avoid heat shock, prewarm all solutions to 37 °C.
5. At least three technical replicates (independent measurements of each condition per one plate, e.g., if you analyze the action of compounds or mutations) are used to be certain of the results. We discard data if there is more than 20% of variation.
6. Cultivation medium (proliferation and differentiation) evaporates rapidly; therefore add 100 μL when renewing medium. Take into consideration marginal effect; that is, medium from the wells of edge rows evaporates faster than from the center.

7. An alternative solution to commercially available XF Assay Medium can be prepared as follows: Separately, dissolve DMEM Base in 500 mL dH_2O and dissolve 1.85 g NaCl in 500 mL dH_2O. Combine both solutions and remove 20 mL from the mixture. To the remaining 980 mL, add 10 mL 200 mM l-glutamine, 1.1 g sodium pyruvate, and 15 mg phenol red. Add 1.8 g glucose for a final concentration of 10 mM. Warm medium to 37 °C and adjust a 7.3 pH using NaOH. pH is temperature sensitive; therefore assay medium needs to be prewarmed ahead of the pH adjustment.
8. An alternative solution to commercially available stock compounds can be prepared as follows: Dissolve oligomycin A in DMSO to make a stock solution of 1 mM; dissolve FCCP in DMSO to make a stock solution of 2 mM; dissolve rotenone in DMSO to make a stock solution of 10 mM; dissolve antimycin A in ethanol to make a stock solution of 10 mM. Store all inhibitors at −20 °C for up to 1 month. All reagents are purchased from Sigma. These agents can be acutely toxic. Wear gloves, protective clothing, face/eye shields, and respiratory protection during preparation. Inhibitors may be light sensitive; prepare them in dark tubes and store them in the dark (Table 1 describes how to prepare stock compounds to load the cartridge) [7].
9. Working concentration of compounds should be titrated for selected cell line and can range from 0.5 to 2.0 for oligomycin and from 0.125 to 2.0 for FCCP. Seahorse recommends using 1 μM of oligomycin for most cell lines.
10. There are two approaches to loading the injection ports of the sensor cartridge: constant loading volume/variable compound concentration and variable loading volume/constant compound concentration. In this chapter, first approach is applied, where constant volume of compound loaded in each injection port requires that each compound be prepared at a different concentration.
11. Normally data are expressed as pmol O_2/min/mkg protein. However, in this protocol satellite cells seeding on the Geltrex-covered surface hindered the measurement of protein content. Geltrex made a considerable contribution to the final amount of protein in the wells; therefore accurate protein concentrations of the tissue cannot be measured (Fig. 2).

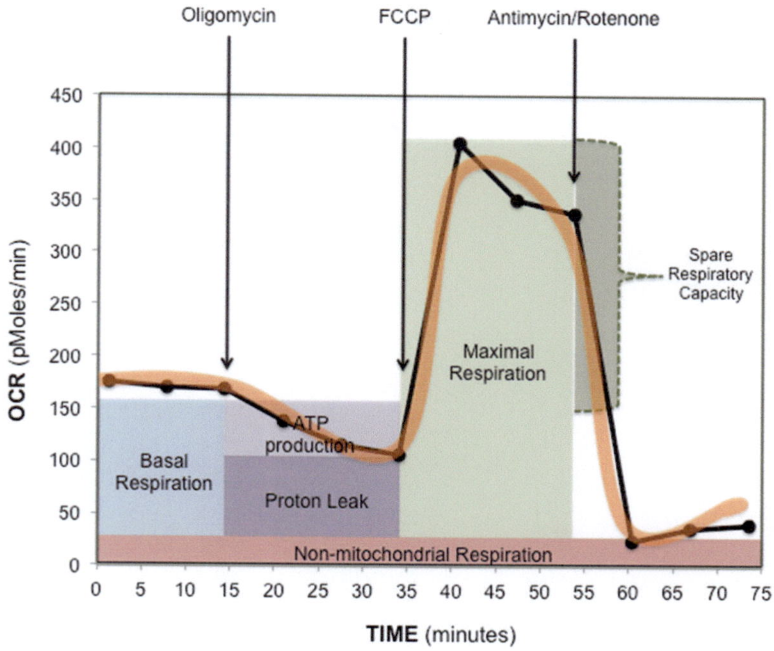

Fig. 2 Typical example of oxygen consumption rates in primary myotubes in the absence or presence of mitochondrial agents (*the orange line*). Data points are obtained from a single plate where oxygen consumption of 16 wells is averaged (OCR) (pMoles/min). *Black dots* indicate the mean values under each condition

Acknowledgments

The authors thank Dr. Vladimir Gogvadze and Ganna Oliynyk for providing access to the equipment and help in setting up the initial protocol. This research was supported by Stiftelsen Samariten and Fredrik och Ingrid Thurings stiftelse.

References

1. Soubannier V, McBride HM (2009) Positioning mitochondrial plasticity within cellular signaling cascades. Biochim Biophys Acta 1793: 154–170
2. Hoppeler H, Fluck M (2003) Plasticity of skeletal muscle mitochondria: structure and function. Med Sci Sports Exerc 35:95–104
3. Sharov VG, Todor AV, Silverman N et al (2000) Abnormal mitochondrial respiration in failed human myocardium. J Mol Cell Cardiol 32: 2361–2367
4. Yadava N, Nicholls DG (2007) Spare respiratory capacity rather than oxidative stress regulates glutamate excitotoxicity after partial respiratory inhibition of mitochondrial complex I with rotenone. J Neurosci 27:7310–7317
5. Mogensen M, Sahlin K, Fernström M et al (2007) Mitochondrial respiration is decreased in skeletal muscle of patients with type 2. Diabetes 56:1592–1599
6. Tigchelaar W, De Jong AM, van Gilst WH et al (2016) In EXOG-depleted cardiomyocytes cell death is marked by a decreased mitochondrial reserve capacity of the electron transport chain. Cell 1:134–143
7. Zhang J, Nuebel E, Wisidagama DRR et al. (2012) Measuring energy metabolism in cultured cells, including human pluripotent stem cells and differentiated cells. Nature Protocols, 7(6).

Chapter 8

An ATP-Based Luciferase Viability Assay for Animal African Trypanosomes Using a 96-Well Plate

Keisuke Suganuma, Nthatisi Innocentia Molefe, and Noboru Inoue

Abstract

Cell viability assays using multi-well cell culture plates are frequently used for in vitro drug screening. We herein describe an ATP-based luciferase viability assay for animal African trypanosomes using a 96-well plate. This assay could be further applied to the screening of novel compounds for the treatment of animal African trypanosomiasis.

Key words Animal African trypanosomiasis, Drug screening, Luciferase assay

1 Introduction

High throughput screening (HTS) assays of large compound libraries have been undertaken to discover novel therapeutic candidates for several parasites [1]. These assays have been performed for a number of parasitic diseases such as American trypanosomiasis (Chagas' disease), leishmaniasis, and malaria [2–5]. The HTS approach has been proven to have the potential to identify new drug candidates with novel modes of action. HTS approaches using 96-well cell culture plates have been well established for members of the *Trypanozoon* subgenus, including *Trypanosoma brucei brucei*, *T. b. gambiense*, *T. b. rhodesiense*, and *T. evansi* [6–10]. Recently, a luciferase-based viability assay to measure the concentration of ATP in cells has been adapted to HTS for the estimation of viable cell numbers of *T. b. brucei* and *T. b. gambiense* [7, 9]. Since, the luciferase-based viability assay method has been shown to offer greater sensitivity and be less time-consuming than the Alamar Blue™ assay system [5, 7]. This assay could be utilized for the HTS of trypanocidal compounds against animal African trypanosome species.

Thus, we have established an ATP-based luciferase viability assay for three species of animal African trypanosomes, namely *T. b. brucei*,

Daniel F. Gilbert and Oliver Friedrich (eds.), *Cell Viability Assays: Methods and Protocols*, Methods in Molecular Biology, vol. 1601, DOI 10.1007/978-1-4939-6960-9_8, © Springer Science+Business Media LLC 2017

T. evansi, and *T. congolense*, using the two available trypanocidal drugs (diminazene aceturate and pentamidine) as reference compounds [11].

2 Materials

2.1 Daily Trypanosome Culture

1. Hirumi's modified Iscove's medium (HMI-9) [12]: Iscove's modified Dulbecco's medium supplemented with 20% heat-inactivated fetal bovine serum (HI-FBS), 60 mM HEPES, 1 mM pyruvic acid sodium salt, 0.1 mM bathocuproine, 0.1 mM hypoxanthine and 16 μM thymidine (HT supplement), 10 μg/L insulin, 5.5 μg/L transferrin and 6.7 ng/L sodium selenite (ITS-X), 0.0001% 2-mercaptoethanol, 0.4 g/L BSA and 2 mM l-cysteine. The contents are mixed and the pH is adjusted to 7.2 with the addition of 10 M NaOH. Finally, the medium (HMI-9) is sterilized using a 0.45 μm filter.
2. Cell culture flask: e.g., Nunc™ cell culture EasYFlask™.
3. Transfer pipette.
4. CO_2 incubator: 5% CO_2, 37 °C (*T. b. brucei* and *T. evansi*), 33 °C (*T. congolense*).
5. Phase-contrast microscope.

2.2 ATP-Based Luciferase Viability Assay in a 96-Well Plate

1. Trypanosomes from daily culture.
2. Phosphate-buffered saline with 10% glucose (PSG).
3. Hemocytometer.
4. Multichannel pipette.
5. Reagent reservoir for the multi-channel pipette (sterilized).
6. Luminometer for the 96-well cell culture plate.
7. A 96-well cell culture plate suitable for the analysis of luminescence: e.g., Nunc™ 96-well optical bottom plate (*see* **Note 1**).
8. Plate shaker.
9. CellTiter-Glo®: Store CellTiter-Glo buffer and CellTiter-Glo substrate at −30 °C. Prior to use, thaw the buffer and the substrate at room temperature. Mix the buffer and substrate completely (CellTiter-Glo reagent). The CellTiter-Glo reagent is aliquoted into 1.5 mL tubes and kept at −30 °C until use.
10. Dimethyl sulfoxide (DMSO): Cell culture grade.
11. Reference compounds: Pentamidine and diminazene aceturate (200 μg/mL in DMSO). Using DMSO as a solvent, prepare 10 mg/mL stock solutions. Subsequently, adjust the concentration of the compounds to 200 μg/mL by 50× dilution with DMSO. The reference compounds are kept at −30 °C until use.

3 Methods

The trypanosome cultures are maintained by replacing the entire culture supernatant with fresh medium every other day (*see* **Note 2**). With the exception of the luminescence measurement and 50% inhibitory concentration (IC_{50}) calculation, all of the steps should be performed under a safety cabinet.

3.1 Preparation of the Trypanosomes

1. Collect trypanosomes from the cell culture flask in the logarithmic phase.
2. Count the trypanosomes using a hemocytometer. If the number of trypanosomes is too high to count using a hemocytometer, PSG can be used to dilute the concentration to a countable value.
3. Adjust the trypanosome concentration to 2×10^5 cells/mL (*T. congolense*) or 1×10^4 cells/mL (*T. b. brucei, T. evansi*) by HMI-9 (*see* **Note 3**).
4. In an incubator, maintain the *T. b. brucei* and *T. evansi* at 37 °C, and *T. congolense* at 33 °C until use.

3.2 Serial Dilution of the Reference Compounds

1. Prepare the 0.5% DMSO-HMI-9 by adding 50 μL of absolute DMSO into 10 mL of HMI-9. Aliquot 500 μL of the 0.5% DMSO-HMI-9 into 7 of 1.5 mL tubes.
2. Thaw the 200 μg/mL of the reference compounds at room temperature. Five microliters of the reference compounds are added into 1 mL of HMI-9 in a new 1.5 mL tube; the concentration of the reference compounds is 1000 ng/mL in 0.5% DMSO-HMI-9.
3. Mix 500 μL of the reference compounds (1000 ng/mL) prepared in **step 2** of Subheading **3.2** with 500 μL of the 0.5% DMSO-HMI-9 prepared in **step 1** of Subheading **3.2**. Subsequently, transfer 500 μL of diluted reference compounds to the next tube for the serial dilution of the reference compounds. After the repetition of 2× serial dilution, the concentrations of the reference compounds will range from 1000 to 3.9 ng/mL.

3.3 Plate Settings for the Cell Viability Assay

1. Add 50 μL of 0.5% DMSO-HMI-9 to the negative control wells (N) and blank wells (B), and 50 μL of serially diluted reference compounds to the test wells (P and D) (Fig. 1).
2. Dispense 50 μL of HMI-9 into the blank wells (Fig. 1).
3. Dispense 50 μL of the trypanosomes prepared in **step 3** of Subheading **3.1** to the negative control wells and test wells, and mix by gently pipetting (Fig. 1).

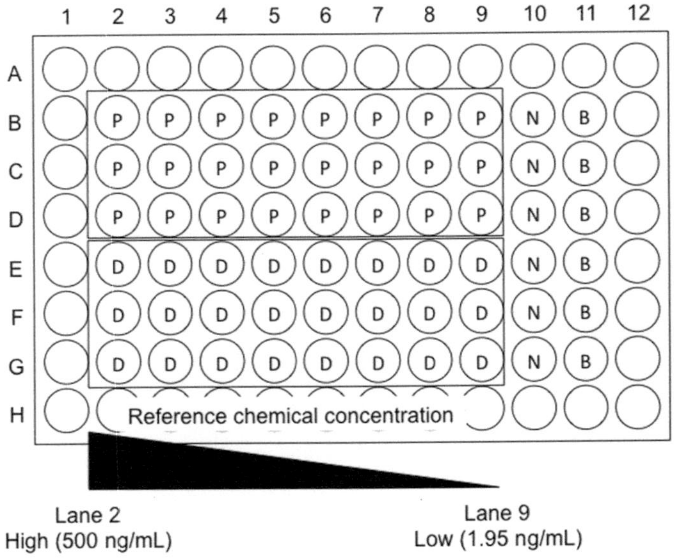

Fig. 1 Example of the plate setting. Example of a 96-well cell culture plate setting. Fifty microliters of serially diluted reference compounds aliquoted from high concentration wells (lane 2) to low concentration wells (lane 9). *P* Pentamidine test wells, *D* Diminazene aceturate test wells, *N* Negative control wells, *B* Blank wells

4. The final parasite cell concentrations that are used in the assay are 1×10^5 cells/mL (*T. congolense*) and 5×10^3 cells/mL (*T. b. brucei* and *T. evansi*), with 0.25% of DMSO in 100 μL of HMI-9 per well. The final concentrations of the reference compounds are 500–1.95 ng/mL. The plates are incubated for 72 h at 33 °C (*T. congolense*) or 37 °C (*T. b. brucei* and *T. evansi*).

3.4 Measurement of Luminescence

1. Observe the trypanosome cultures under a phase-contrast microscope (*see* **Note 4**).
2. Thaw and equilibrate CellTiter-Glo reagent at room temperature.
3. Add 100 μL of CellTiter-Glo reagent to each well and mix gently with a multi-channel pipette (*see* **Note 5**).
4. Gently shake the cell culture plate using a plate shaker to promote cell lysis.
5. Measure the luminescence values using a luminometer.

3.5 IC_{50} Value Calculation

1. Subtract of the mean luminescence value of the blank wells from each well.
2. Calculate the survival rate (%) by dividing the luminescence values of each well by the mean luminescence value of the negative control wells.

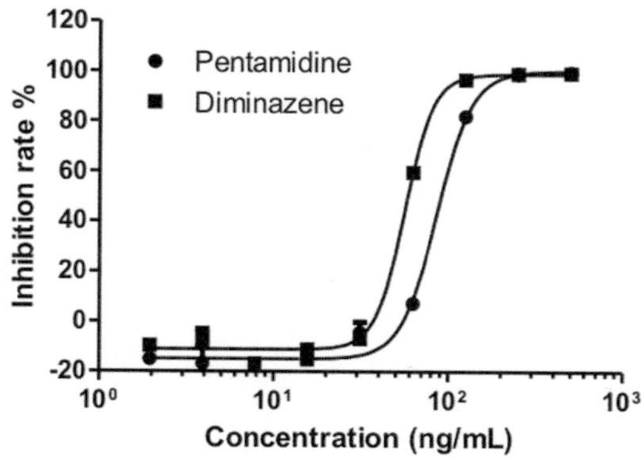

Fig. 2 IC_{50} values of pentamidine and diminazene aceturate in *T. congolense*. The IC_{50} values of diminazene aceturate and pentamidine in *T. congolense* were evaluated using the ATP-based luciferase viability assay. The rate of inhibition is relative to that of the control well (where no compound compounds were added). The sigmoidal dose response curve graphs were plotted using the GraphPad PRISM 5 software program. The IC_{50} values of pentamidine and diminazene aceturate were calculated as 100.45 ± 26.08 ng/mL (169.48 ± 44.00 nM) and 55.98 ± 13.15 ng/mL (108.65 ± 25.25 nM), respectively (Reproduced from ref. [11] with permission from The Japanese Society of Veterinary Science)

3. Calculate the inhibition rate (%) by subtracting the survival rate from 100.
4. The IC_{50} value of the serially diluted reference compounds is calculated by plotting the inhibition rate (0% inhibition = the luminescence of negative control wells) against log in a software program (e.g., GraphPad Prism) (Fig. 2).

3.6 Further Application of This Assay System

This system can be applied to the screening of trypanocidal compounds from drug libraries and to calculate the IC_{50} values of novel trypanocidal compounds by the serial dilution of compounds [13].

4 Notes

1. Non-transparent white cell culture plates (e.g., 96-well LumiNunc™ White, sterile) can be used for the measurement of luminescence in this assay. However, we recommend the use of a transparent bottom plate (e.g., Nunc™ optical bottom plate) because the condition of the trypanosomes in the 96-well culture plate and the trypanocidal effects are observed by phase-contrast microscopy.
2. *T. congolense* BSFs proliferate on the bottom of cell culture flask in an adhesion-dependent manner. Thus, *T. congolense* BSFs are detached from the culture flask using a transfer pipette and *T. congolense* BSFs are discarded with the supernatant in the daily maintenance of the culture. *T. b. brucei* and *T. evansi*

BSFs proliferate in the medium as a suspension culture. The logarithmic-phase trypanosomes are suitable for the cell viability assay; thus, in the trypanosome cultures should be assayed the day after the replacement of the medium.

3. A growth curve was plotted to show the growth kinetics of cells and calculate the doubling time (Td). To optimize the cell culture conditions, 100 μL of trypanosomes were incubated in a 96-well plate at 37 °C (*T. b. brucei* and *T. evansi*) or at 33 °C (*T. congolense*). Using a hemocytometer, the trypanosomes were counted every 24 h after cultivation until post-cultivation day 6 after appropriate dilution with PSG. Td was calculated by counting the cells in the log phase of growth and using the equation: Td = (t2–t1) × log (2)/log (q2/q1). Two measurements were made: the initial count (q1) at time (t1) and the resultant density following 24 h of incubation (q2, t2) [5, 8]. The trypanosome concentrations were optimized for our using trypanosome strains (*T. congolense* IL3000, *T. evansi* Tansui, and *T. b. brucei* GUTat3.1 strains); thus, it might be necessary to optimize the trypanosome concentrations for each of the trypanosome strains that are used.
4. The trypanocidal activities of the reference compounds are visually evaluated by phase-contrast microscopy.
5. The manufacturers' manual of the CellTiter-Glo recommends the use of 100 μL of CellTiter-Glo reagents for the 96-well cell culture plate assay. However, we can decrease the amount of CellTiter-Glo reagents to 25 μL/well without affecting the luminescence [11].

Acknowledgments

This study was financially supported by the Japan Society for the Promotion of Science (JSPS), Grants-in-Aid for Scientific Research from the Ministry of Education, Culture, Sports, Science and Technology (MEXT), and the AMED/JICA, SATREPS.

References

1. Pecoul B (2004) New drugs for neglected diseases: from pipeline to patients. PLoS Med 1:e6
2. Lang T, Goyard S, Lebastard M, Milon G (2005) Bioluminescent *Leishmania* expressing luciferase for rapid and high throughput screening of drugs acting on amastigote-harbouring macrophages and for quantitative real-time monitoring of parasitism features in living mice. Cell Microbiol 7:383–392
3. Engel JC, Ang KK, Chen S, Arkin MR, McKerrow JH, Dolyle PS (2010) Image-based high-throughput drug screening targeting the intracellular stage of *Trypanosoma cruzi*, the agent of Chagas' disease. Antimicrob Agents Chemother 54:3326–3334
4. Smilkstein M, Sriwilaijaroen N, Kelly JX, Wilairat P, Riscoe M (2004) Simple and inexpensive fluorescence-based technique for high-throughput antimalarial drug screening. Antimicrob Agents Chemother 48:1803–1806
5. Sykes ML, Baell JB, Kaiser M, Chatelain E, Moawad SR, Ganame D et al (2012)

Identification of compounds with antiproliferative activity against *Trypanosoma brucei brucei* strain 427 by a whole cell viability based HTS campaign. PLoS Negl Trop Dis 6:e1896

6. Räz B, Iten M, Grether-Bühler Y, Kaminsky R, Brun R (1997) The Alamar Blue assay to determine drug sensitivity of African trypanosomes *(T. b. rhodesiense and T. b. gambiense) in vitro.* Acta Trop 68:139–147
7. Sykes ML, Avery VM (2009) A luciferase based viability assay for ATP detection in 384-well format for high throughput whole cell screening of *Trypanosoma brucei brucei* bloodstream form strain 427. Parasit Vectors 2:54
8. Sykes ML, Avery VM (2009) Development of an Alamar Blue viability assay in 384-well format for high throughput whole cell screening of *Trypanosoma brucei brucei* bloodstream form strain 427. Am J Trop Med Hyg 81:665–674
9. Van Reet N, Pyana P, Rogé S, Claes F, Büscher P (2013) Luminescent multiplex viability assay for *Trypanosoma brucei gambiense.* Parasit Vectors 6:207
10. Gillingwater K, Büscher P, Brun R (2007) Establishment of a panel of reference *Trypanosoma evansi* and *Trypanosoma equiperdum* strains for drug screening. Vet Parasitol 148:114–121
11. Suganuma K, Allamanda P, Hakimi H, Zhou M, Angeles JM, Kawazu S et al (2014) Establishment of ATP-based luciferase viability assay in 96-well plate for *Trypanosoma congolense.* J Vet Med Sci 76:1437–1441
12. Hirumi H, Hirumi K (1991) *In vitro* cultivation of *Trypanosoma congolense* bloodstream forms in the absence of feeder cell layers. Parasitology 102(Pt 2):225–236
13. Suganuma K, Sarwono AE, Mitsuhashi S, Jąkalski M, Okada T, Nthatisi M et al (2016) Mycophenolic acid and its derivatives as potential chemotherapeutic agents targeting inosine monophosphate dehydrogenase in *Trypanosoma congolense.* Antimicrob Agents Chemother 60(7):4391–4393

Chapter 9

SYBR® Green I-Based Fluorescence Assay to Assess Cell Viability of Malaria Parasites for Routine Use in Compound Screening

Maria Leidenberger, Cornelia Voigtländer, Nina Simon, and Barbara Kappes

Abstract

Owing to its fast and reliable assessment of parasite growth, the SYBR® Green I-based fluorescence assay is widely used to monitor drug susceptibility of malaria parasites. Its particular advantages are that it is a simple, one-step procedure and very cost-effective making it especially suited for high through put screening of newly developed drugs and drug combinations. Here we describe a SYBR® Green I-based fluorescence assay protocol to be used for routine screening of compounds and extracts in a research laboratory environment. A variation of the standard protocol is also provided allowing to address stage-specific effects of fast-acting drugs.

Key words Fluorescence assay, SYBR Green, Growth inhibition assay, Malaria, *Plasmodium falciparum*, Compound screening

1 Introduction

All in vitro antimalarial drug sensitivity assays are based on the in vitro culture technique developed by Trager and Jensen (1976) allowing for a continuous culture of *Plasmodium falciparum* in a blood medium-mixture under controlled atmospheric conditions (*see* Subheading 3.3) [1]. This continuous culture technique enabled the development of several antimalarial drug sensitivity assay systems that are inter alia used (a) to screen for novel or improved chemical entities with antimalarial activity, (b) to examine the interaction of drug combinations, (c) to monitor the emergence of drug-resistant parasite strains, and (d) to determine the baseline sensitivity to new drugs before their market launch in a country.

The [^{3}H]hypoxanthine incorporation assay developed in 1979 by Desjardins and coworkers has been the gold standard for antimalarial drug screening over decades [2]. Numerous

Daniel F. Gilbert and Oliver Friedrich (eds.), *Cell Viability Assays: Methods and Protocols*, Methods in Molecular Biology, vol. 1601, DOI 10.1007/978-1-4939-6960-9_9,

antimalarial drug sensitivity microtiter plate assays have been developed since then omitting the use of radioactivity for the detection of parasite growth inhibition [3–10]. A detailed description of a number of these assays is provided in a comprehensive book by Basco [11]. Therefore only two of these assays, which are widely used in the community, will be mentioned here. In 1993, Makler et al. established a novel colorimetric screening technique based on the activity of the parasite lactate dehydrogenase (pLDH) and the commercially available Malstat™ reagent [12]. In brief, 3-acetylpyridine adenine dinucleotide (APAD), a nicotinamide adenine dinucleotide (NAD) analogon, is reduced to APADH by pLDH, which in a subsequent reaction is used to reduce the yellow tetrazolium salt, nitroblue tetrazolium, to the blue formazan product in the presence of phenazine ethosulfate. Reaction progress is detected by measuring the absorbance at 650 nm in a plate reader. In 2002, Noedl and coworkers adapted a commercially available enzyme-linked immunosorbent assay (ELISA) based on monoclonal antibodies directed against *P. falciparum* histidine-rich protein II to malaria drug sensitivity testing [7]. Although widely used both assays are, however, not advisable for high-throughput screening (HTS), since they are too expensive or involve multistep procedure [13, 14]. This problem was resolved with the introduction of the SYBR green I dye-based fluorescence assay that has been developed by Smilkstein and coworkers in 2004 [9]. This assay is suited for HTS and has been reported to be as sensitive as the [^{3}H]hypoxanthine incorporation assay [13]. Since its implementation, the SYBR green I fluorescence assay has been extensively validated and compared with existing methods and is nowadays for most applications the method of choice among the in vitro assays for testing the sensitivity of the human malaria parasites to drugs [13–17].

The SYBR green I fluorescence assay relies on the binding of SYBR green I to parasite DNA. It preferentially binds to double-stranded versus single-stranded DNA or RNA [18]. The fluorescence intensity therefore reflects the amount of double-stranded DNA in an individual sample. Mature erythrocytes are acaryotes lacking a nucleus and accordingly DNA and RNA. Thus the fluorescence signal detected in parasite cultures originates from the binding of the dye to parasite DNA in any of the erythrocytic stages of the parasite [4]. The fluorescence intensity can be determined using a minifluorometer, a fluorescence spectrophotometer, or a HTS versatile fluorescence-activated microplate reader.

2 Materials

Prepare all solutions with ultrapure water from a Milli-Q water filtration station (sensitivity of 18 MΩ at 25 °C) and analytical grade reagents. Unless otherwise indicated reagents are prepared and kept at room temperature (RT). Cell culture medium and solutions for the treatment of parasite cultures have to be sterile. Therefore, autoclave or sterile filter (0.2 μm filter) all solutions that are used. Prepare solutions consisting of multiple components from sterile stock solutions under a clean bench or sterile-filter it after preparation. In general, use media and solutions prewarmed to 37 °C for culturing parasites or to treat a parasite culture.

1. AlbuMAX® I: 10% in ultrapure H_2O. Filter the solution through a 0.2 μm Steritop filter unit (for the preparation of AlbuMAX® I *see* Subheading 3.1).
2. Complete medium: RPMI 1640 containing 2 mM l-glutamine, 25 mM N-2-hydroxyethylpiperazine-N-2-ethanesulfonic acid (HEPES) and 24 mM $NaHCO_3$ supplemented with 0.1 mM hypoxanthine, 25 μg/mL gentamicin and 0.5% AlbuMAX® I to constitute complete medium. Store at 4 °C.
3. Incomplete medium: RPMI 1640 w/o AlbuMAX® I and w/o hypoxanthine. Store at 4 °C (*see* **Note 1**).
4. Washed type A positive human erythrocytes with a hematocrit of 50% (*see* Subheading 3.2).
5. D-sorbitol: 5% in ultrapure H_2O. Filter through a 0.2 μm Steritop filter unit.
6. Cell culture dishes for suspension cell lines (10 cm × 22 mm).
7. Black 96 well microtiter plate for suspension cultures with cover plate.
8. 75 cm^2 cell culture flasks.
9. 12-channel pipette.
10. Antimalaria drugs: Prepare 10 mM stocks in cell culture-tested dimethyl sulfoxide (DMSO). Store at −80 °C.
11. SYBR® Green I nucleic acid stain. Store at −20 °C (*see* **Note 2**).
12. Lysis buffer: 40 mM Tris–HCl, pH 7.5, 10 mM EDTA, pH 8.0, 0.02% saponin, 0.08% Triton X-100, SYBR® Green I 1:10,000 (*see* **Note 3**).
13. Chloroquine diphosphate salt: 10 mM stock in ultrapure H_2O. Store at −80 °C.
14. 100% Methanol.
15. Giemsa solution.
16. Easycoll.

3 Methods

3.1 Preparation of AlbuMAX® I

1. Weigh out 50 g AlbuMAX® I in a 500 mL glass cylinder and add ultrapure H_2O to 400 mL.
2. Mix vigorously. It takes about 30 min until AlbuMAX® I is dissolved.
3. Prior to further processing, check carefully that AlbuMAX® I is, indeed, completely dissolved. Add ultrapure H_2O to 500 mL, mix thoroughly and filter the solution through a 0.2 μm Steritop filter unit.
4. Make aliquots of 25 mL and freeze them at –20 °C for long-term storage.

3.2 Preparation of Erythrocytes from Human Blood Donations

1. Use a 500 mL bag of fresh human red blood cells (usually 300–400 mL) from one donor.
2. Make 30 mL aliquots in 50 mL sterile conical tubes.
3. Pellet red blood cells by centrifugation at 2000 × *g* for 10 min at RT (acceleration 9, deceleration 7).
4. Discard the supernatant and wash the red blood cell pellet with incomplete medium.
5. Repeat this washing step.
6. Determine the precise volume of the erythrocyte pellet and add the same volume of incomplete medium to resuspend the erythrocyte pellet.
7. The red blood cell solution has now a hematocrit of 50% and can be stored for up to 4–6 weeks at 4 °C.

3.3 Parasite Culturing

1. Maintain *P. falciparum* in a continuous culture in human erythrocytes in RPMI 1640 supplemented with AlbuMAX® I as serum substitute at 37 °C under controlled atmospheric conditions of 5% CO_2, 3% O_2, and 92% N_2 [1].
2. Cultivate the infected red blood cells in 10 cm petri dishes with a hematocrit of 5% (*see* **Note 4**). Parasitemia should be maintained between 3 and 5% (*see* Subheading 3.4 for details on the determination of parasitemia).
3. Dilute parasite cultures every 48 h to a parasitemia of 0.4%. Use prewarmed complete medium and uninfected erythrocytes.

3.4 Determination of Parasitemia via Thin Blood Film

The determination of parasitemia via a thin blood film, which is described below, is schematically depicted in Fig. 1.

1. Remove the culture supernatant and resuspend the infected red blood cells in 10 mL of prewarmed complete medium.
2. Transfer 100 μL of the suspension into a 1.5 mL microcentrifuge tube and centrifuge for 1 min at 800 × *g*.

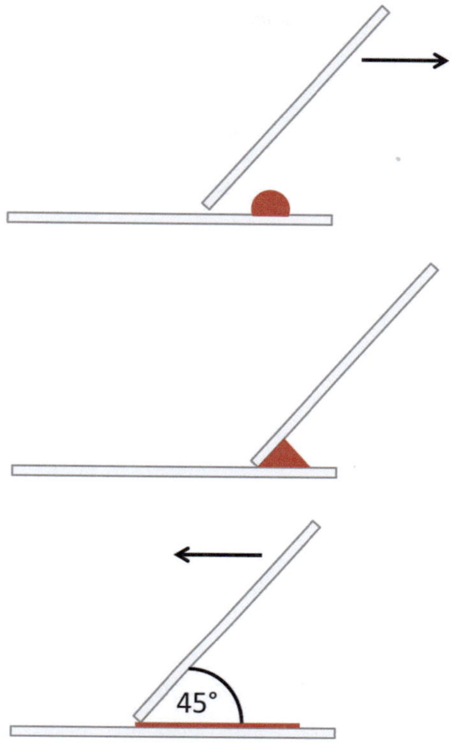

Fig. 1 *Preparation of a thin blood film*: Schematic representation

3. Remove 70 μL of the supernatant and resuspend the remaining cell pellet.
4. Transfer 5 μL of the suspension to a microscope slide.
5. Approach a spreader slide to the suspension.
6. Hold the spreader slide at a 30–45° angle and push it forward smoothly.
7. Wait until the smear is air dried.
8. Dip the slide in 100% methanol in order to fix the cells to the slide.
9. Wait until the smear is air dried again.
10. Incubate the slide for 15 min in Giemsa staining solution (dilution: 1:15 in ultrapure H_2O) in a coplin jar.
11. Remove the slide from the coplin jar, rinse it briefly in a jet of ultrapure H_2O, and let it air dry.
12. Analyze the slide with a microscope at 1000× magnification in oil immersion.
13. Count the number of infected erythrocytes in 1000 erythrocytes and determine the parasitemia by applying the formula below.
14. $$parasitemia\ [\%] = 100 \times \frac{number\ of\ infected\ erythrocytes}{1000}.$$

3.5 Parasite Synchronization by Sorbitol Treatment

Parasite cultures are synchronized by sorbitol treatment according to Lambros and Vanderberg [19]. In our hands and with our culture conditions, the 3D7 strain has an intraerythrocytic growth cycle of 44 h. The time values given below refer to this cycle lengths. If a given parasite strain has a shortened or prolonged intraerythrocytic development, these values have to be adapted accordingly.

1. Synchronize the parasite culture 44 h before the planned assay. Parasitemia should not exceed 5%.
2. Transfer the parasite culture into a 50 mL conical tube.
3. Centrifuge the culture at RT for 2 min at 800 × *g* (acceleration 9, deceleration 7).
4. Remove the supernatant and resuspend the pellet in 10 mL of prewarmed 5% D-sorbitol.
5. Incubate the suspension for 15 min in a water bath at 37 °C and invert the falcon tube gently after 7 min.
6. Centrifuge the suspension for 2 min at 800 × *g* (acceleration 9, deceleration 7) and RT.
7. Remove the supernatant and wash the erythrocytes in 20 mL of prewarmed incomplete medium.
8. Resuspend the red blood cells in 10 mL of prewarmed complete medium.
9. Determine the parasitemia (*see* Subheading 3.4).
10. Dilute the culture with the appropriate volume of uninfected erythrocytes and complete medium to a hematocrit of 5% and a parasitemia of 2%. One culture dish will be sufficient.
11. Incubate parasite culture at 37 °C and the above-described atmospheric conditions for further 44 h before starting the assay (*see* **Note 5**). Change culture medium after 20–28 h.

3.6 Assay Preparation of the Cell Proliferation Assay (Standard Protocol)

The SYBR® Green I drug sensitivity assay is adapted from Smilkstein et al. [9]. This assay is performed with a cell suspension containing synchronized ring-stage parasites at a parasitemia of 0.2% and a hematocrit of 2%. Each drug is screened in triplicates in a given assay and in three independent assays (*see* **Note 6**). Standard antimalarial drugs such as chloroquine, artesunate, or dihydroartemisinin (DHA) serve as positive controls for growth inhibition.

1. Use a black 96 well microtiter plate (*see* **Note 7**).
2. Add 50 μL of complete medium to each well.
3. Dedicate 8 wells for the negative control (erythrocytes w/o parasites) and 16 wells for positive control (erythrocytes with parasites w/o drug). An example on how a microtiter plate is routinely loaded is shown in Fig. 2.

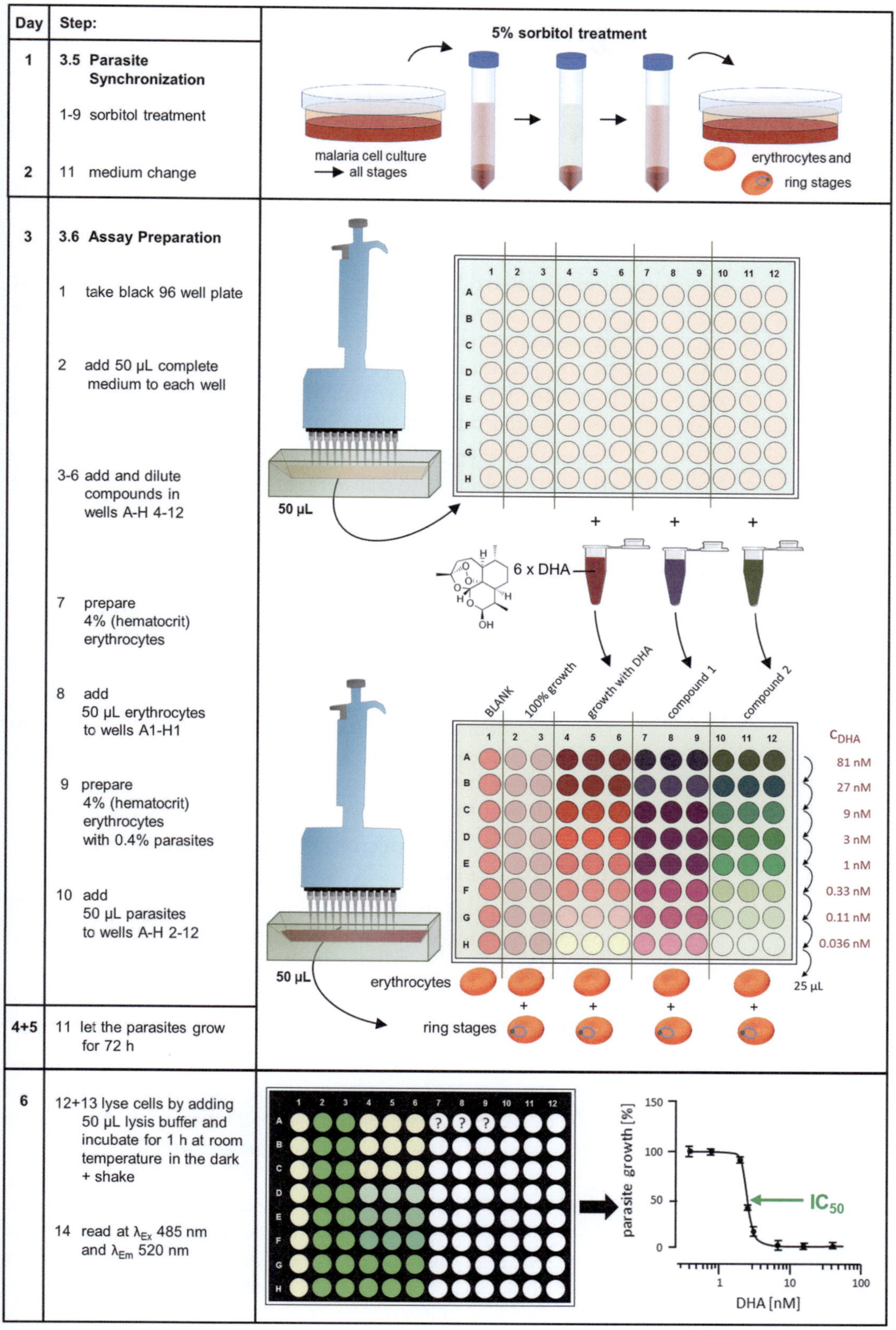

Fig. 2 *Fluorescence-based in vitro antimalarial drug sensitivity assay*: Schematic representation of experimental steps in their chronological order

4. Prepare drug dilutions in complete medium. Dilute the drugs to their sixfold starting concentration in a 1.5 mL microcentrifuge tube. Mix properly. The sixfold starting concentration of chloroquine and artesunate is 1458 nM and of DHA 486 nM.
5. Add 25 μL of the sixfold starting concentration of the drug to the first row of the microtiter plate (*see* Fig. 2). Mix carefully by pipetting.
6. Prepare serial dilutions of the drugs (1:3) by transferring 25 μL from row 1 to row 2 with a 12-channel pipette. Mix carefully. Again transfer 25 μL from row 2 to row 3 using a 12-channel pipette. Continue this procedure until all dilutions have been prepared. Discard the remaining 25 μL from the wells of the last row after dilution has been made.
7. Prepare an erythrocyte solution with a hematocrit of 4% in complete medium and transfer it into a reagent reservoir. 6 mL is sufficient for loading of a complete microtiter plate.
8. Add 50 μL of this solution to each well of the negative control (*see* Fig. 2).
9. Add a highly synchronized ring-stage culture to the erythrocyte solution in the reagent reservoir so that the final parasitemia will be 0.4%. The final hematocrit of this parasite solution has to be adjusted to 4%. Mix carefully avoiding air bubbles.
10. Use a 12-channel pipette to immediately transfer 50 μL of the parasite solution to each well of positive control and of the drug dilutions. Start with the positive control and the lowest drug dilution to avoid cross-contamination.
11. Incubate the loaded microtiter plate under controlled atmospheric conditions at 37 °C for 72 h (*see* Subheading 3.3).
12. Prepare 6 mL lysis buffer per plate in a reagent reservoir and add 50 μL to each well using a 12-channel pipette (*see* **Note 3**).
13. Lyse the parasites by incubating them in lysis buffer at RT for 1 h in the dark with constant shaking.
14. Detect the fluorescence signal using a fluorescence-activated microplate reader (top reading). Use the instrument settings with an excitation wavelength of 485 nm and an emission wavelength of 520 nm (exposure time 0.1 or 0.2 s).

3.7 Experimental Modification: Stage-Specific Evaluation of Drug Activity

In order to determine the efficacy of a drug on a defined asexual blood stage of *P. falciparum*, malaria parasites have to be tightly synchronized to a cycle window of ± 2 h. To determine the effect of a drug on a given parasite stage, parasites have to be synchronized twice within an interval of 16 h (*see* Fig. 3).

To perform this tight synchronization, the precise duration of the intraerythrocytic growth cycle of the respective parasite strain has to be known (**Note 8**). Since the drug assay is only performed

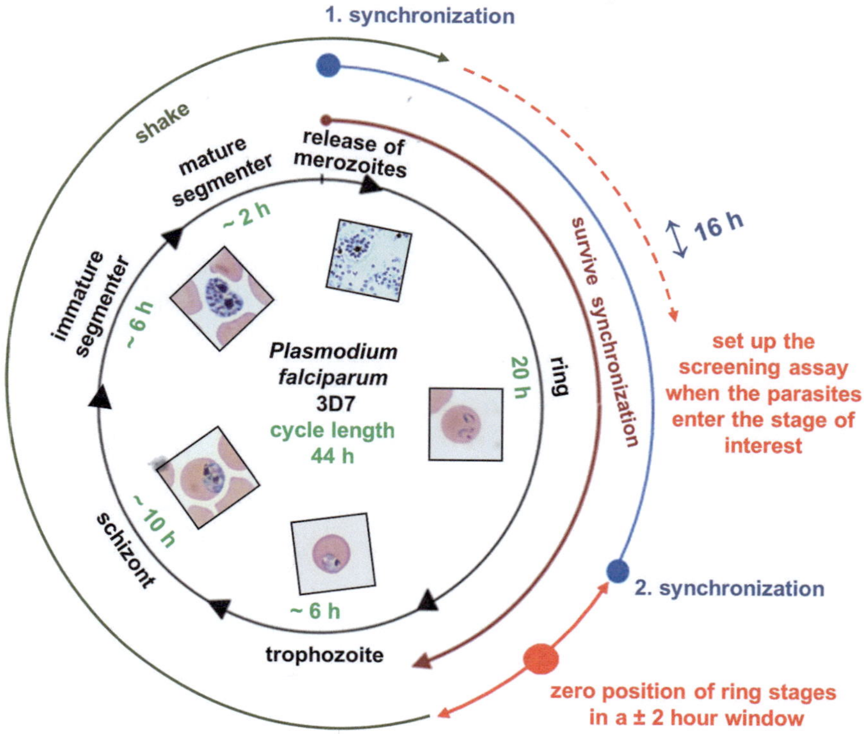

Fig. 3 *Time frame of the individual asexual blood stages of the P. falciparum 3D7 strain*: The 44 h intraerythrocytic growth cycle can divided into four stages based on morphological and biochemical criteria. During almost half of the intraerythrocytic development (ring stage (1–20 h)) the parasite displays only little measurable metabolic activity. Protein and nucleic acid synthesis are negligible. The following stage, the trophozoite (20–26 h) is characterized by an extensive RNA and protein biosynthesis. RNA synthesis peaks during this phase of parasite development. Rings and trophozoites are mononucleated cells. DNA synthesis is initiated in trophozoites. In the subsequent schizont stage (26–36 h) 4–5 rounds of mitotic divisions take place that result in a single syncytial cell containing 8–32 nuclei. In the final segmenter stage (36–44 h) individual merozoites are formed by budding of nuclei and cytoplasmic organelles from the main parasite mass. Merozoites (about 44 h) are released, which infect fresh erythrocytes and thereby initiate a new intraerythrocytic cycle. The time spans given relate to the 44 h growth cycle of the 3D7 strain. The synchronization scheme to obtain synchronous parasite stages with a cycle window of ±2 h is depicted in the figure

for the duration of the respective parasite stage, it has to be stressed that the assay outlined below is only applicable for fast-acting drugs or compounds (*see* also **Note 9**).

1. Start with a parasite culture with a high initial parasitemia (10–15%) that mainly consists of ring-stage parasites. Ideally change medium 5–8 h prior to the first synchronization.
2. Synchronize the culture twice within an interval of 16 h (*see* Subheading 3.5).
3. Separate surviving ring forms and erythrocytes from parasite debris on an Easycoll step gradient after the second synchronization. The step gradient is made as follows. Pour a 3 mL layer

of 75% Easycoll in incomplete medium in a 15 mL conical tube. Resuspend the cells of a culture plate in 3 mL incomplete medium and carefully put these on top of the Easycoll layer avoiding any disturbances of this layer.

4. Centrifuge the gradient at RT for 10 min at 1450 × *g* in a swing-out rotor centrifuge (acceleration 7 and deceleration 5). Parasite debris (parasites that died due to the sorbitol treatment) accumulates in the interphase between the incomplete medium and Easycoll layer. Erythrocytes and ring stages are found at the bottom of the gradient (*see* Fig. 4).
5. Aspirate the cell pellet from the bottom of the gradient with a 1 mL pipette tip. To do so move the tip through the upper layers until it touches the pellet surface. Then start with the aspiration of the pellet.
6. Transfer the obtained cell pellet into a new 15 mL conical tube and wash it 3 × with 10 mL incomplete medium to remove the remaining Easycoll.
7. Resuspend the cells in 50–60 mL complete medium and transfer them into a 75 cm^2 culture flask (Parasitemia can be checked at this point (*see* Subheading 3.4). It should be around 1–2%).
8. Transfer the flask to a cell culture incubator and fix it on a slowly moving shaker with its bottleneck upwards until the newly formed merozoites have completed their invasion of erythrocytes and developed into ring stages. This step is required to maintain synchronicity, to achieve a higher parasitemia, and to obtain erythrocytes infected with a single parasite.
9. Change culture medium 30–35 h after the second synchronization step. Do not replace it earlier to avoid suction of released merozoites.

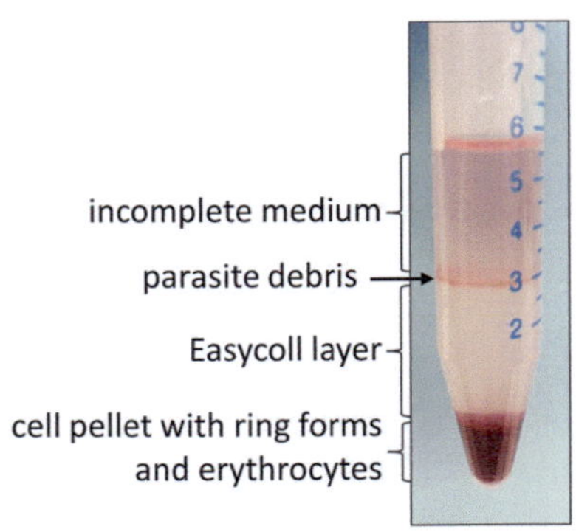

Fig. 4 *Easycoll gradient to separate ring stages from parasite debris after centrifugation:* The cell pellet at the bottom contains the desired ring-stage parasites. Parasite debris is found in the interphase between the Easycoll layer and the incomplete medium

10. Start with the preparations for the compound screening assay in advance (roughly 1–2 h). The assay has to be ready before the parasites at the zero position (0 ± 2 h) are entering the stage of interest (*see* Fig. 3 for a precise definition of the zero position).
11. According to the time schedule presented in Fig. 3, the zero position corresponds to ~46 h for trophozoite stages, ~52 h for schizonts, ~62 h for segmenter stages, and ~68 h for merozoites after the second synchronization step (Table 1).
12. To analyze for stage-specific effects of a given drug, the routine protocol described under Subheading 3.6 has to be slightly modified. Parasites have to be plated with a parasitemia of about 1%.
13. Fluorescence will be measured at the end of a given stage except for merozoites where the fluorescence will be determined 4 h after the setup of the assay (Table 1).

3.8 Calculation and Validation of the Experimental Datasets

1. To determine the percentage of growth inhibition, export the raw data from the plate reader to Excel.
2. Arithmetically average the growth values of the positive control (parasite culture w/o drugs) and determine the mean values per well.
3. Subtract the averaged values of the negative control from the mean values of the positive control and determine the standard deviation.
4. Use triplicates to calculate mean values of a given drug concentration. Subtract the mean value of the negative control and determine the percentage of growth inhibition.
5. To calculate the half maximal inhibitory concentration (IC_{50}) of a given compound required to inhibit parasite growth by 50%, import the above data into Sigma plot. Use sigmoidal equation category and hill function to determine the precise IC_{50} value [20].

Table 1

Overview on time periods between the different parasite treatments

Stage of interest	Time period between the two synchronizations steps	Time span between the second synchronization and start of the assay	Time between assay set up and read out
Trophozoite	16 h	46 h (44 h + 2 h)	6 h
Schizont	16 h	52 h (44 h + 2 h + 6 h)	10 h
Segmenter	16 h	62 h (44 h + 2 h + 6 h + 10 h)	8 h
Merozoite	16 h	68 h (44 h + 2 h + 6 h + 10 h + 6 h)	4 h

4 Notes

1. For some wash steps it is not necessary to use complete medium. Therefore AlbuMAX® I and hypoxanthine are usually omitted.
2. SYBR® Green I is light sensitive. The solution has therefore to be protected from light by aliquoting the purchased stock solution in light safe microcentrifuge tubes, which are stored at −20 °C. The working solution of SYBR® Green I is a 1:10,000 dilution in lysis buffer.
3. Lysis buffer is prepared freshly from the following stock solutions: 1 M Tris–HCl, pH 7.5, 0.5 M EDTA, pH 8.0, 10% saponin (stored at −20 °C; only thaw once), 20% Triton X-100. Six millilitre lysis buffer is sufficient for one microplate. Lysis buffer containing SYBR® Green I needs to be used up within 24 h.
4. All *P. falciparum* lab strains are originally derived from human infections. The 3D7 clone, which is used for routine screening was derived from NF54 by limiting dilution [21]. NF54 was collected from a patient living close to Schipol Airport, Amsterdam, who had never left the Netherlands and acquired the so-called airport malaria [22].
5. Check the physical constitution of the parasite culture via a Giemsa smear. If the parasites are in good shape, determine the parasitemia of the synchronized culture precisely 44 h after synchronization. Typically a parasitemia of 10% ±1.5% is expected with 99% of the parasites being ring stages. Immediately start with the drug assay.
6. Note that each drug has to be screened in three independent assays to obtain three biological replicates.
7. Black plates absorb light. They are therefore used to minimize the fluorescence background and crosstalk. For this reason black plates are usually used for fluorescence-based assays.
8. When the growth cycle of a given parasite strain differs from the 44 h cycle of 3D7, the two synchronizations steps have to be adjusted according to the actual length of the ring phase of the respective parasite strain. To put it another way, if the intraerythrocytic growth cycle of a parasite strain is slower than the one of 3D7, the time point for the second synchronization step has to be postponed. If the cycle is, however, faster than the one of 3D7, the time point for the second synchronization has to be antedated. A shortened or extended intraerythrocytic growth cycle will further effect the duration of the individual parasite stages, which will also be either shortened or prolonged. Thus the precise start and duration of the respective parasite stages have to be established first, since the drug has to be administered at the beginning and for the duration stage of the stage of choice.

9. Note that this assay is although not applicable for most of the drugs interfering with individual steps in cell cycle progression, since checkpoints controlling the correct order of events during the nuclear division cycles are absent [23].

Acknowledgment

Cornelia Voigtländer gratefully acknowledges funding of the Erlangen Graduate School in Advanced Optical Technologies (SAOT) by the German Research Foundation (DFG) in the framework of the German excellence initiative. This work was supported by the project preparatory fund of the University of Erlangen-Nürnberg.

References

1. Trager W, Jensen JB (1976) Human malaria parasites in continuous culture. Science 193:673–675
2. Desjardins RE, Canfield CJ, Haynes JD et al (1979) Quantitative assessment of antimalarial activity in vitro by a semiautomated microdilution technique. Antimicrob Agents Chemother 16:710–718
3. Baniecki ML, Wirth DF, Clardy J (2007) High-throughput Plasmodium falciparum growth assay for malaria drug discovery. Antimicrob Agents Chemother 51:716–723
4. Bennett TN, Paguio M, Gligorijevic B et al (2004) Novel, rapid, and inexpensive cell-based quantification of antimalarial drug efficacy. Antimicrob Agents Chemother 48:1807–1810
5. Corbett Y, Herrera L, Gonzalez J et al (2004) A novel DNA-based microfluorimetric method to evaluate antimalarial drug activity. Am J Trop Med Hyg 70:119–124
6. Druilhe P, Moreno A, Blanc C et al (2001) A colorimetric in vitro drug sensitivity assay for Plasmodium falciparum based on a highly sensitive double-site lactate dehydrogenase antigen-capture enzyme-linked immunosorbent assay. Am J Trop Med Hyg 64:233–241
7. Noedl H, Wernsdorfer WH, Miller RS et al (2002) Histidine-rich protein II: a novel approach to malaria drug sensitivity testing. Antimicrob Agents Chemother 46:1658–1664
8. Quashie NB, De Koning HP, Ranford-Cartwright LC (2006) An improved and highly sensitive microfluorimetric method for assessing susceptibility of Plasmodium falciparum to antimalarial drugs in vitro. Malar J 5:95
9. Smilkstein M, Sriwilaijaroen N, Kelly JX et al (2004) Simple and inexpensive fluorescence-based technique for high-throughput antimalarial drug screening. Antimicrob Agents Chemother 48:1803–1806
10. Weisman JL, Liou AP, Shelat AA et al (2006) Searching for new antimalarial therapeutics amongst known drugs. Chem Biol Drug Des 67:409–416
11. Basco K (2007) Field application of in vitro assays for the sensitivity of human malaria parasites to antimalarial drugs. In: World Health Organization
12. Makler MT, Ries JM, Williams JA et al (1993) Parasite lactate dehydrogenase as an assay for Plasmodium falciparum drug sensitivity. Am J Trop Med Hyg 48:739–741
13. Abiodun OO, Gbotosho GO, Ajaiyeoba EO et al (2010) Comparison of SYBR Green I-, PicoGreen-, and [3H]-hypoxanthine-based assays for in vitro antimalarial screening of plants from Nigerian ethnomedicine. Parasitol Res 106:933–939
14. Bacon DJ, Latour C, Lucas C et al (2007) Comparison of a SYBR green I-based assay with a histidine-rich protein II enzyme-linked immunosorbent assay for in vitro antimalarial drug efficacy testing and application to clinical isolates. Antimicrob Agents Chemother 51:1172–1178
15. Co EM, Dennull RA, Reinbold DD et al (2009) Assessment of malaria in vitro drug combination screening and mixed-strain infections using the malaria Sybr green I-based fluorescence assay. Antimicrob Agents Chemother 53:2557–2563
16. Johnson JD, Dennull RA, Gerena L et al (2007) Assessment and continued validation

of the malaria SYBR green I-based fluorescence assay for use in malaria drug screening. Antimicrob Agents Chemother 51:1926–1933

17. Vossen MG, Pferschy S, Chiba P et al (2010) The SYBR Green I malaria drug sensitivity assay: performance in low parasitemia samples. Am J Trop Med Hyg 82:398–401
18. Zipper H, Brunner H, Bernhagen J et al (2004) Investigations on DNA intercalation and surface binding by SYBR Green I, its structure determination and methodological implications. Nucleic Acids Res 32:e103
19. Lambros C, Vanderberg JP (1979) Synchronization of Plasmodium falciparum erythrocytic stages in culture. J Parasitol 65:418–420
20. Beez D, Sanchez CP, Stein WD et al (2011) Genetic predisposition favors the acquisition of stable artemisinin resistance in malaria parasites. Antimicrob Agents Chemother 55:50–55
21. Walliker D, Quakyi IA, Wellems TE et al (1987) Genetic analysis of the human malaria parasite Plasmodium falciparum. Science 236:1661–1666
22. Ponnudurai T, Leeuwenberg AD, Meuwissen JH (1981) Chloroquine sensitivity of isolates of Plasmodium falciparum adapted to in vitro culture. Trop Geogr Med 33:50–54
23. Doerig C, Chakrabarti D, Kappes B et al (2000) The cell cycle in protozoan parasites. Prog Cell Cycle Res 4:163–183

Chapter 10

Screening Applications to Test Cellular Fitness in Transwell® Models After Nanoparticle Treatment

Bastian Christ*, Christina Fey*, Alevtina Cubukova, Heike Walles, Sofia Dembski, and Marco Metzger

Abstract

Nanoparticles (NPs) in biotechnology hold great promise for revolutionizing medical treatments and therapies. In order to bring NPs into clinical application there is a number of preclinical in vitro and in vivo tests, which have to be applied before. The initial in vitro evaluation includes a detailed physicochemical characterization as well as biocompatibility tests, among others. For determination of biocompatibility at the cellular level, the correct choice of the in vitro assay as well as NP pretreatment is absolutely essential. There are a variety of assay technologies available that use standard plate readers to measure metabolic markers to estimate the number of viable cells in culture. Each cell viability assay has its own set of advantages and disadvantages. Regardless of the assay method chosen, the major factors critical for reproducibility and success include: (1) choosing the right assay after comparing optical NP properties with the read-out method of the assay, (2) verifying colloidal stability of NPs in cell culture media, (3) preparing a sterile and stable NP dispersion in cell culture media used in the assay, (4) using a tightly controlled and consistent cell model allowing appropriate characterization of NPs. This chapter will briefly summarize these different critical points, which can occur during biocompatibility screening applications of NPs.

Key words Nanoparticles, Cytotoxicity assay, Impedance spectroscopy, Colloidal dispersion of nanoparticles

1 Introduction

Well-tailored multifunctional nanoparticles (NPs) play a major role in the development of future-oriented advanced functional materials for life science applications. The surface engineering and application of NPs in biotechnology holds great promise for revolutionizing medical treatments and therapies in areas such as imaging, faster diagnostics, drug delivery, and tissue regeneration. Because of the NP size, which is similar to subcellular structures such as proteins or DNA, they can interact with each other and

*Bastian Christ and Christina Fey contributed equally to this work.

Daniel F. Gilbert and Oliver Friedrich (eds.), *Cell Viability Assays: Methods and Protocols*, Methods in Molecular Biology, vol. 1601, DOI 10.1007/978-1-4939-6960-9_10, © Springer Science+Business Media LLC 2017

initiate various processes at this level. Despite the already existing diversity of R&D work on the NP field, there is still reinforced need to study NP clinical relevant properties and to bring them into the application. One of the biggest challenges here is the correlation of physicochemical NP properties such as size, morphology, and surface quality and their in vivo interactions [1].

More and more new smart NP systems are synthesized. There are numerous methods to investigate physicochemical properties of novel NP. The NP composition (qualitative and quantitative) can be determined using wet chemical analysis, mass spectrometry with inductively coupled plasma (ICP), and X-ray fluorescence (XRF). The dynamic light scattering (DLS) and the Fraunhofer diffraction studies allow the determination of particle size and polydispersity. More information on the particle size and morphology can be obtained by the transmission (TEM) and scanning electron microscopy (SEM). Another method, applied in the characterization of NPs, is X-ray diffraction (XRD). Thus, crystal structure and size can be determined. Usually, there are different spectroscopic methods such as nuclear magnetic resonance (NMR), infrared (IR) and Raman spectroscopy, as well as determination of the surface charge (Zeta-potential) available for surface characterization. The biggest challenge is the quantitative determination of the surface features. NP surface can be characterized by thermogravimetry (TGA), elemental analysis (CHNS), X-ray photoelectron spectroscopy (XPS), and potentiometric and photometric techniques. These methods, established in the surface analysis as standard, should be adapted and calibrated for the NP characterization.

The methods of NP biocompatibility estimating are also imperfect. For a reliable in vitro determination of pathophysiological effects at the cellular level, the correct choice of the method of examination as well as NP pretreatment is absolutely essential. The method must be selected so that there are no artifacts, otherwise false-positive or -negative results will be achieved. Different tests can be performed to evaluate the cytotoxicity of NP (cell growth and cell survival): the NADH/NADPH-dependent MTT test [3-(4.5-dimethyl-thiazol-2-yl)-2.5-diphenyl-tetrazolium bromide], WST-1 test [2-(4-iodophenyl)-3-(4-nitrophenyl)-5-(2.4-disulphophenyl)2H-tetrazolium], and eluation test according to DIN EN ISO 10993-5 or luminescent ATP-dependent tests such as CellTiter-Glo® from Promega. Before examination, it is important to choose the appropriate test depending on the material system and further NP application. For example, it is crucial to note that the read-out of some methods is based on luminescence (e.g., CellTiter-Glo®). In the case of fluorescence dye-doped NP systems, the influence of NP absorption and emission properties on the detection and following results must be excluded. A quite promising and complementary nondestructive possibility to determine cell fitness after NP application is the impedance spectroscopy or transepithelial electrical resistance (TEER) measurement. In contrast to standard testing procedures, these methods allow comparison of individual tissue models before and after a treatment [2]. Interestingly, also mild effects, e.g., by repeated washing steps with PBS already led to reduced electrical resistance in skin models, which later recovered indicating the high sensitivity of this technique. Furthermore, MTT-based assays, for instance, assess only the viable cells, thus nonviable cell layers such as present in epidermal models are not considered, however these cells pose the major part of the skin barrier. In conclusion, impedance spectroscopy could be useful to identify sub-irritative (e.g., stinging, burning, or itching sensation) or mild chemical effects on barrier integrity, for instance.

With respect to NP applications, the next important aspect is the NP stability under examination conditions. Without appropriate surface modification, NPs tend to agglomerate and sediment in physiological solutions due to their pH and salt content [3–5]. This needs to be addressed in order to ensure a consistent assessment of the therapeutic and diagnostic potential of NPs in cell culture experiments or animal testing: studies concerning particle concentrations are only reliable when carried out with stable dispersions that guarantee an accurate dosing. The effect of different sized particles on cells can only be studied if the particles do not form large agglomerates [3, 6, 7]. The agglomeration leads to an unwanted reduction of surface area and therefore falsifies results [3, 8]. Consequently, choosing the right stabilizing agent is of significant importance. However, the efficiency of a stabilizer depends not only on the NP composition but also on its surface properties and the chosen cell culture medium [5]. Therefore, since there is no general solution, the best surfactant has to be identified in each individual particle case and medium [9].

2 Materials

2.1 Reagents

1. 70% Ethanol solution.
2. Sterile deionized water.
3. Cells and cell-specific culture media.
4. MTT [3-(4.5-dimethyl-thiazol-2-yl)-2.5-diphenyl-tetrazolium bromide] reagent.

Preparation of MTT stock solution: Solve 60 mg MTT in 20 mL VE-water (3 mg/mL final concentration), filter solution with a 0.2 μm sterile filter and store in light-protected containers as aliquots at −20 °C until use.

5. WST-1 reagent.

6. CellTiter-Glo®-Kit.
7. 10% Sodium dodecyl sulfate (SDS) stock solution in cell culture medium.
8. Isopropanol.
9. DPBS+/− (Dulbecco's Phosphate Buffered Saline with/without $MgCl_2$ and $CaCl_2$).
10. Cell-specific coating solutions, e.g., rat tail collagen type I (100 μg/mL in 0.1% acetic acid).
11. Transwell® inserts.

2.2 Tools and Instruments

1. Sterile hood.
2. Centrifuge.
3. Spectrophotometer.
4. Ultrasonic bath.
5. Vortex mixer.
6. Vials, caps.
7. Well plates.
8. Micropipettes with pipette tips.
9. Incubator.
10. Water bath.
11. Impedance spectrometer LCR HiTESTER 3522-50 (HIOKI E.E. Corp.) and custom-made user interface, programmed in LabVIEW (National Instruments).
12. Millicell ERS-2 (Millipore) in combination with the electrode type STX3 (World Precision Instruments); storage in sterile DPBS− with 1% antimycotic–antibiotic solution.

3 Methods

3.1 Choosing the Right Assay After Comparing Optical NP Properties with the Read-Out Method of the Assay

This section describes how to find the right assay to determine cytotoxicity of NPs.

1. Record an absorption spectrum of the NPs in the solvent used while reading-out the assay (*see* **Note 1**).
2. Compare the absorption properties of NPs with the read-out wavelength of cytotoxicity assay.

   Assay examples (compare Subheading 3.4):

   WST-1 assay: The read-out wavelength is 450 nm. This means that NPs must not absorb light at 450 nm.

   MTT assay: The read-out wavelength is 570 nm. This means that NPs must not absorb light at 570 nm.

CellTiter-Glo® assay: Luciferase emits light in the wavelength range from 460 to 670 nm. The testing NPs may not absorb light in this wavelength range (*see* **Note 2**).

3. Choose an assay with no interactions between optical NP properties and the read-out method in the assay. An example is shown in Fig. 1: absorption and emission scans of (a) fluorescein isothiocyanate (FITC) dye-doped silica NPs and (b) ATTO 647N dye-doped NPs in water with a concentration of 3 mg/mL are shown. The FITC-doped NPs absorb light in the range of 410–550 nm; the ATTO 647N dye-doped NPs in the range of 550–700 nm. Accordingly, the read-out wavelength of the assay must not be situated within these wavelength ranges for each NP system.

3.2 Verifying Colloidal Stability of NPs in Cell Culture Media

This section describes how to proof whether NPs are well dispersed in cell culture media and do not form agglomerates.

1. Transfer an appropriate amount of NPs to be tested in a cytotoxicity assay into a vessel (*see* **Note 3**) and centrifuge the NPs to a pellet (*see* **Note 4**).
2. Discard the supernatant and redisperse the NP pellet in cell culture media you need for your assay by treating the vessel in an ultrasonic bath until all NPs are well dispersed (*see* **Note 5**).
3. Prepare a second vessel with the same volume of cell culture media incl. additives as a reference system (*see* **Note 6**).
4. Select a wavelength that shows no absorbance in the spectrum of NP dispersion (*see* **Note 7**).
5. Vortex the vessels gently and pipet 200 μL of the prepared NP dispersion and of pure cell culture media in a 96-well plate and

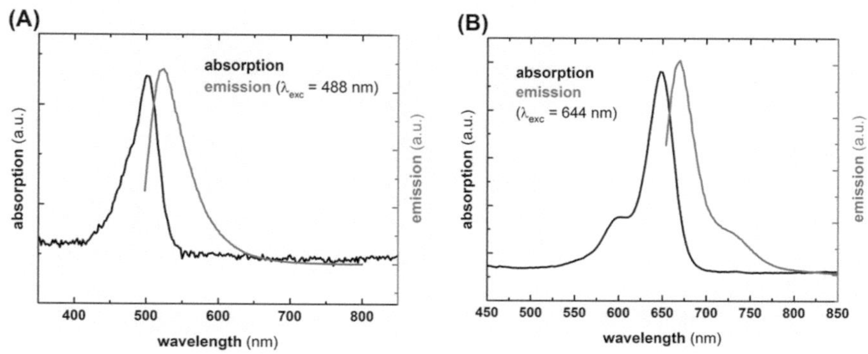

Fig. 1 In order to choose the right viability assay interactions between optical NP properties and the read-out method in the assay has to be proven. *Example*: different absorption (*black*) and emission spectra (grey, λ_{exc} = 488 nm) of (**a**) FITC dye-doped silica NPs and (**b**) ATTO647N dye-doped silica NPs in water with a NP diameter of 60 nm and a concentration of 3 mg/mL

measure the absorption at the selected wavelength with a plate reader directly after preparing the solutions. Proof, if both values in absorption measurements are the same.

6. Incubate the well plate and the vessels in an incubator as long as the cells are treated with NP dispersion in the chosen cytotoxicity assay.
7. Visually evaluate comparatively the two vessels and particularly ensure, if there are sedimented or floating NPs or not. Figure 2 shows in cell culture media sedimented NPs (right) and stable dispersed NPs (left).
8. Vortex the vessels gently and pipet 200 μL of the samples in a well plate. The absorption at the selected wavelength should be measured again. If the NPs are well dispersed both values in absorbance should be the same.

3.3 Preparing a Sterile and Stable NP Dispersion in Cell Culture Media Used in the Assay

This section describes how to prepare a sterile particle dispersion with a defined concentration in cell culture media.

1. Calculate for each well insert in your assay 200 μL of NP dispersion with the desired NP concentration.
2. Transfer the appropriate amount of NPs in a sterile hood into a sterile cap (*see* **Note 3**) and centrifuge the NPs to a pellet (*see* **Note 4**).
3. Discard the supernatant and redisperse the pellet in 1 mL of a 70% ethanol solution by shortly vortexing the cap and treating in an ultrasonic bath until the pellet is fully dispersed.

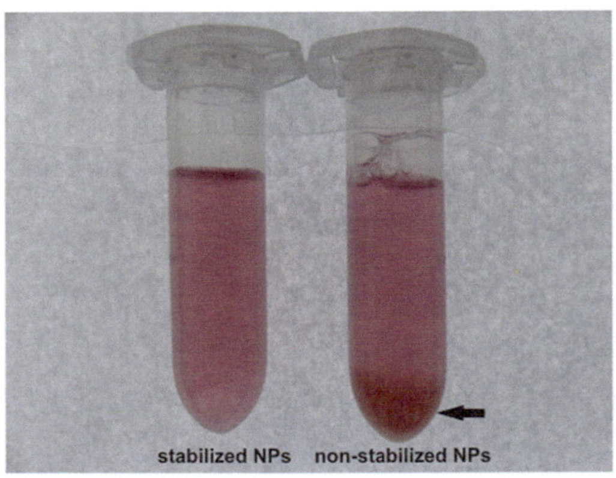

Fig. 2 The colloidal stability of NPs in cell culture media needs to be considered. *Example*: calcium fluoride NPs with a diameter of ca. 10 nm dispersed and incubated in cell culture media with 10% FCS. The NPs in the right vial are non-stabilized on its surface and consequently sedimented (*arrow*). In the left vial the NPs are stabilized on its surface and are still well dispersed in cell culture media after extended incubation

4. Centrifuge the NPs to a pellet and wash the NPs for at least two times with sterile water to remove all traces of ethanol. One washing step includes redispersing and centrifugation of the NPs.
5. If the NPs need to be stabilized, apply the stabilizer to the NPs under sterile conditions, wash the NPs when necessary, and centrifuge the NPs to a pellet.
6. Redisperse the NPs in a defined volume of cell culture media used in cytotoxicity assay to obtain the required NP concentration.

3.4 Cellular Application of NPs and Examples of Suitable Cytotoxicity Assays

Cytotoxicity tests are usually done on monolayer cultures of target cells in multiwell-format (e.g., 96-well). However, in some cases it might be also necessary to determine cell toxicity in 3D Transwell®-like test systems. This section summarizes different quantitative assays to determine cell fitness and viability after NP application on Transwell®-like test systems.

3.4.1 Preparing Transwell®-like Test Systems (e.g., Caco-2 Colorectal Cancer Cells)

1. Place inserts in 24-well plate and coat with collagen I (200 μL/insert) for 15 min (lid closed).
2. Let coated inserts dry for 15 min with open lid under the hood (*see* **Note 8**).
3. Prepare at least technical duplicates for each test substance and biological experiment, also for the positive/negative control (1% SDS/without NPs) and the blank (no cells).
4. Seed cells at a density of $8 \times 10^4/cm^2$ (e.g., Caco-2, *see* **Note 9**).
5. Culture cells for 7–21 days in the incubator at 37 °C.
6. Tightness of cultures needs to be monitored before NPs can be applied. This can be non-invasively done via TEER measurement or impedance spectroscopy (*see* **Note 10** and (D)).
7. Change medium three times per week (200 μL in apical and 1 mL in basolateral chamber).

3.4.2 Application of NPs

1. Shortly vortex prepared NP dispersion (*see* Subheading 3.2), sonicate for ~5 min, and finally vortex before application on cells.
2. Aspirate medium, directly add 200 μL NP dispersion per condition and well into apical compartment, add 1 mL cell-specific medium into basolateral compartment (*see* **Note 11**).
3. Incubate for up to 24 h in the incubator at 37 °C and document the applied NPs microscopically, if applicable.

3.4.3 Quantitative Cell Fitness/Vitality Assays

MTT assay: The detection of cell vitality via the MTT test is based on the reduction of the yellow, water-soluble dye MTT to the blue-violet, water-insoluble formazan salt. MTT is positively charged and readily penetrates viable eukaryotic cells. The subsequent NADH/

NADPH-dependent, partial reduction of MTT by succinate in the mitochondria is dependent on the enzyme succinate dehydrogenase (mitochondrial complex II). The amount of absorbance correlates well with viable cell number, although only cell metabolism is detected. The tetrazolium reduction assay offers a less expensive option that may achieve adequate performance depending on experimental design. The procedure is as follows:

1. Thaw prepared MTT reagent (3 mg/mL) and vortex (*see* **Note 12**).
2. Mix MTT reagent 1:2 with DPBS+ in centrifuge tubes (1.5 mg/mL final concentration).
3. Aspirate medium from cells and wash gently five times with DPBS+ (*see* **Note 13**).
4. Add MTT mixture to cells that they are completely coated (e.g., 200 μL per 24-well).
5. Incubate cells for 1–4 h in the incubator at 37 °C (cover plate with aluminum foil).
6. Aspirate reagent and gently wash cell layer three times with DPBS+.
7. Document the purple cell staining with the camera, if applicable.
8. MTT can be resolved from cells, e.g., with 300 μL isopropanol and by shaking the 24-well plate overnight at 4 °C (*see* **Note 14**).
9. Transfer 200 μL of each well into a transparent 96-well microplate.
10. Quantify with a spectrophotometer at 500–600 nm; subtract blank measurement without cells (*see* **Note 15**).

WST-1 assay: The test principle of WST-1 test is similar to the MTT assay, although WST-1 is negatively charged and does not readily penetrate cells. The assay also measures the formation of formazan based on the water-soluble tetrazolium salt (WST-1). Since the WST-1 dye is not cell invasive cells can be continued to culture again after WST-1 test, although cell might have suffered from the procedure. The procedure is as follows:

1. Thaw WST-1 reagent, vortex, and mix 1:10 with DPBS+ in centrifuge tubes.
2. Aspirate medium from cells and wash gently five times with DPBS+.
3. Add WST-1 mixture to cells that they are just coated (e.g., 200 μL per 24-well).
4. Incubate cells for 30 min in the incubator at 37 °C and transfer 100 μL of each well into a 96-well microplate.

5. Quantification with a spectrophotometer at 450 nm; reference wavelength 620 nm.
6. Aspirate reagent and wash gently three times with DPBS+.
7. Add specific cell culture medium apical and basolateral if cells need to be cultured again.

CellTiter-Glo® assay: The detection of cytotoxicity via CellTiter-Glo® test kit is based on the detection of cellular ATP as an indicator of metabolically active cells. After adding the reagent (CellTiter-Glo®) it results in cell lysis and generation of a luminescent signal, which is proportional to the present ATP and respective cell number. The ATP assay is by far the most sensitive method of measuring viable cells using a plate reader, with typical sensitivity that is two orders of magnitude better than the MTT-based assays (*see* Fig. 3). Therefore, the ATP detection assay is by far the most sensitive, has fewer steps, is the fastest to perform, and has the least amount of interference. The procedure is as follows:

1. Thaw and mix CellTiter-Glo® reagents according to manufacturers' recommendation.
2. Aspirate cell culture medium and gently wash cells five times with DPBS+ (*see* **Note 16**).
3. Add 100 μL cell-specific medium and afterwards 100 μL CellTiter-Glo® mix per well and shake for 2 min on a shaker.
4. Incubate for 10 min in the dark at room temperature and transfer 100 μL of each well into a 96-well lumitrac microplate.
5. Quantify luminescence with a spectrophotometer; subtract blank measurement of the specific cell medium (*see* **Note 17**).

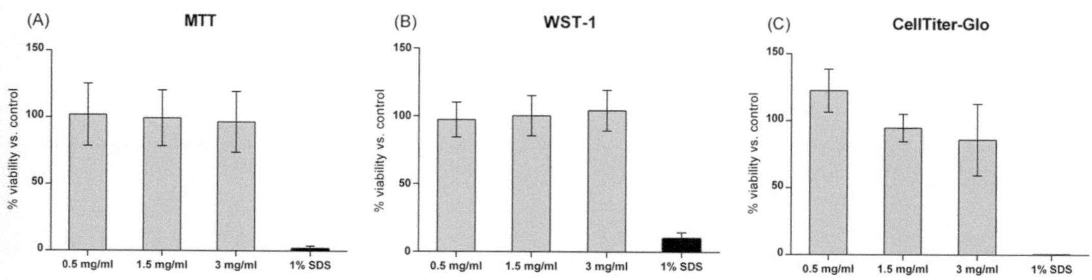

Fig. 3 There are a variety of assay technologies available that measure metabolic markers to estimate the number of viable cells in culture. *Example*: representative results of dose-dependent analysis of cell fitness using MTT (**a**), WST-1 (**b**), and CellTiter-Glo® (**c**) assays. Prior assay application three different concentrations (0.5, 1.5, and 3 mg/mL) of CaF_2-NPs were applied on tight Caco-2 Transwell® cultures (barrier integrity proven via TEER measurement) for 24 h. Within the tested concentrations no significant toxicity was observed for this particular NP type, however the CellTiter-Glo® assays appears more sensitive and suggests mild effects on cell viability (i.e., ~80% viability vs. control) with the highest concentration applied. Application of 1% SDS served as positive control

TEER measurement and impedance spectroscopy: The standard output reading for electrical characterization of cells, cell layers, and tissues is the TEER. When measuring at different frequencies, a generated impedance spectrum can be depicted in a Bode plot composed of the amplitude and phase angle, which allow one to identify electrical elements of the sample and subsequently may correlate it to cell fitness. The electrical characteristics can be quantified by employing an equivalent circuit and a mathematical model (*see* [2]). The procedure is as follows:

TEER measurement:

1. Calibration should be done once a day: The meter should display 1000 Ω, otherwise it needs to be adjusted accordingly.
2. Disinfection/Equilibration: Immerse the electrode tips in 70% ethanol for 15 min. Allow to air dry for 15 s and equilibrate the electrode in cell culture media for 15 min.
3. Measurement: Place the electrode steady in that way that the shorter tip is in the insert and the longer tip is in the outer well of the Transwell® system. Wait fluctuations and take the value (*see* **Note 18**).

Impedance spectroscopy:

1. Wash the electrodes with VE-water; spray with 70% ethanol and allow to dry completely.
2. Transfer the Transwells into appropriate 24-well insert system (e.g., BRANDplate).
3. Change cell culture medium and measure the impedance spectra in a frequency ranging between 1 Hz and 100 kHz using impedance spectrometer (e.g., LCR HiTESTER 3522-50 and appropriate custom-made user interface including LabVIEW software).

For representative results of quantitative assays for nontoxic NPs, *see* Fig. 3.

4 Notes

1. If the read-out method uses cell culture media or salt buffer solutions with pH indicators, you can also use water as a solvent for recording the absorption spectrum.
2. In the case of fluorescent NPs, you have to make sure that luminescence of luciferase does not excite NPs to fluorescence.
3. NP samples are mostly kept in dispersion. Determine the particle concentration in mg/mL by drying a defined volume of NP dispersion in an oven and determining afterwards the mass of the NPs after drying.

4. You can calculate the centrifugation parameters by knowing the NP diameter, NP density, medium density, viscosity of medium, and gravitational force.
5. It is important to use the same cell culture media with all additives as used in cytotoxicity assay.
6. For instance, NP can be electrostatically stabilized by adsorbing biopolymers, proteins, organic acids, etc. on NP surface.
7. For instance, in the case of FITC-doped silica NPs in Fig. 1a select a wavelength higher than 550 nm. For the ATTO 647N dye-doped NPs (Fig. 1b) select a wavelength below 550 nm.
8. The complete evaporation of remaining collagen solution is essential to eliminate any acetic acid solvent.
9. Individual cell concentration needs to be tested dependent to the cell type.
10. TEER values should be >200 Ω *cm^2 (Caco-2). Tightness can be validated in transport studies using appropriate reference compounds such as FITC-dextrane (e.g., 40 kDa FITC-dextrane transport to basolateral compartment should be less than 1% after 1 h).
11. The amount of serum might influence the NP properties, cellular uptake, and transport speed. Usually, 0–10% serum or BSA in buffer or medium is used.
12. Before applying MTT-based assays appropriate controls (e.g., MTT assay reagent and test compound, e.g., NPs in culture medium without cells present) are necessary to check to proof for assay chemistry interference with NPs. Examples of the literature include ascorbic acid, vitamin A, sulfhydryl-containing compounds including reduced glutathione, coenzyme A, dithiothreitol, and chemicals that uncouple electron transport from oxidative phosphorylation of ATP.
13. If cells are not gently washed they tend to detach easily.
14. Wrap plates with parafilm to avoid loss of volume.
15. The amount of signal generated is dependent on several parameters including: the concentration of MTT, the length of the incubation period, the number of viable cells, and their metabolic activity. All of these parameters should be considered when optimizing the assay conditions to generate a sufficient amount of product that can be detected above background.
16. Wash gently to avoid cell detachment.
17. As for the MTT assay, here it is equally important to include a dose-dependent ATP control experiment (i.e., defined ATP with NP and without cells) since NPs could also react with ATP and therefore falsify the results.

18. The electrode must be sufficiently covered with medium, and medium level inside and outside the Transwell® must be equal (e.g., 300 μL inside and 1.3 mL outside for 24-wells). Average at least three random measurement points per well.

Acknowledgments

The authors kindly thank the Fraunhofer Society and the Bavarian State ministry for economy and media, energy and technology (Az.:VI/3-6622/453/12) for financially supporting the work.

References

1. Moore TL, Rodriguez-Lorenzo L, Hirsch V et al (2015) Nanoparticle colloidal stability in cell culture media and impact on cellular interactions. Chem Soc Rev 44(17):6287–6305
2. Groeber F, Engelhardt L, Egger S et al (2015) Impedance spectroscopy for the non-destructive evaluation of in vitro epidermal models. Pharmaceut Res 32(5):1845–1854
3. Allouni ZE, Cimpan MR, Høl PJ et al (2009) Agglomeration and sedimentation of TiO2 nanoparticles in cell culture medium. Colloid Surf B: Biointerfaces 68(1):83–87
4. Ji Z, Jin X, George S et al (2010) Dispersion and stability pptimization of TiO_2 nanoparticles in cell culture media. Environ Sci Technol 44(19):7309–7314
5. Bihari P, Vippola M, Schultes S et al (2008) Optimized dispersion of nanoparticles for biological in vitro and in vivo studies. Particle Fibre Toxicol 5(1):1–14
6. Cho EC, Zhang Q, Xia Y (2011) The effect of sedimentation and diffusion on cellular uptake of gold nanoparticles. Nat Nano 6(6):385–391
7. Guiot C, Spalla O (2013) Stabilization of TiO2 nanoparticles in complex medium through a pH adjustment protocol. Environ Sci Technol 47(2):1057–1064
8. Lakshminarasimhan N, Kim W, Choi W (2008) Effect of the agglomerated state on the photocatalytic hydrogen production with in situ agglomeration of colloidal TiO_2 nanoparticles. J Phys Chem C 112(51):20451–20457
9. Koch S, Kessler M, Mandel K et al (2016) Polycarboxylate ethers: The key towards non-toxic TiO_2 nanoparticle stabilisation in physiological solutions. Colloid Surf B: Biointerfaces 143:7–14

Chapter 11

Assays for Analyzing the Role of Transport Proteins in the Uptake and the Vectorial Transport of Substances Affecting Cell Viability

Emir Taghikhani, Martin F. Fromm, and Jörg König

Abstract

Endogenous compounds, drugs, or other xenobiotics may affect cell viability. A prerequisite for intracellular cell damage is the uptake of such substances across the plasma membrane into cells. Furthermore, the subsequent transporter-mediated export out of cells may influence cell viability. Therefore, transport proteins mediating the uptake (uptake transporter) or export (export pumps) of substances in and out of cells are important determinants of cell viability. Uptake transporters mostly belong to the superfamily of solute carriers (SLC transporters), whereas export pumps are members of the ABC-transporter superfamily (ATP-binding cassette). Cell systems recombinantly overexpressing uptake transporters (single transfectants) or multiple-transfected cell models expressing simultaneously an uptake transporter together with an export pump (double transfectants) are important in vitro tools for analyzing protein-mediated transport of potentially cell toxic compounds.

Here we describe different in vitro transport assays for the functional analysis of transport proteins. Using single-transfected HEK293 cells stably overexpressing an uptake transporter, substances can be tested as potential substrates (uptake assay) or potential transport inhibitors (inhibition assay) for the respective transport protein. Vectorial transport of substances with the uptake across the basolateral plasma membrane and the export across the apical membrane of polarized grown MDCKII cells can be analyzed using double-transfected cell models with the simultaneous overexpression of an uptake transport and an export pump in vectorial transport assays, thereby mimicking physiological transport processes, e.g., in liver or kidney.

Key words Uptake assay, Vectorial transport, Uptake transporter, Export pump, Drug transport

1 Introduction

Uptake transporters and efflux pumps determine plasma and tissue concentrations of a broad range of drugs, xenobiotics, or endogenous compounds. Furthermore, protein-mediated uptake is a prerequisite for the subsequent intracellular action or metabolism of substances. Therefore, the intracellular concentration of potential cytotoxic compounds depends on uptake transporters in the plasma membrane of cells. In addition, also the transporter-mediated

Daniel F. Gilbert and Oliver Friedrich (eds.), *Cell Viability Assays: Methods and Protocols*, Methods in Molecular Biology, vol. 1601, DOI 10.1007/978-1-4939-6960-9_11, © Springer Science+Business Media LLC 2017

export of substances (e.g., drugs) or their metabolites may affect this intracellular concentration. One well-known drug affecting cell viability is cisplatin. Cisplatin is widely used for chemotherapy but this therapy often is limited by severe side effects such as nephrotoxicity and ototoxicity [1]. In kidney, cisplatin is taken up into cells of the terminal proximal tubule and of the distal nephron where it causes apoptosis or necrosis. The uptake transporter OCT2 (gene symbol *SLC22A2*) mediates this cisplatin uptake, and it has been demonstrated that OCT2-mediated uptake plays a pivotal role in the development of cisplatin-induced oto- and nephrotoxicity [2]. Therefore, transport proteins mediating the uptake or the export of substances are important for cell viability and it is important to characterize substances as transporter substrates or transport inhibitors.

Functionally, transport proteins can be classified into transporters mediating the uptake of substances (e.g., drugs) into cells and transporters mediating the export of substances (e.g., drug metabolites) out of cells [3]. Uptake transporters mostly belong to the superfamily of SLC transporters (SLC = solute carriers). Today, this superfamily consists of 52 SLC families with more than 400 identified transport proteins. Particularly important for the transport of drugs are the families SLC21/SLCO and SLC22. The SLC21/SLCO family encodes for OATP proteins (OATP = Organic Anion Transporting Polypeptides) and members of this family are expressed in nearly all human tissues and cells investigated. Important family members for drug treatment are OATP1B1 (gene symbol *SLCO1B1*) and OATP1B3 (*SLCO1B3*), both expressed in human liver mediating the uptake of endogenous substances and drugs (e.g., statins or antibiotics) from blood into hepatocytes. The SLC22 family comprises organic cation transporters (OCTs) and organic anion transporters (OATs), both of which are important for drug and xenobiotic transport.

Export transporters mostly belong to the superfamily of ABC transporters (ABC = ATP-binding cassette) mediating the export of substances out of cells against a concentration gradient driven by ATP hydrolysis. The ABC superfamily comprises seven families (ABCA–ABCG) with members of the ABCB family (e.g., P-glycoprotein), the ABCC family (e.g., multidrug resistance associated protein 2–MRP2), and the ABCG family (breast cancer resistance protein–BCRP) as important for the transport of drugs, drug conjugates, and xenobiotics. In addition to ABC transporters, also family members of the SLC47 family are important for mediating the export out of cells. MATE (multidrug and toxin extrusion proteins) proteins are members of this family and important for the excretion of drugs and drug metabolites into bile and urine [3].

Taken the known substrate spectrum of all transporters important for drug and xenobiotic transport into account, it is obvious that in vitro cell systems for the standardized analysis of uptake transporters and export pumps are important tools for investigating substances potentially affecting cell viability. Uptake transporters are investigated routinely by so-called single-transfected cells recombinantly overexpressing the respective transporter. They can be used in normal uptake assays or in inhibition assays, in which the investigated substance is added as possible transport modulator to an uptake assay using a prototypic substrate for the respective transport protein.

Because it is much more difficult analyzing export pumps in normal transport assays, double- and multiple-transfected cell lines with the simultaneous expression of an uptake transporter and an export pump are important in vitro tools for analyzing the vectorial transport of substances. Such cell models are mostly based on MDCKII cells that grow in a polarized fashion separating a basolateral and an apical compartment. Uptake transporters are localized in the basolateral membrane of polarized grown MDCKII cells mediating the uptake of substances from the basolateral compartment into cells. The export is mediated by proteins in the apical membrane of MDCKII cells transporting the substance into the apical compartment. In recent years several double- and multiple-transfected MDCKII cell models have been established expressing different combinations of uptake transporters and export proteins [4–10].

In this chapter, we describe three different experimental setups for the analysis of transport proteins. Based on single-transfected HEK293 cells overexpressing solely an uptake transporter, substances can be identified as substrates in simple uptake assays (uptake assay). Prerequisite for such analyses are substances that are available in radiolabeled form or for which an LC/MS-MS analysis can be performed after the uptake assay. If this is not the case, substances can be added as potential transport inhibitors by using the single-transfected HEK293 cells with a known transport substrate (inhibition assay). In addition, vectorial transport studies using double-transfected MDCKII cells will be described (vectorial transport studies).

2 Materials

2.1 Cell Culture

1. **Phosphate buffered saline (PBS):** commercially available as a ready-to-use solution. Store at 2–8 °C.
2. **Minimum essential medium (MEM)** containing 10% heat inactivated fetal calf serum and 1% Penicilline-streptomycin; selection antibiotic needs to be added (e.g., Geneticin or Hygromycin). Store at 2–8 °C.
3. **Poly-D-lysine:** stock solution of 10 mg/mL is commercially available (stored at −20 °C) and 0.1 mg/mL working solution (diluted in PBS, sterile filtered, stored at −20 °C)

4. **Sodium butyrate stock solution500 mM:** Weigh 2.75 g sodium butyrate and dissolve in 50 mL PBS; sterile filtration is necessary. Store at 2–8 °C.
5. **Trypsin solution:** commercially available as a ready-to-use solution. Store at −20 °C.
6. **Membrane inserts: ThinCerts®**, pore size 0.4 μM, diameter: 14 mm (for vectorial transport studies in particular).

2.2 Transport Assay

1. **Uptake buffer:** Weigh 8.3 g NaCl (142 mM), 0.37 g KCl (5 mM), 0.17 g K_2HPO_4 (1 mM), 0.29 g $MgSO_4 \times 7\ H_2O$ (1.2 mM), 0.22 g $CaCl_2$ (1.5 mM), 0.99 g Glucose (5 mM), and 2.97 g HEPES (12.5 mM). Transfer all the compounds into the cylinder and add water to a volume of 900 mL. Mix and adjust the pH with diluted NaOH to pH 7.3. Make up to 1 L with water. Sterile filtration of the buffer is necessary. Store the filtered uptake buffer at 4 °C.
2. **0.2% SDS solution:** Weigh 2 g sodium-dodecylsulfate and transfer to a 1 L graduated cylinder containing about 100 mL of water. Gently dissolve the SDS by using a magnetic stir bar. When the compound is totally dissolved, make up to 1 L with water. Store at room temperature.
3. **Donor solution**: contains the substrate of interest (*see* **Note 1**).
4. **Ultima Gold XR** or similar liquid scintillation cocktail (*see* **Note 2**).

2.3 Protein Measurement

For the normalization of the transport data to the respective protein amount we routinely use a bicinchoninic acid assay (BCA Protein Assay Kit).

2.4 Data Analysis

Microsoft Excel and GraphPad Prism 5.01.

3 Methods

3.1 Uptake Assay

A routinely performed uptake assay (Fig. 1) using HEK-OATP1B1 cells and the respective HEK-Co cells is described. (*See* **Note 3** for further information about the cell culture conditions).

Cell lines are obtained by transfecting parental HEK293 cells with an expression vector [e.g., pcDNA3.1(+)] containing the *SLCO1B1* cDNA encoding OATP1B1 [11]. HEK293 cells transfected with the empty expression vector [e.g., pcDNA3.1(+)] and selected under the same experimental conditions serve as control.

3.1.1 Seeding of HEK-OATP1B1 and HEK-Co Cells

1. Remove MEM from cellular monolayer by aspiration; wash the cells by gently applying 3 mL of PBS.
2. Remove PBS and add 1.5 mL of Trypsin solution to detach cells.

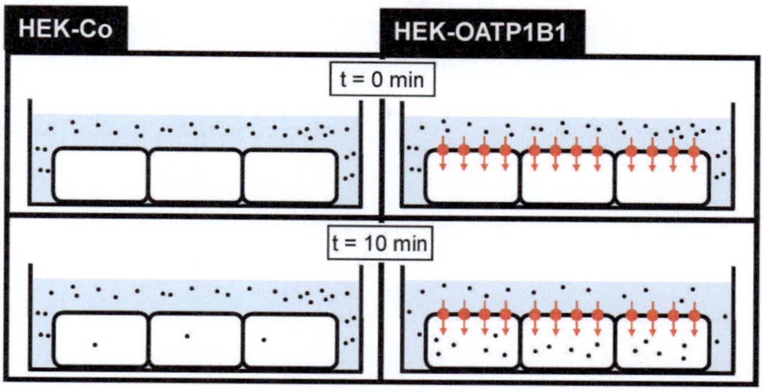

Fig. 1 Uptake assay. Setup for analyzing the function of uptake transporters expressed in stably transfected HEK293 cells. As an example HEK-OATP1B1 cells expressing the human hepatic drug uptake transporter OATP1B1 (gene symbol SLCO1B1) are shown together with the respective HEK-Co (control) cells. At time point 0 min, the donor solution (containing the radiolabeled substance—black circles) is applied onto the cell layer and after 10 min the uptake assay is terminated and the cell-associated radioactivity can be determined. In this example, the substance is transported by the OATP1B1 protein resulting in more substance intracellularly compared to the HEK-Co cells

3. Incubate the cells at 37 °C, 5% CO_2 for 3 min.
4. Use 5 mL MEM to collect the dissolved cells; transfer the suspension into a 50 mL-Tube.
5. Centrifuge the suspension for 3 min at 50 × *g*.
6. Remove the MEM on top of the cellular pellet; dissolve the pellet into a single cell suspension by using 15 mL of fresh MEM; frequently pipetting up and down helps to separate cells from another.
7. Dilute the single cell suspension in a ratio of 1:4; count the number of cells in 10 μL of the 1:4-dilution by using a hemocytometer (*see* **Note 4**).
8. Dilute the cell suspension in a suitable way to get a cell amount of 7×10^5 cells/0.8 mL.
9. Coat the cavities of the 12-well plate with poly-D-lysine by pipetting 200 μL of a 0.1 mg/mL poly-D-lysine solution into each cavity; incubate the 12-well plate for 30 min at room temperature.
10. Remove residual poly-D-lysine solution; dry the 12-well plate for another 30 min at room temperature (*see* **Note 5**).
11. Pipette 0.8 mL of the cell suspension into the wells of the 12-well plate; incubate the seeded cells at 37 °C, 5% CO_2 for 24 h in the incubator.

3.1.2 Induction of Cells

1. To increase the amount of the recombinantly expressed transport protein, cells were routinely induced with sodium butyrate [12]. For induction, dilute sodium butyrate in fresh medium to get a final concentration of 10 mM (*see* **Note 6**).
2. Aspirate MEM on cells; gently apply 0.8 mL of the freshly prepared induction medium into the cavities of the 12-well plate.
3. Incubate the induced cells at 37 °C, 5% CO_2 for additional 24 h.

3.1.3 Transport Assay

1. Warm up the prepared donor solutions to a temperature of 37 °C; additionally warm up a large aliquot of uptake buffer.
2. Remove MEM from cells; wash the cells by gently applying 1 mL prewarmed uptake buffer into every cavity.
3. Remove the uptake buffer; the cells are now ready for the uptake experiment.
4. Apply 300 μL of donor solutions (containing the labeled substrate) onto the cells; incubate the cells at 37 °C, 5% CO_2 for desired time points [(e.g., 10 min) *see* **Note 7**].
5. After incubation put the cells on ice to stop uptake.
6. Quickly aspirate the donor solution from the cavities; subsequently wash each cavity three times by applying 1 mL of ice-cold uptake buffer (*see* **Note 8**).
7. For cell lysis apply 800 μL of 0.2% SDS; incubate the cells at room temperature on a shaker for 15 min.
8. Homogenate the cells by gently pipetting up and down (*see* **Note 9**).
9. Measure the radioactivity in the donor solution in each cell homogenate by using a scintillation counter. Use 500 μL of every lysate and of the donor solution for the measurement (*see* **Note 2**).
10. Determine the protein amount in each homogenate by a BCA assay.

3.2 Inhibition Assay

In principle, the inhibition assay is carried out like the uptake assay with donor solutions containing in addition to the radioactive-labeled transport substrate also desired concentrations of the substance tested as possible transport inhibitor (Fig. 2, *see* **Note 10**).

3.3 Vectorial Transport Assay

An assay for analyzing uptake and subsequent export of substances in one cell system by using polarized grown MDCKII cells simultaneously overexpressing an uptake transporter and an export pump (Fig. 3). Here we describe an assay using MDCK-OATP1B3-MRP2 cells expressing the human uptake transporter OATP1B3 together with the export pump MRP2. Single-transfected MDCK-OATP1B3, MDCK-MRP2, and MDCK-Co cells served

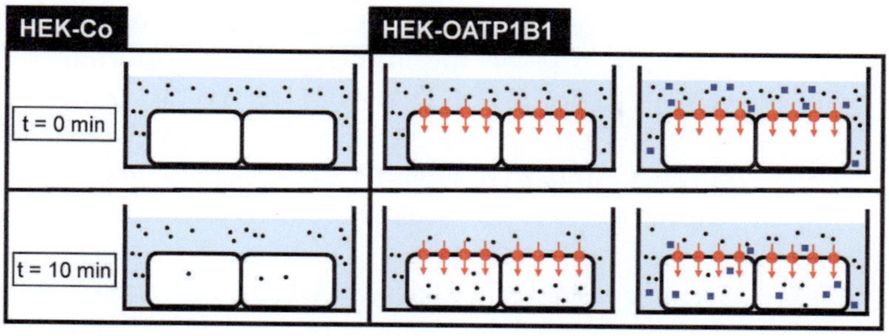

Fig. 2 Inhibition assay. Setup for an inhibition assay. As for the uptake assay, HEK-OATP1B1 and HEK-Co cells are shown. In addition to an assay using a labeled prototypic substrate (black circles) an assay with an added possible transport inhibitor (blue squares) is shown. After 10 min of incubation, reduced uptake of the prototypic substrate can be measured in the inhibition assay compared to the assay without added transport inhibitor

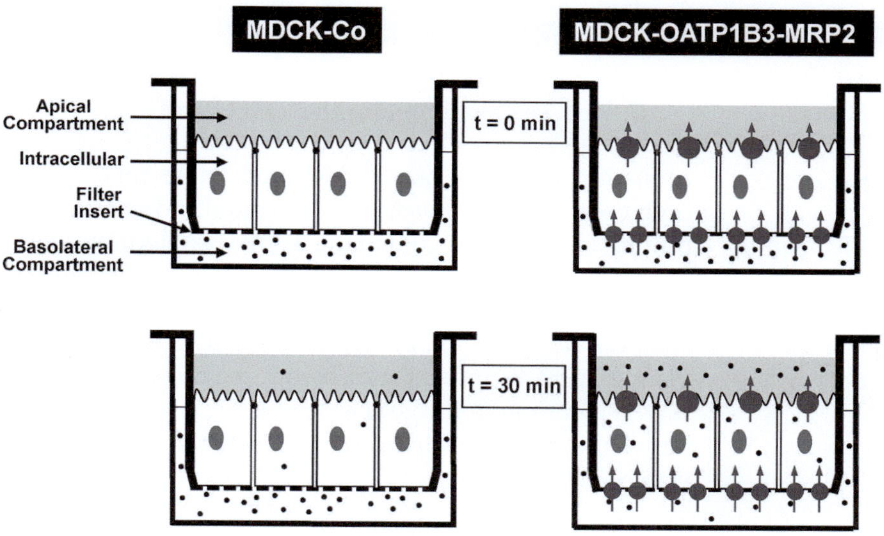

Fig. 3 Vectorial transport assay. Setup for a vectorial transport assay. MDCKII cells (MDCK-Co cells and MDCK-OATP1B3-MRP2 cells recombinantly overexpressing the human hepatic uptake transporter OATP1B3 and the apically localized export pump MRP2) were cultured on ThinCert® filter inserts separating a basolateral from an apical compartment. A substrate of both transporters (black circles) was added to the basolateral compartment, and after 30 min, the radioactivity in the apical compartment can be measured demonstrating the vectorial transport of the investigated substance mediated by the uptake into the cells and the subsequent export across the apical membrane out of the cells

as control (*see* **Note 3** for further information about the cell culture conditions).

The MDCK-OATP1B3-MRP2 cell line was established by transfecting parental MDCKII cells with two different expression vectors (e.g., pcDNA3.1(+)–geniticin resistance and pcDNA3.1/Hygro–hygromycin resistance), each vector containing a cDNA (*SLCO1B1* and *ABCC2*) encoding for the respective transport protein [4].

Single-transfected MDCKII cells recombinantly overexpressing only one of the investigated transporters (e.g., MDCK-OATP1B1 or MDCK-MRP2) served as control. In addition, MDCKII cells transfected with an empty expression vector and selected under the same experimental conditions as single-transfected MDCKII cells were used as control cells (MDCK-Co).

3.3.1 Seeding of MDCKII Cells

Seeding of stably transfected MDCKII cells is performed as described (Subheading 3.1.1: "Seeding of HEK-OATP1B1/HEK-Co cells") with the exception, that MDCKII cells are seeded onto ThinCert® membrane inserts (Greiner Bio—One International), which are placed inside the cavities, separating a basolateral (downside) from an apical compartment (topside). The following aspects have to be taken into account during the procedure:

1. Prepare the 12-well plate for the MDCKII cells by putting a ThinCert® into each cavity. Apply 0.8 mL of MEM into the basolateral compartment.
2. Apply an amount of 5×10^5 MDCKII cells (*see* **Note 11**) in 0.8 mL cell medium into the ThinCerts®.

3.3.2 Induction of Cells

Induction is carried out as described. Note that the medium in both compartments needs to be aspirated and restored with induction medium (MEM containing 10 mM sodium butyrate).

3.3.3 Vectorial Transport Assay

Before the transport assay, a 12-well plate is prepared with donor solution (0.8 mL per cavity). This plate is called Plate A. Store Plate A at 37 °C immediately before the transport assay.

As mentioned, the cells are prepared in a 12-well plate seeded into ThinCerts®, which are placed inside the cavities of the 12-well plate = Plate B.

1. Warm up a large aliquot of uptake buffer to a temperature of 37 °C.
2. Remove MEM in the basolateral and in the apical compartment of "Plate B." Wash the cells with 1 mL prewarmed uptake buffer.
3. Remove uptake buffer and add 8 mL of fresh, prewarmed uptake buffer into the basolateral and apical compartment.
4. To start the transport assay, transfer every ThinCert® from Plate B into the cavities of Plate A containing the donor solution and incubate the cells at 37 °C, 5% CO_2 for desired time points.
5. At desired time points (e.g., after 5, 10, 20, 30 min), remove 50 μL of uptake buffer from the apical compartment. Replace the removed volume with 50 μL of fresh, prewarmed uptake buffer every time.

6. At the final time point (e.g., after 30 min) quickly transfer the ThinCerts® back into the cavities of Plate B, which is stored on ice. Remove buffer and wash the cells three times by applying 1 mL of ice-cold uptake buffer.
7. For cell lysis, cut out the membrane of each ThinCert® carrying the cells. Transfer each membrane into a 2.0 mL reaction tube and add 600 μL of 0.2% SDS. Incubate the cells at room temperature on a shaker for 15 min for cell lysis.
8. Homogenate the cells by gently pipetting up and down (*see* **Note 9**).
9. Determine the radioactivity in the donor solutions, the cell homogenate, and in each sample taken from the apical compartment by scintillation counting (*see* **Note 2**). For this use 400 μL donor solution, 400 μL homogenate, and 50 μL of each apical sample.
10. Determine the protein concentration in each homogenate by a BCA assay.

3.4 Data Analysis

3.4.1 Uptake Assays

The intracellular accumulation of the substrate can be calculated as follows (units are added in square brackets):

$$\frac{c(\text{substrate})[\mu M] \times cpm_{\text{substrate}} \times 10^6}{\text{protein content}\left[\frac{\mu g}{mL}\right] \times cpm_{\text{donor solution}} \times \text{incubation time}[\text{min}]}$$

The factor 10^6 is a consequence of the mathematical conversion, which leads to a final expression of the intracellular accumulation in the unit pmol × mg protein^{-1} × min^{-1}.

For transport assays with single-transfected HEK-Co and HEK-OATP1B1 cells, respectively, the net uptake (uptake, mediated by the transport protein of interest) is calculated as follows:

$$\begin{aligned} &\textit{Net}\ \text{uptake}[\text{pmol} \times mg\ \text{protein}^{-1} \times \text{min}^{-1}] \\ &= (\text{Uptake into } HEK-\text{OATP1B1 cells}) \\ &\quad -(\text{Uptake into } HEK-Co \text{ cells}) \end{aligned}$$

Figure 4a shows the analysis with blotting the uptake values of HEK-OATP1B1 and HEK-Co cells and the resulting net uptake against the substrate concentration. This kind of analysis can be used to determine kinetic constants such as K_m and $V_{\max}$ values for the transporter and the respective substrate.

3.4.2 Inhibition Assays

Using inhibition assays, kinetic values such as IC_{50} values or K_i values can be determined. Figure 4b shows the determination of an IC_{50} value. The net uptake is calculated as described. The average net uptake without the inhibitor will be set to 100%. The uptake in the presence of the inhibitor is shown as percentage of

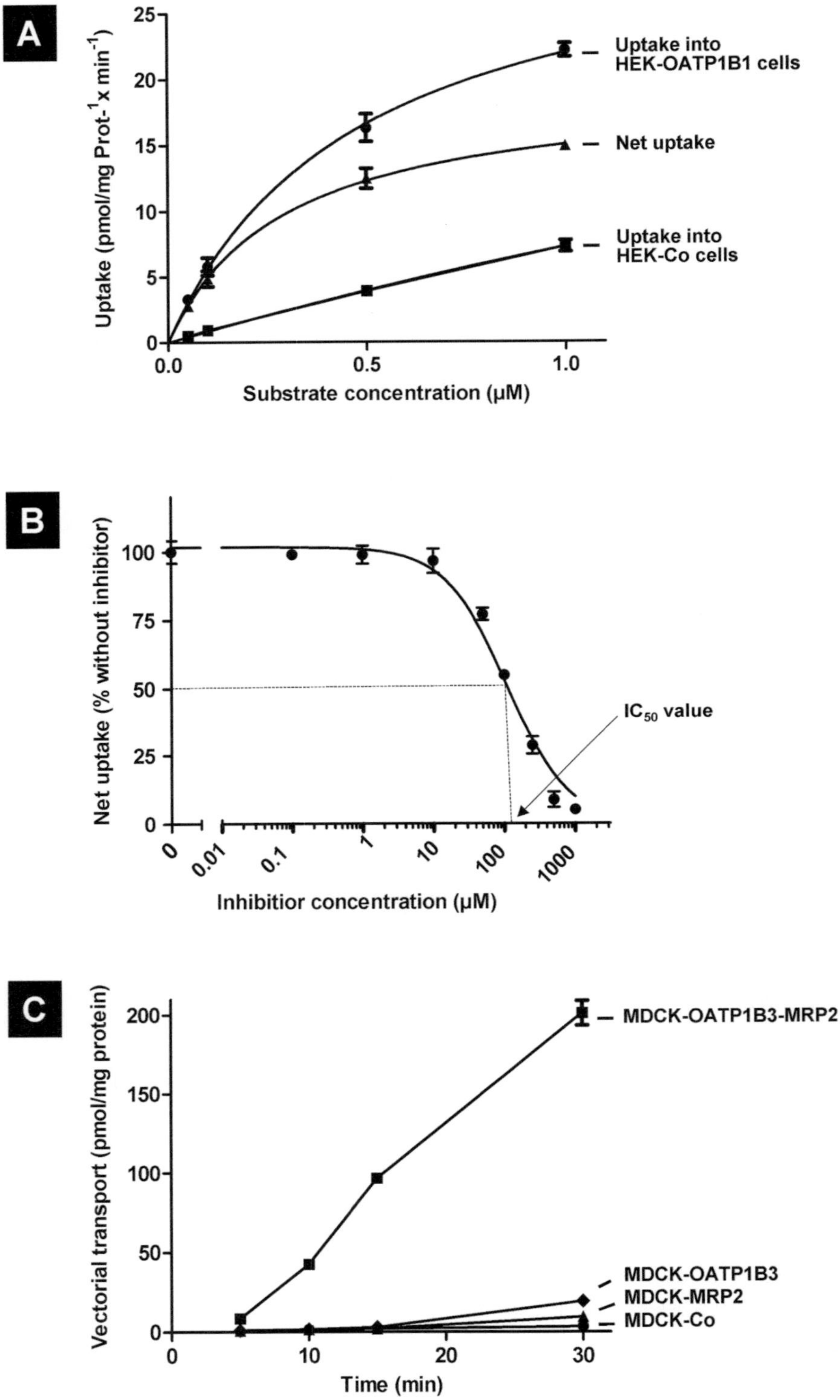

Fig. 4 Data analysis. (**a**) Analysis of an uptake experiment showing the concentration-dependent uptake of a substance into HEK-Co (control) cells and into HEK-OATP1B1 cells recombinantly overexpressing the hepatic uptake transporter OATP1B1. The resulting net uptake can be used to determine kinetic constants for the substrate and the respective transporter. (**b**) Analysis of an inhibition experiment showing the net uptake of a prototypic substrate

the net uptake without the inhibitor. This will be calculated as follows:

$$\textit{Net}\ \text{uptake}\,(\%\ \text{without inhibitor}) = \frac{\text{Average}\ \textit{net}\ \text{uptake}\,(\text{inhibitor in a certain concentration})}{\text{Average}\ \textit{net}\ \text{uptake}\,(\textit{no}\ \text{inhibitor})} \times 100$$

IC_{50} values can automatically be calculated by using, e.g., the GraphPad Prism software package.

3.4.3 Vectorial Transport Assays

Vectorial transport data can be determined as substrate-dependency or time-dependency analysis. Figure 4c shows the analysis of a time-dependency study. For this, at desired time points (5/10/15/30 min) samples have been taken from the apical compartment as described. The amount of radioactivity in the apical compartment represents the amount of substrate taken up from the basolateral compartment into the cells and exported out of the cells into the apical compartment and is therefore a measure for both transport processes (uptake and export).

Vectorial transport will be calculated as follows:

$$\frac{c(\text{substrate})[\mu M] \times cpm_{\text{substrate}} \times 10^6}{\text{protein content}\left[\frac{\mu g}{mL}\right] \times cpm_{\text{donor solution}}}$$

4 Notes

1. The donor solution primarily contains the compound of interest in a radiolabeled form (if available; if not—see inhibition assay). Unlabeled substance is supplemented to reach a certain substrate concentration. Uptake buffer is used for the preparation of the donor solution.

   Donor solutions for inhibition assays contain a known radiolabeled standard substrate of the transport protein (e.g., bromosulfophthalein for OATP1B1). Unlabeled substrate can be added to reach a certain substrate concentration. Uptake buffer is used for the preparation of the solutions. The concentration of the standard substrate is equal in all donor solutions.

Fig. 4 (continued) without and with added transport inhibitor. The net uptake is given in percent uptake without added inhibitor. In this example, the added substance has an IC_{50} value of 125 μM. (**c**) Analysis of a vectorial transport experiment. MDCK-OATP1B3-MRP2 double transfectants and the respective control cells (MDCK-OATP1B1, MDCK-MRP2, and MDCK-Co) were used in a time-dependency analysis. Aliquots were taken from the apical compartment after 5, 10, 15, and 30 min demonstrating that sufficient vectorial transport can be measured only when both transport proteins are present in the cell model (data adapted from [4])

The tested unlabeled compound is added into this donor solution in different concentrations (e.g., 1/5/10 μM, etc.).

2. All samples are measured in scintillation vials, 4 mL of Ultima Gold XR is added to a suitable amount of cell homogenate. The necessary volume of sample depends on the assay and is described in detail in the respective sections of the method section in this manuscript. The radioactivity of the samples is measured by a scintillation counter.
3. Cells for transport assays need to be passaged every 3–4 days, depending on the state of confluence. Cell culture is carried out in an atmosphere of 5% CO_2 at 37 °C in cell culture flasks (75 cm^2); seeding of the cells is carried out in 12-well plates. It is important to pay attention to the passage number of the recombinant HEK293 and MDCKII cells because with higher passage numbers subclones may be selected not reflecting the original clones used immediately after culturing the cells. Therefore, only passage numbers between 10 (freshly thawed) and 40 will be used for transport experiments.
4. This is an important step because all transport data will be normalized to the respective protein amount and the protein amount strongly depends on the cell number determined. The number of cells in original mixture will be counted as follows:

$$\left(\frac{\text{number of cells counted}}{(\text{proportion of chamber counted})(\text{volume of squares counted})}\right)$$
$$\left(\frac{\text{volume of diluted sample}}{\text{volume of original mixture in sample}}\right)$$

5. Because transfected HEK293 cells will be manipulated by several washing steps during uptake or inhibition studies, carefully coating the cavities will prevent the detaching of the cells.
6. Higher sodium butyrate concentrations may be toxic for the cells.
7. The optimal time point for uptake experiments have to be elucidated in time-dependency experiments. For each substrate, a time point in the initial uptake rate (linear uptake) should be selected. Therefore, time points between 1 and 10 min are routinely used.
8. Be careful when washing the cells to avoid detaching.
9. Also be careful at this step because of the SDS in the lysing solution foaming may be a problem.
10. For the first inhibition experiments, a wide concentration range of the substance should be used (e.g., 0.1/1/10/100 μM). After the initial experiment, additional measuring points can be included for the determination of a valid IC_{50} value.

11. The correct cell number for seeding MDCKII cells on ThinCerts® has to be elucidated for each cell line. It is important that the cells will form a tight monolayer after being cultured and right before the vectorial transport experiments. If the monolayer is leaky, this will result in radioactivity in the apical compartment due to diffusion and not due to transport processes. Leakiness needs to be regularly checked using non-permeable markers (e.g., inulin) or by measuring transepithelial resistance.

References

1. Karasawa T, Steyger PS (2015) An integrated view of cisplatin-induced nephrotoxicity and ototoxicity. Toxicol Lett 237:219–227
2. Ciarimboli G, Deuster D, Knief A et al (2010) Organic cation transporter 2 mediates cisplatin-induced oto- and nephrotoxicity and is a target for protective interventions. Am J Pathol 176:1169–1180
3. König J, Müller F, Fromm MF (2013) Transporters and drug-drug interactions: important determinants of drug disposition and effects. Pharmacol Rev 65:944–966
4. Cui Y, König J, Keppler D (2001) Vectorial transport by double-transfected cells expressing the human uptake transporter SLC21A8 and the apical export pump ABCC2. Mol Pharmacol 60:934–943
5. Ishiguro N, Maeda K, Saito A et al (2008) Establishment of a set of double transfectants coexpressing organic anion transporting polypeptide 1B3 and hepatic efflux transporters for the characterization of the hepatobiliary transport of telmisartan acylglucuronide. Drug Metab Dispos 36:796–805
6. König J, Zolk O, Singer K et al (2011) Double-transfected MDCK cells expressing human OCT1/MATE1 or OCT2/MATE1: determinants of uptake and transcellular translocation of organic cations. Br J Pharmacol 163:546–555
7. Letschert K, Komatsu M, Hummel-Eisenbeiss J et al (2005) Vectorial transport of the peptide CCK-8 by double-transfected MDCKII cells stably expressing the organic anion transporter OATP1B3 (OATP8) and the export pump ABCC2. J Pharmacol Exp Ther 313:549–556
8. Mita S, Suzuki H, Akita H et al (2005) Vectorial transport of bile salts across MDCK cells expressing both rat Na+-taurocholate cotransporting polypeptide and rat bile salt export pump. Am J Physiol Gastrointest Liver Physiol 288:G159–G167
9. Nies AT, Herrmann E, Brom M et al (2008) Vectorial transport of the plant alkaloid berberine by double-transfected cells expressing the human organic cation transporter 1 (OCT1, SLC22A1) and the efflux pump MDR1 P-glycoprotein (ABCB1). Naunyn Schmiedebergs Arch Pharmacol 376:449–461
10. Kopplow K, Letschert K, König J et al (2005) Human hepatobiliary transport of organic anions analyzed by quadruple-transfected cells. Mol Pharmacol 68:1031–1038
11. Seithel A, Eberl S, Singer K et al (2007) The influence of macrolide antibiotics on the uptake of organic anions and drugs mediated by OATP1B1 and OATP1B3. Drug Metab Dispos 35:779–786
12. Cui Y, König J, Buchholz JK et al (1999) Drug resistance and ATP-dependent conjugate transport mediated by the apical multidrug resistance protein, MRP2, permanently expressed in human and canine cells. Mol Pharmacol 55:929–937

Chapter 12

Metabolite Profiling of Mammalian Cell Culture Processes to Evaluate Cellular Viability

Isobelle M. Evie, Alan J. Dickson, and Mark Elvin

Abstract

Metabolite profiling allows for the identification of metabolites that become limiting during cell culture and/or for finding bottlenecks in metabolic pathways that limit culture growth and proliferation. Here we describe one protocol with two different sampling methodologies for GC–MS-based metabolite profiling. We also highlight an example of the types of datasets that are attainable and how such datasets can be evaluated to identify factors related to cell viability. We also demonstrate, via the same methodology, the accurate quantification of a number of metabolites of interest.

Key words Gas chromatography–mass spectrometry (GC–MS), Metabolites, Extracellular profile, Intracellular profile, Metabolomics, Metabolite profiling, Mammalian cells

1 Introduction

In mammalian cell culture maintaining optimum viability is essential in extending culture duration. Therefore assessment of viability throughout culture is an essential process. In order to maintain cellular proliferation and cell viability throughout culture, mammalian cells require a supply of specific nutrients to meet their metabolic demands [1], with metabolic activity seen as an indicator of cell health [2]. While successful attempts have been made to extend and maintain cell viability through optimization of commercial cell culture medium [3], metabolic bottlenecks remain. The depletion and exhaustion of metabolites limits cellular viability, particularly throughout long-term culture [4]. Such issues highlight the need for further understanding of the effects of culture conditions on cell viability. While assays can inform of cell death (e.g., vital dye exclusion [5]), metabolite profiling enables for an earlier indication of progression towards a poor viability.

Most commonly, mammalian cell culture is characterized by a highly glycolytic state paralleled by high rates of waste production [6]. When nutrients become limiting throughout culture,

Daniel F. Gilbert and Oliver Friedrich (eds.), *Cell Viability Assays: Methods and Protocols*, Methods in Molecular Biology, vol. 1601, DOI 10.1007/978-1-4939-6960-9_12,

cells cannot generate energy required for critical cellular processes. Ultimately, nutrient limitations lead to a restricted ATP supply to the cells [4], limiting cellular growth and viability. Having predictive early metabolic indicators would allow timely intervention to prevent loss of viability.

With the development of Omics-based technologies [7], stringent methodologies are now available to assess the link between metabolic function and cell viability, with metabolic profiles found to favorably shift under optimized culture conditions [8]. Cells grown in optimized conditions are known to lead to increased ATP production [9], demonstrating the importance of metabolic status for cellular viability.

Nutrient starvation and by-product accumulation can trigger cell death through apoptosis, limiting cell viability [10]. To understand and address such metabolic consequences, metabolite profiling can aid in understanding the pattern of metabolism that occurs in different cell culture regimes [11]. Metabolite profiling strategies have been used to improve cell viability through understanding the metabolic demands of specific cell lines [1], identify apoptosis-inducing metabolites [12], and to identify cell-engineering targets [13].

Analytical mass spectrometry platforms such as gas chromatography–mass spectrometry (GC–MS) and liquid chromatography–mass spectrometry (LC–MS) are widely utilized for metabolite profiling [14, 15]. Such profiles allow for identification of metabolites, through comparison of individual peaks of the fragmentation pattern, to metabolomic databases and mass spectral libraries [16, 17]. In this protocol, we describe and demonstrate through exemplar data, a validated adapted method [18], enabling for parallel extracellular (footprint) and intracellular (fingerprint) metabolite profiling of metabolites from suspension-cultured mammalian cells.

This method involves the recovery of metabolites by direct sampling from spent medium (extracellular) and through quenching of cells (intracellular) to stop cellular metabolism before separation of the cells from the medium [18, 19]. Sampling is followed by methanol and water extractions of the metabolites from the quenched cells [18, 20]. The metabolite samples generated through this method are amenable to analysis by mass spectrometry. Initially samples are derivatized after extraction, allowing for compounds of poor volatility, polarity, and stability to gain properties more amendable to mass spectrometry analysis, enabling the detection of a wider range of metabolites [21].

Here we also demonstrate and describe through the use of GC–MS, an example of the types of datasets that are attainable. Datasets received through the application of mass-spectrometry provide a broad semiquantitative profile of a range of metabolites. To allow for determination of the concentration of metabolites of interest, we also demonstrate the use of standards in tandem with the metabolic samples analyzed, allowing for accurate quantification of metabolites.

The generated metabolic profiles enable the identification of metabolites that relate to the transition in changes to cell viability throughout different phases of cell culture. Such data can be used to identify factors related to the status of cell viability.

2 Materials

Prepare all solutions using ultrapure water (prepared by purifying deionized water, to attain a sensitivity of 18 MΩ-cm at room temperature) and use analytical grade reagents where possible. Prepare and store all reagents at room temperature (unless indicated otherwise).

2.1 Extracellular (Footprint) Sampling

1. Analytical grade methanol (*see* **Note 1**).
2. Analytical grade water (*see* **Note 2**).
3. Analytical grade isopropanol (*see* **Note 3**).
4. Water/methanol/isopropanol solution (v/v/v, 2:5:2) (*see* **Note 4**).
5. Solution of internal standard in a ratio of 2:5:2 water/methanol/isopropanol (v/v/v) (*see* **Note 5**).

2.2 Intracellular (Fingerprint) Sampling

1. Ice-cold analytical grade water (*see* **Note 6**).
2. Analytical grade methanol stored at −80 °C (*see* **Note 7**).
3. Analytical grade absolute ethanol.
4. 8.5% (w/v) of ammonium bicarbonate (AMBIC) in water. Adjust the pH to 7.4 with 1 M hydrochloric acid.
5. Quenching solution: 60% (v/v) analytical grade methanol, 0.85% (w/v) AMBIC. Prepare quenching solution by combining 60 mL analytical grade methanol, 10 mL of 8.5% (w/v) AMBIC, and 29 mL of analytical grade water. Adjust the pH to 7.4 with 1 M hydrochloric acid and store at 4 °C (*see* **Note 8**).

2.3 GC–MS Sample Preparation

1. 40 mg/mL methoxyamine hydrochloride (MOX) in pyridine in a 1.5 mL micro-centrifuge tube—vortex until clear (*see* **Note 9**). This should always be prepared fresh on the day of GC–MS sample preparation.
2. *N*-Methyl-*N*-(trimethylsilyl) trifluoroacetamide (MSTFA) containing 1% (v/v) trichloromethylsilane (TCMS) stored at 4 °C.

3 Methods

Here we describe one protocol with two different sampling methodologies for GC–MS-based metabolite profiling, and this protocol could also be used with other mass spectrometry-based technologies such as LC–MS. It is important to collect samples for both footprint and fingerprint analysis at the same time (*see* **Note 10**). We carry out all the following procedures at room temperature unless otherwise specified.

3.1 Extracellular Metabolite Sample Preparation (Footprint Sampling)

Refer to Fig. 1 for an illustrative diagram of footprint sampling.

Collect medium samples for extracellular metabolite analysis throughout batch culture by taking 200 μL of medium and removing the cells via centrifugation (500 × *g* for 5 min) in a bench top micro-centrifuge. This clarified medium is then stored at −80 °C until assays are performed. Once all the samples have been collected we then proceed to follow the method outlined below, so we process all the samples from the same batch culture at once (*see* **Note 11**).

1. Take 20 μL of growth medium supernatant (with cells removed—*see* **Note 12**) and mix with 200 μL analytical grade methanol (*see* **Note 1**) and 5 μL internal standard (if required—*see* **Note 5**) in a 1.5 mL micro-centrifuge tube.
2. Vortex briefly and centrifuge (12,000 × *g* for 2 min) using a bench top micro-centrifuge.
3. Remove supernatant to a fresh 1.5 mL micro-centrifuge tube.
4. Lyophilize the extracellular samples using a rotary centrifugal evaporator at 30 °C until a dry metabolite pellet is observed (*see* **Note 13**). Larger volume samples may take longer to dry.
5. Store lyophilized samples at −80 °C until ready for next stages of assessment via GC–MS (*see* Subheading 3.3).

3.2 Intracellular Metabolite Sample Preparation (Fingerprint Sampling)

This sampling method is a two-step process involving quenching of cells (Subheading 3.2.1), followed by extraction of metabolites from the quenched cells (Subheading 3.2.2). The quenching of cells (Subheading 3.2.1) is time sensitive and therefore if preparing more than one sample it is recommended that two researchers work together, or that sampling is carried out in single batches.

3.2.1 Quenching of Cells

Refer to Fig. 2 for an illustrative diagram of quenching of cells.

1. Using a hemocytometer or an automated counting system, calculate the total culture volume required for 1×10^7 viable cells (one sample volume).
2. Pipette five sample volumes of the quenching solution into a 50 mL conical tube (*see* **Note 14**).
3. Cool the quenching solution to −40 °C using a dry ice ethanol bath or a cryostat (*see* **Note 15**).
4. Pipette one sample volume of cells into the middle of the −40 °C quenching solution in the 50 mL falcon tube (*see* **Note 16**).

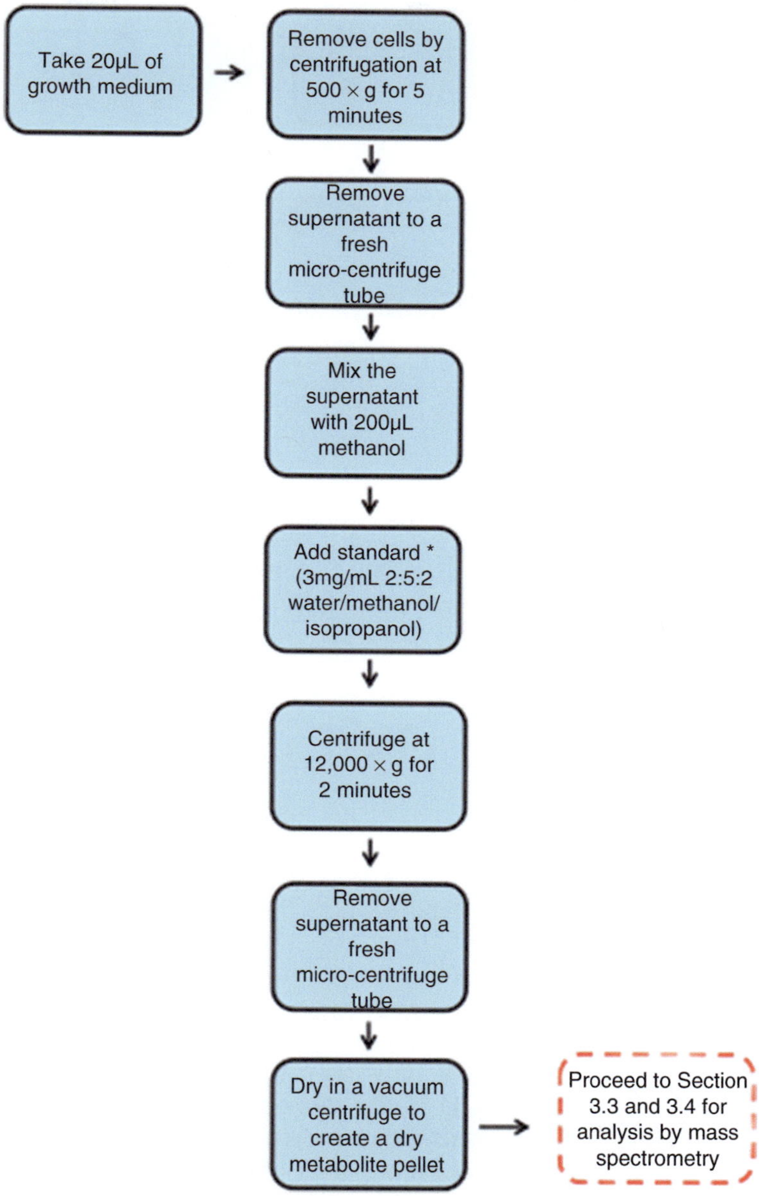

Fig. 1 Extracellular (footprint) sampling from suspension cultured cells. 20 μL of growth medium is mixed with 200 μL methanol and GC-MS standard (if required) in 2:5:2 water/methanol/isopropanol. The samples are centrifuged at 12,000 × *g* for 2 min before being removed to a fresh micro-centrifuge tube. The samples are dried down in a vacuum centrifuge to create a dry metabolite pellet. *Asterisk*: Refer to **Note 5**

5. Immediately after adding the sample volume, gently mix the sample by a single inversion of the tube.
6. Pellet the cells by centrifugation at 1000 × *g* for 1 min at −20 °C.
7. Remove the supernatant using an aspirator and a vacuum line (*see* **Note 17**).

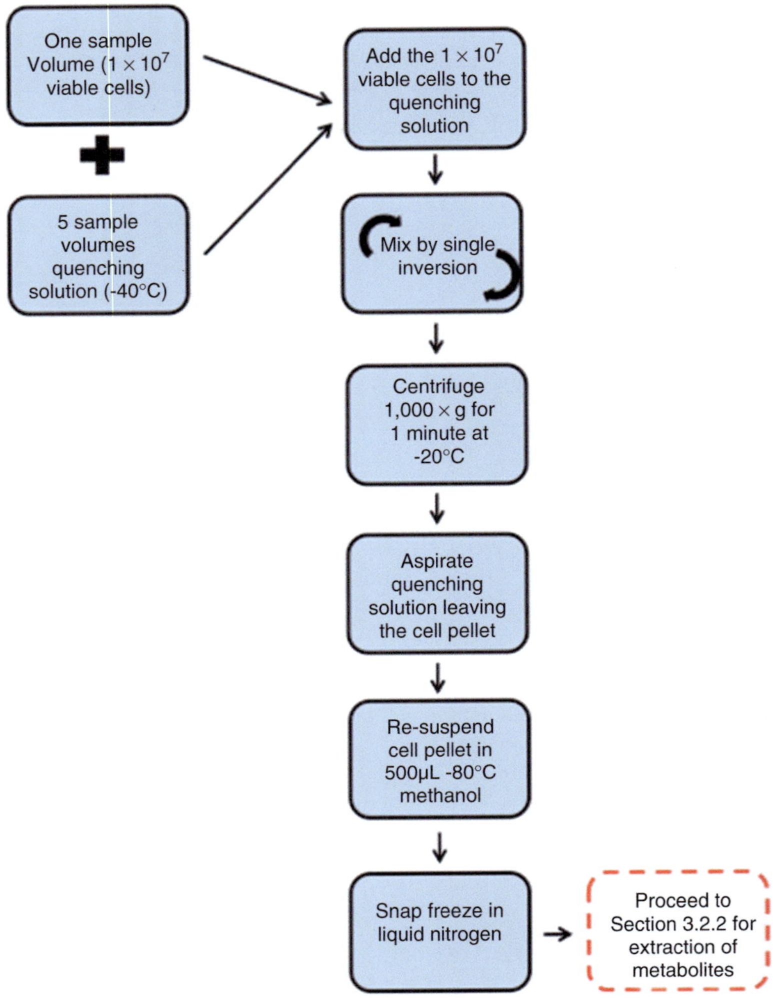

Fig. 2 Quenching of cells from suspension culture. One sample volume of cells (1×10^7 viable cells) is quenched in five volumes of quenching solution at −40 °C. The cells are pelleted by centrifugation and an aspirator is used to remove the quenching solution. The cell pellet is resuspended in −80 °C methanol and snap frozen in liquid nitrogen. The metabolites are then extracted from the quenched cells as described in Subheading 3.2.2 and Fig. 3

8. Resuspend the cell pellet in 500 µL –80 °C 100% methanol and transfer the supernatant into a fresh 1.5 mL microcentrifuge tube.
9. Snap-freeze the sample in liquid nitrogen.

After completion of the quenching of cells (Subheading 3.2.1), the samples can be frozen before extraction of the metabolites from the quenched cells (Subheading 3.2.2), if preferred. This allows for Subheading 3.2.2 to be carried out when all the required sampling and quenching is completed and allows for the same batch of standard (if required for experimental design) to be used across all samples.

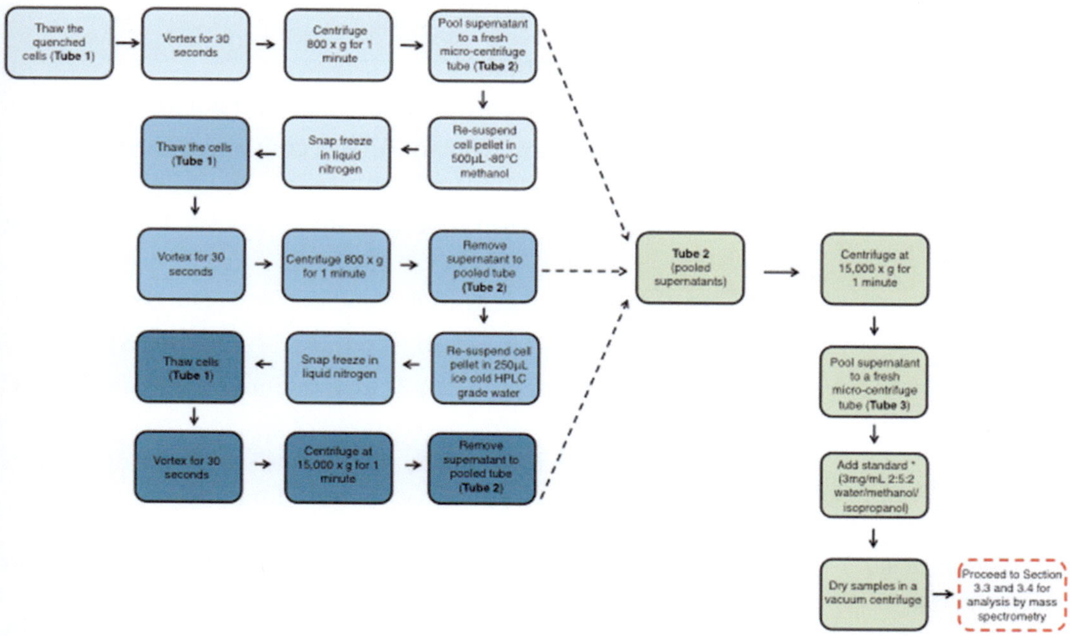

Fig. 3 Extraction of metabolites from the quenched cells. The metabolites are extracted using two 100% methanol and one water extraction. The quenched frozen cells in 100% methanol (Fig. 2) are thawed, vortexed, and pelleted by centrifugation. The supernatant is pooled into a new tube and the cell pellet is resuspended in 100% methanol and snap frozen in liquid nitrogen. The freeze thaw cycle is repeated and the cells are vortexed and pelleted by centrifugation and the supernatant is pooled with the previous methanol extraction. The cell pellet is resuspended in HPLC grade water and snap frozen in liquid nitrogen. The freeze thaw cycle is repeated for the last time and the cells are vortexed and pelleted by centrifugation. The supernatant is removed and pooled with the previous two methanol extractions. The pooled extracts are centrifuged to remove any potential remaining cell debris. The supernatant is removed to a fresh centrifuge tube and dried down in a vacuum centrifuge to create a dry metabolite pellet. *Asterisk*: Refer to **Note 5**

3.2.2 Extraction of Metabolites

Refer to Fig. 3 for an illustrative diagram of extraction of metabolites.

1. Thaw the quenched cells on dry ice (**step 9**, Subheading 3.2.1) (*see* **Note 18**).
2. Vortex the cells for 30 s.
3. Pellet the cells by centrifugation at 800 × *g* for 1 min.
4. Transfer the supernatant to a fresh 1.5 mL micro-centrifuge tube (Tube 2) stored on dry ice.
5. Resuspend the cell pellet in 500 μL –80 °C 100% methanol.
6. Snap-freeze in liquid nitrogen.
7. Repeat **steps 2–6** for Tube 1 and remove the supernatant to the previous pooled extractions in the tube on dry ice (Tube 2).
8. Resuspend the cell pellet (Tube 1) in 250 μL ice-cold water (*see* **Note 19**).

9. Snap-freeze in liquid nitrogen.
10. Repeat **steps 1** and **2**.
11. Pellet the cells by centrifugation at 15,000 × *g* for 1 min.
12. Transfer the supernatant to Tube 2 containing the two previous pooled 100% methanol extractions.
13. Centrifuge the pooled supernatant (Tube 2) at 15,000 × *g* for 1 min.
14. Transfer to a fresh centrifuge tube (Tube 3) (*see* **Note 20**).
15. Add 5 μL internal standard (if required—*see* **Note 5**).
16. Dry the supernatant (Tube 3) in a vacuum centrifuge at 30 °C (*see* **Note 21**).
17. Store lyophilized samples at −80 °C until ready for next stages of assessment via GC–MS (*see* Subheading 3.3).

3.3 Derivatization of Extracellular and Intracellular Metabolite Preparations

The lyophilized metabolite extracts (from both extracellular (Subheading 3.1) and intracellular (Subheading 3.2) metabolite preparations stored at −80 °C) are prepared for GC–MS analysis by a two-stage derivatization procedure (methoxyamination followed by trimethylsilylation) outlined below:

1. After removal of samples from −80 °C, the samples are briefly lyophilized again (30 °C for 10 min) using a rotary centrifugal evaporator (*see* **Note 22**).
2. After 10 min the samples are methoxyaminated by addition of 10 μL methoxyamine hydrochloride (MOX, 40 mg/mL in pyridine) to the lyophilized pellet.
3. The samples are then incubated for 90 min at 30 °C shaking at approximately 195 rpm in a temperature-controlled orbital shaking incubator.
4. After 90 min the methoxyaminated samples are then trimethylsilylated by adding 90 μL *N*-methyl-*N*-(trimethylsilyl) trifluoroacetamide (MSTFA) containing 1% (v/v) trichloromethylsilane (TCMS) in a fume hood.
5. Incubate samples for 30 min at 37 °C.
6. After 30 min the derivatized samples are centrifuged (at 12,000 × *g* for 2 min) using a bench top micro-centrifuge and the resulting supernatants (*see* **Note 23)** are transferred to silanized glass GC–MS vials immediately prior to analysis on the GC–MS instrument (*see* **Note 24**).

3.4 Gas Chromatography–Mass Spectrometry

Extracellular and intracellular metabolite samples are randomized (*see* **Note 25**) before being injected into the GC–MS instrument. For GC–MS we routinely use a 7890A GC system coupled to a 5975C Inert XL MSD mass spectrometer with a triple axis detector and fitted with a DB-5MS + DG column (250 μm × 30 m,

Table 1
Gas chromatography (GC) settings

Gas chromatograph settings	Value/units
Carrier gas	Helium (mL/min)
Inlet temperature	250 °C
Oven temperature	60 °C
Temperature ramp 1	325 °C at 10 °C/min

Table 2
Mass spectrometry (MS) settings

Mass spectrometer settings	Value/units
Source temperature	230 °C
Quadrupole temperature	150 °C
Solvent delay	5.9 (min)
Scan range	50–600 (M/Z)

0.25 μm film thickness with 10 m Duraguard). Helium (1.2 mL/min) is used as the carrier gas. Our GC–MS method is retention time locked (RTL) to our internal standard myristic acid-d_{27} (*see* **Note 26**). The GC–MS settings used for a metabolite run are described in Tables 1 and 2.

Briefly, components are separated by isothermal chromatography for 1 min at 60 °C, followed by an increase to 325 °C at a rate of 10 °C/min then 10 min at 325 °C. Mass spectra are acquired in positive ion mode using electron impact ionization at 70 eV. The injector, MS source, and MS quad temperatures are set at 250, 230, and 150 °C, respectively. Mass spectra are scanned from 50 to 600 mass units.

3.5 Data Processing/Analysis

Metabolite peaks in the raw chromatograms are identified using a program called MSD ChemStation (Agilent Technologies) to search the Fiehn GC–MS Metabolomics RTL Library [16]. The Fiehn mass spectral library contains mass spectra of more than 1000 metabolite derivatives originating from more than 700 metabolites [16]. Derivatization is necessary to restrict the potential number of compounds formed from individual metabolites to a manageable number, and to enable the volatilization of metabolites with high boiling points for effective separation in the GC. Compounds originating from the same parental metabolite are annotated in the Fiehn mass spectral library using a numbering system (e.g., Glucose 1 or Glucose 2). Metabolite identifications from GC–MS datasets are based on retention times and specific fragmentation patterns (*see* **Note 27**).

3.6 Exemplar Results

To maintain viability cells require nutrients to meet their metabolic demands. Here we show a validated metabolic dataset gained through intracellular and extracellular metabolite profiling throughout cell culture for the identification of bottlenecks that limit cellular viability (Fig. 4). The metabolic profile allowed for the development of a simple feeding regime to alleviate metabolic limitations to cell viability. Extracellular metabolite profiling alone allows for the identification of the exhaustion and accumulation of metabolites of interest, but when used in tandem with intracellular metabolite profiling, allows for a greater understanding of changes to external metabolites on the flux of TCA cycle intermediates.

The complete exhaustion of the key metabolite glucose by day 5 of cell culture has led to a transition in cellular metabolism from production to use of lactate (Fig. 4b). The depletion of glucose affects maintenance of cellular viability due to a lack of TCA cycle intermediates for important cellular functions. From the metabolic profiles received, a developed feeding regime where glucose is fed back into the culture on day 4 of culture up to the initial concentration for day 0 of culture enables for an increase in cell viability (Fig. 4a). The effect of an altered nutrient environment on cellular metabolism can be identified through an increased flux of TCA cycle intermediates (Fig. 4b). The metabolite profiles also enable for an understanding of the effect of an altered nutrient environment and the associated increase in cellular viability, on the consumption of associated metabolites (alanine, glutamate) with an increase in cell viability leading to the exhaustion of extracellular metabolites in the cellular environment (Fig. 4b).

3.6.1 Metabolite Standards for Accurate Quantification

Metabolite profiling through mass spectrometry allows for a semi-quantitative approach for determination of the pattern on metabolism occurring throughout cell culture. Therefore quantification of metabolites of interest requires the use of metabolic assays in conjunction with the metabolic profiles received. Such approaches, while accurate, require optimization for detection of individual metabolites of interest.

The use of a “metabolic cocktail” of standards run in tandem with the samples on the mass spectrometer, allows for quantification of metabolites in the samples through the use of standard curves. The standards can be made up of specific metabolites of choice in a range of concentrations, allowing for quantification of several metabolites at once.

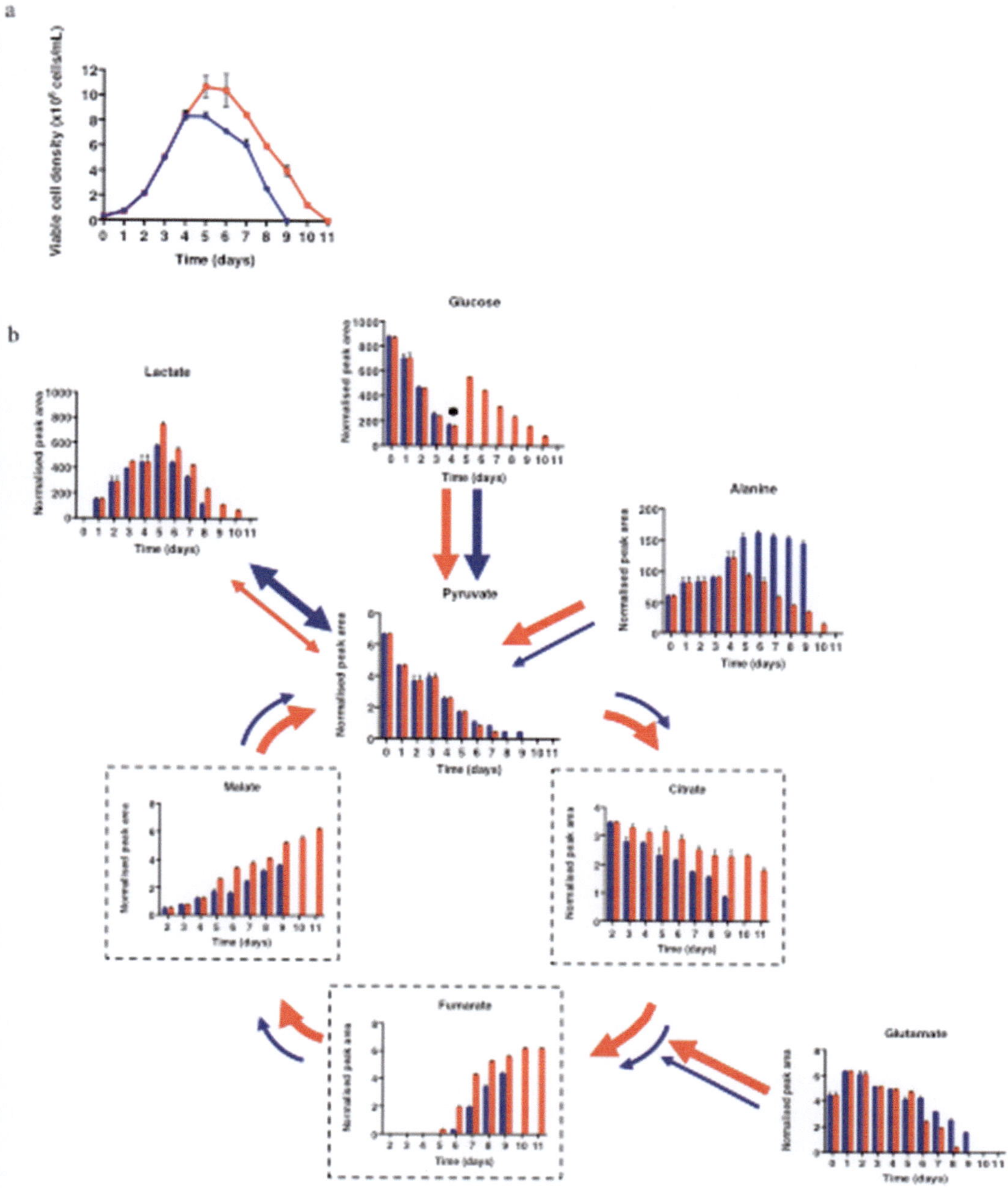

Fig. 4 Intracellular and extracellular metabolite profiles to improve cell viability. (**a**) Fingerprint and footprint samples were obtained throughout batch culture for fed (*red*) and un-fed cells (*blue*). (**b**) The samples were analyzed by GC–MS for metabolites of interest. The bar charts mapped onto the metabolic pathways represent the abundance of metabolites throughout culture. The bars are colored as indicated by the growth profile with blue bars representing control (un-fed) cells and red bars represent fed-batch cultures. Graphically data represents three biological replicates with error bars representing the standard error of the mean. The *arrows* indicate the direction of the reaction with *double arrowheads* indicating the reaction is reversible. The thickness of the *arrows* indicates an increased flux of metabolites. Graphs in *dashed lined boxes* represent intracellular metabolites and (*asterisk*) indicates the time of feed

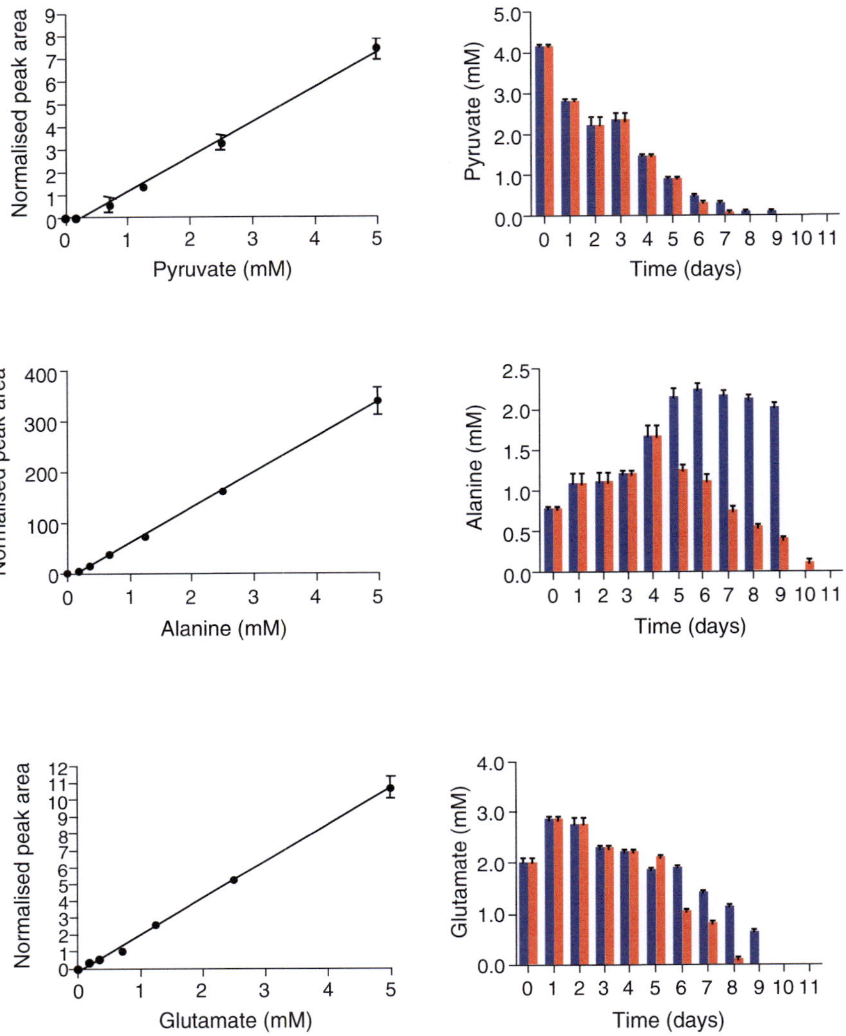

Fig. 5 Metabolite standards for quantification of metabolites of interest. Standard curves were generated for the metabolites alanine, pyruvate, and glutamate by determining the normalized peak area for a range of concentrations from 0 to 5 mM. Standard curves represent three technical replicates with standard deviation error bars. The standard curves were used for quantification of the metabolite profiles given by GC-MS (Fig. 4b) for accurate quantification of the metabolites pyruvate, alanine, and glutamate. *Blue bars* represent control (un-fed) cells and *red bars* represent fed-batch cultures. Data represents three biological replicates with error bars representing the standard error of the mean

Here we show the results obtained for standard curves for the metabolites pyruvate, alanine, and glutamate (Fig. 5). The standards were run in tandem with the GC–MS samples (Fig. 4b) and the curves were used for accurate quantification of the metabolites for fed and un-fed cells (Fig. 5).

The outcome of such approaches to quantification could allow for established metabolite datasets to be assembled into models composed of metabolic reactions in relation to the cellular environment. Such datasets could allow for the emergence of predictive models relating the state of cellular metabolism on cell viability.

4 Notes

1. We use LC–MS Ultra CHROMASOLV® methanol.
2. We use LC–MS CHROMASOLV® water.
3. We use Optima® 2-Propanol.
4. We find it best to prepare this solution fresh each time.
5. The standard of choice (if required) is dependent on experimental design and will vary based on the choice of library used for detection of metabolites. We use a 3 mg/mL solution of myristic acid-d_{27} in a ratio of 2:5:2 water/methanol/isopropanol (v/v/v) as our GC–MS internal standard, and it is **IMPORTANT** that this is prepared fresh each time.
6. We use LC–MS CHROMASOLV® water. It is important that this is kept ice-cold.
7. We use LC–MS Ultra CHROMASOLV® methanol. It is important that this is kept at –80 °C to prevent turnover of metabolites.
8. We use LC–MS Ultra CHROMASOLV® methanol, LC–MS CHROMASOLV® water, and Ammonium Bicarbonate (AMBIC) to make the quenching solution. It's **CRITICAL** that fresh quenching solution must be prepared for each intracellular preparation because of pH drift of the solution over time.
9. We find it best to pre-warm the pyridine first before adding the MOX as this will aid in dissolving the MOX into solution.
10. Extracellular and intracellular sample collection for metabolite analyses should be run in parallel (i.e., collected at the same time on the same day) so that direct comparisons of the metabolite data can be made. This is a very **IMPORTANT** step and should always be done.
11. It is important to collect all samples for extracellular metabolite analysis before proceeding to the methanol precipitation so as to use the same batch of internal standard for all samples. This means that there should be less variability when normalizing the data.
12. We routinely take 200 μL of medium and remove the cells via centrifugation (at 500 × *g* for 5 min) in a bench top micro-centrifuge.

13. We use an Eppendorf *Concentrator* Plus set to aqueous liquid mode at 30 °C (pellet times may vary between the size of samples to lyophilize).
14. For efficient manipulation of the sample in the 50 mL falcon tube, the maximum sample volume cannot exceed 8 mL. If a larger volume is required for 1×10^7 viable cells, split the sample equally amongst two or more falcon tubes as required and recombine for extraction (Subheading 3.2.2).
15. Check the temperature of the quenching solution with a thermometer. If cooling more than one falcon tube, check the temperature across all tubes to ensure that they are all the same temperature.
16. It is **CRITICAL** that the quenching volume is at –40 °C when the cells are added for efficient quenching of cellular metabolism. Temperatures below –40 °C will freeze the cells (resulting in white material being observed when the cells are added) due to damage of the cell membranes. Temperatures above –40 °C will not quench the cells.
17. Keep the aspirator in place for a few seconds after the supernatant is removed. This is to ensure that all of the supernatant is removed to prevent medium contamination in your samples.
18. If you stored your samples in the –80 °C freezer after completing quenching of cells (Subheading 3.2.1), please note your samples will already be thawed at –80 °C.
19. Needs to be HPLC grade water.
20. This step is important for removing any remaining cell debris from the sample.
21. If preferred the samples can be stored at this point at –80 °C until future derivatization (Subheading 3.3) and/or analysis by mass spectrometry (Subheading 3.4).
22. This step is done to remove any possible trace of moisture that may have accumulated during storage at –80 °C.
23. There is usually some debris present in the micro-centrifuge tubes after spinning them down. It is very **IMPORTANT** to only put clear supernatant into the silanized GC–MS vials.
24. Derivatized samples have an expected stability of up to 24 h, so larger batches of samples are split into smaller batches to ensure reproducibility between GC–MS runs.
25. We load our GC–MS samples in a randomized order so to minimize bias and not introduce any artifacts into our metabolite datasets.
26. We use myristic acid-d_{27} as our internal GC–MS standard (other standards are available). We add this to all our extracellular and intracellular sample preparations as our GC–MS

method is retention time locked to this compound. This means that we get comparable chromatograms between every run.

27. Metabolite identification is based on the retention times closely matching and the ratios of a primary (quantifying) fragment peak area with two secondary (qualifying) fragment peak areas.

Acknowledgment

We gratefully acknowledge the financial support of UK BBSRC and TSB. We are also grateful to Dr. Chris Sellick from MedImmune, who was instrumental in developing the methodology for this GC–MS-based technique.

References

1. Sellick CA, Croxford AS, Maqsood AR, Stephens G, Westerhoff HV, Goodacre R, Dickson AJ (2011) Metabolite profiling of recombinant CHO cells: designing tailored feeding regimes that enhance recombinant antibody production. Biotechnol Bioeng 108(12):3025–3031
2. Browne SM, Al-Rubeai M (2011) Defining viability in mammalian cell cultures. Biotechnol Lett 33:1745–1749
3. Jayapal KP, Wlaschin KF, Hu W-S, Yap MGS (2007) Recombinant protein therapeutics from CHO cells-20 years and counting. Chem Eng Prog 103(10):40–47
4. Dickson AJ, Pogson CI (1977) The metabolic integrity of hepatocytes in sustained incubations. FEBS Lett 83(1):27–32
5. Stoddart MJ (2011) Cell viability assays: introduction. Methods Mol Biol. 740:1–6. doi:10.1007/978-1-61779-108-6_1
6. Ahn WS, Antoniewicz MR (2012) Towards dynamic metabolic flux analysis in CHO cell cultures. Biotechnol J 7(1):61–74
7. Kildegaard HF, Baycin-Hizal D, Lewis NE, Betenbaugh MJ (2013) The emerging CHO systems biology era: harnessing the omics revoloution for biotechnology. Current Opin Biotech 24(6):1–6
8. Luo J, Vijayasankaran N, Austen J, Santuray R, Hudson T, Amanullah A, Li F (2012) Comparative metabolite analysis to understand lactate metabolism shift in Chinese hamster ovary cell culture process. Biotechnol Bioeng 109(1):146–156
9. Chong WPK, Reddy SG, Yusufi FNZ, Lee DY, NSC W, Heng CK, MGS Y, Ho YS (2010) Metabolomics driven approach for the improvement of Chinese hamster ovary cell growth: overexpression of malate dehydrogenase II. J Biotechnol 147(2):116–121
10. Braasch K, Nikolic-Jaric M, Cabel T, Salimi E, Brdiges GE, Thompson DJ, Butler M (2013) The changing dielectric properties of CHO cells can be used to determine early apoptotic events in a bioprocess. Biotechnol Bioeng 110(11):2902–2914
11. Dickson AJ (2014) Enhancement of production of protein biopharmaceuticals by mammalian cell cultures: the metabolomics perspective. Current Opin Biotech 30:73–79
12. Chong WPK, Yusufi FNK, Lee D-Y, Reddy SG, Wong NSC, Heng CK, Yap MGS, Ho YS (2011) Metabolomics-based identification of apoptosis-inducing metabolites in recombinant fed-batch CHO culture media. J Biotechnol 151(2):218–224
13. Majors BS, Betenbaugh MJ, Pederson NE, Chiang GC (2008) Enhancement of transient gene expression and culture viability using Chinese hamster ovary cells overexpressing Bcl-x(L). Biotech Bioeng 101(3):567–578
14. Chong WPK, Thng SH, Hiu AP, Lee D-Y, Chan ECY, Ho YS (2012) LC-MS-based metabolic characterization of high monoclonal antibody-producing chinese hamster ovary cells. Biotech Bioeng 109(12):3103–3111
15. Dunn WB, Broadhurst D, Begley P, Zelena E, Francis-McIntyre S, Anderson N, Brown M, Knowles JD, Halsall A, Nicholls AW, Wilson ID, Douglas BK, Goodacre R, and The Human Serum Metabolome (HUSERMET) Consortium (2011) Procedures for large scale metabolic profiling of serum and plasma using gas

chromatography and liquid chromatography coupled to mass spectrometry. Nature 6(7): 1060–1083

16. Kind T, Wohlgemuth G, Lee DY, Lu Y, Palazoglu M, Shahbaz S, Fiehn O (2009) FiehnLib-mass spectral and retention index libraries for metabolomics based on quadrupole and time-of-flight gas chromatography/mass spectrometry. Anal Chem 81(24):10038–10048
17. Kanehisa M, Goto S, Sato Y, Furumichi M, Tanabe M (2012) KEGG for integration and interpretation of large-scale molecular data sets. NAR 40(1):109–114
18. Sellick CA, Hansen R, Stephens GM, Goodacre R, Dickson AJ (2011) Metabolite extraction from suspension-cultured mammalian cells for global metabolite profiling. Nat Protoc 8(6): 1241–1249
19. Sellick CA, Knight D, Croxford AS, Maqsood AR, Stephens GM, Goodacre R, Dickson AJ (2010) Evaluation of extraction processes for intracellular metabolite profiling of mammalian cells: matching extraction approaches to cell type and metabolite targets. Metabolomics 6:427–438
20. Sellick CA, Hansen R, Maqsood AR, Dunn WB, Stephens GM, Goodacre R, Dickson AJ (2009) Effective quenching processes for physiologically valid metabolite profiling of suspension cultured mammalian cells. Anal Chem 8:174–183
21. Agilent Technologies (2007) Considerations for selecting GC/MS or LC/MS for metabolomics [pdf]. USA: Agilent Technologies. https://www.chem.agilent.com/Library/selec5onguide/Public/5989C6328EN.pdf

Chapter 13

Assaying Spontaneous Network Activity and Cellular Viability Using Multi-well Microelectrode Arrays

Jasmine P. Brown, Brittany S. Lynch, Itaevia M. Curry-Chisolm, Timothy J. Shafer, and Jenna D. Strickland

Abstract

Microelectrode array (MEA) technology is a neurophysiological method that allows for the spontaneous measure of activity in neural cultures and determination of drug and chemical effects thereon. Recent introduction of multi-well MEA (mwMEA) formats have dramatically increased the throughput of this technology, allowing more efficient compound screening. Rapid characterization of compounds for neuroactivity or neurotoxicity hazard evaluation following acute, chronic, or developmental exposures ideally would also consider compound effects on cell health, and to do so in the same well requires a multiplexed approach. Procedures describing the multiplexed method to acute and developmental screening are described in this chapter.

Key words Microelectrode array (MEA), Neurophysiological method, mwMEA, Developmental screening

1 Introduction

Multi-well microelectrode array (mwMEA) systems have increased throughput of traditional MEAs, making them an effective in vitro screening tool to prioritize large sets of compounds. This phenotypic approach can be applied to both drug discovery and drug safety screening, as well as acute [1, 2], chronic, and developmental [3–5] screening for cardiac and neurotoxicity. For example, currently, thousands of compounds lack sufficient toxicity data, especially as it relates to the nervous system and its development, and mwMEAs have been proposed as one approach to address this issue [6]. However, discrimination of compound effects on cell function versus cell health remains challenging as some formats of higher throughput mwMEA plates are opaque. Thus, determination of cell health was often completed in "sister" cultures at different densities. This method proved to be both suboptimal and inefficient, because it required the preparation of additional culture

Daniel F. Gilbert and Oliver Friedrich (eds.), *Cell Viability Assays: Methods and Protocols*, Methods in Molecular Biology, vol. 1601, DOI 10.1007/978-1-4939-6960-9_13, © Springer Science+Business Media LLC 2017

materials and dosing solutions. The advancement of a multiplexed approach allowing for simple and rapid characterization of compound effects on both neurophysiological and cellular viability endpoints within the same network provides a method to differentiate between compound-induced changes in neural activity and overall reductions in cell health [7]. The ability to determine if endpoint specific effects of a compound (e.g., changes in network firing rates) and changes in viability occur simultaneously remains an important aspect of in vitro screening for neurotoxicity. The methods demonstrated in this chapter utilize a mixed primary cortical culture comprising both inhibitory and excitatory neurons and glial cells. However, these methods may be adapted for other MEA platforms (i.e., 96-well platforms) or cell types (i.e., cardiac cells).

Using a multiplexed screening approach we demonstrate a simple and rapid method for simultaneous determination of compound effects on both neural network function and cellular health. Here we present comprehensive methods for assessment of compound effects on neural network function following acute or developmental exposure. These methods include culturing, recording, and viability assessment on mwMEA plates. Specifically, to assess changes in function, primary cortical cultures were grown on mwMEA plates (48 wells). Changes in spontaneous network activity in response to treatment were monitored for both acute (~40 min on *day* in vitro 13) and developmental exposures (~15 mins on *days* in vitro 2, 5, 7, 9, and 12). Effects on cell health were assessed from the same well following recording by measuring both lactate dehydrogenase (LDH) release and metabolic activity (CellTiter Blue; CTB). Modifications to each of the viability assays were needed for both acute and developmental exposures (*see* Subheading 3.7 for details). While facilitating rapid determination of both changes in neural network function and viability, these methods also serve to reduce cost, time, and animal use. These methods could easily be adapted to a variety of other cell types and exposure conditions.

2 Materials

The materials listed below are referenced in multiple procedures. Those specific to a certain procedure are listed in their respective materials section. All procedures in this chapter use sterile technique and are done in a laminar flow hood with the exception of the viability assays described later in this chapter.

1. Laminar flow hood.
2. 37 °C water bath.
3. Incubator (37 °C ± 1 °C, 5% CO_2 ± 1%, 95% humidity).
4. Vacuum flasks/trap.

5. Vortex.
6. Multi-well MEA plates (mwMEA, Axion M768-KAP-48) with viable neural cultures.
7. Pipet Aid and disposable pipettes.
8. Single channel pipette.
9. Multichannel pipette.
10. Reagent reservoir (Costar cat #4870 or equivalent).
11. Sterile pipette tips sized to fit pipettes.
12. Neurobasal-A media (NB) supplemented with B-27 (500 mL NB (1×) Gibco, 10 mL B-27 Supplement (50×), 5 mL GlutaMax (100×), 5 mL Pen-Strep) (NB/B27), pH adjusted to 7.4.
13. 70% ethanol.

2.1 Polyethyleneimine (PEI) Preparation and mwMEA Coating

1. pH meter.
2. Stir plate and stir bar.
3. Balance.
4. Beakers (500 mL, 1 L or other appropriate size).
5. Graduated cylinder.
6. Bottle top filter (Cellulose acetate) (0.22 μm) (Corning Cat# 430521).
7. HEPES (FW = 238.3, Sigma H7523 or equivalent).
8. Polyethyleneimine (PEI) (Sigma Cat# P3143).
9. Deionized (dI) water.
10. 5 M NaOH.

2.2 MEA Recording

1. Axion Integrated Studio (AxIS) software v2.0 or later on a computer workstation.
2. Axion Biosystems Maestro 768-channel amplifier and Middleman data acquisition interface.

2.3 Developmental Screening Assay

1. Sterile polypropylene 96-well dosing plate.
2. Stock compounds.
3. Axion Integrated Studio (AxIS) software v2.0 or later.
4. Axion Biosystems Maestro 768-channel amplifier and Middleman data acquisition interface.

2.4 Acute Screening Assay

1. Polypropylene 96-well dosing plate.
2. Stock compounds.
3. Axion Integrated Studio (AxIS) software v2.0 or later.
4. Axion Biosystems Maestro 768-channel amplifier and Middleman data acquisition interface.

2.5 Viability

2.5.1 CellTiter Blue

1. CellTiter Blue® Cell Viability Assay kit (Promega, catalog # G8081).
2. BMG FLUOstar OPTIMA Fluorescence/Luminescence Microplate Reader and Optima v2.20 R11 software or equivalent (excitation and emission wavelengths of 544 and 590 nm, respectively).
3. 96-well white opaque plate.
4. Test material: exposed cultures in mwMEA plates.
5. Plate map.

2.5.2 LDH Release (Acute Screening)

1. CytoTox 96® Non-Radioactive Cytotoxicity Assay kit (Promega, catalog # G1780).
2. Molecular Devices VERSAMax plate reader and SoftMax Pro v5.3 software or equivalent (Absorbance 490 nm).
3. Clear 96-well plates with flat bottom.
4. Test material: exposed cultures in mwMEA plates.
5. Plate map.

2.5.3 LDH Total (Developmental)

1. CytoTox 96® Non-Radioactive Cytotoxicity Assay kit (Promega, catalog # G1780).
2. Molecular Devices VERSAMax plate reader and SoftMax Pro v5.3 software or equivalent (Absorbance 490 nm).
3. Phosphate Buffered Saline (PBS, Gibco cat# 20012043).
4. Clear 96-well plates with flat bottom.
5. Test Material: exposed cultures in mwMEA plates.
6. Plate map.

3 Methods

3.1 Preparation of Polyethyleneimine (PEI) Solution (0.05% PEI in 50 mM HEPES, pH 8.0)

Prior to culturing cells on mwMEAs, the surface of the wells must be coated with a substrate that facilitates attachment of the cells. Our protocols use PEI, but other substrates may be more appropriate depending on the cell type.

1. Determine the volume of solution needed.
2. Weigh out the appropriate amount of PEI (0.5 mg/mL) into the container that the solution is to be prepared (*see* **Note 1**).
3. Measure 90% of the volume of water needed and add to container containing the PEI.
4. Weigh out the appropriate amount of HEPES (11.9 mg/mL) and add to the container.
5. Mix until both are completely dissolved.

Table 1
Chart of volumes per well for plate types

Plate type	PEI	3× wash	Media addition
12-well	100 μL	1 mL	1.5 mL
48-well	**200 μL**	**400 μL**	**500 μL**
96-well	100 μL	200 μL	100 μL

48-well MEA plates are generally used for both the acute and developmental assays. These methods were written with the use of a 48-well MEA plate in mind. Other plate sizes (12-well and 96-well) can be optimized for these procedures

6. Adjust pH of the buffer to 8.0 with 5.0 M NaOH. Add water to final volume.
7. Filter sterilize through 0.22 μm filter unit.
8. Aliquot into desired quantities and store at –20 °C.

3.2 Coating mwMEA Plates with Polyethyleneimine (PEI)

1. Add the appropriate amount of PEI to each well of an mwMEA plate ensuring that the bottom of the well is completely covered (if needed, gently tap the perimeter of the plate for full coverage) (Table 1).
2. Place the plate in the incubator for 1 h.
3. Remove PEI from wells by aspiration and rinse thoroughly with sterile water three times (Table 1) (*see* **Note 2**).
4. Allow plates to air dry in the hood with the lid off. Turn the lids open end up to ensure sterility.
5. Once dry, plates can immediately be used for cell culture or taped closed and stored in the refrigerator until they are ready for use (can be refrigerated up to 7 days after coating).

3.3 Culturing on mwMEA Plates

Primary cortical neural cultures are prepared from the neocortex of Long Evans rat pups (0–24 h old) and is detailed in [1] with modifications. All procedures involving animals were approved by the National Health and Environmental Effects Laboratory Institutional Animal Use and Care Committee. Cortical cells are plated at a seeding density of 1.5×10^5 cells/well on 48-well MEA plates pre-coated with 0.05% polyethylenimine (PEI) (Subheading 3.2: Coating mwMEA Plates with PEI). Cells are administered via a 25 μL media/laminin drop directly onto the microelectrode array (*see* **Note 3**). After 2 h, 475 μL of NB/B27 media is added to each well. Viable cultures form complex neural networks over time and can remain stable for several weeks in culture. Neural firing increases in synchrony and bursting activity as the culture ages, and this activity can be quantified [5]. Plates can either be immediately dosed following seeding and monitored over a period of time (Subheading 3.5) or acutely exposed to compound (Subheading 3.6).

3.4 MEA Recording

1. Turn on the Middle Man and open AxIS.
2. Once the program has connected to the Maestro and initialized, set the heater to 37 °C.
3. Maestro settings should be set as follows: Neural Spikes, 1200× gain, and Median Referencing selected.
4. Add the Digital Filter to the Maestro and use the default settings:
   - High Pass Filter = Butterworth.
   - High Pass Cutoff Frequency = 300 Hz (Acute), 0.1 Hz (Developmental).
   - Low Pass = Butterworth.
   - Low Pass Cutoff Frequency = 5000 Hz (*see* **Note 4**).
5. Add Spike Detector to Digital Filter: Set the standard deviation to 8× and select Adaptive Threshold Crossing method.
6. Add Burst Detection (optional): Use the Interspike Interval Threshold method of detection and make sure that the settings are as follows: Max ISI = 100 ms, Minimum number of spikes = 5, Mean Firing Rate Estimation = 10 s, and Synchrony = 20 ms (*see* **Note 5**).
7. For experimental recordings, "AxIS Raw" files are collected (*see* **Note 6**). Additionally, you may also wish to record other files while collecting the .raw file (*see* **Note** 7).
8. Once the heater has reached 37 °C, place the mwMEA plate to be recorded into the Maestro. Ensure that the plate is in the correct position, with the notched corner on the upper left side.
9. Select play and ground any electrodes per well with noise over 5 μV (*see* **Notes 8** and **9**).
10. Record for the desired length of time. Length of recording time is determined by the investigator. After the time is completed, select the stop icon.

3.5 Developmental Screening Assay

This procedure begins with the dosing of cortical cultures in triplicate mwMEA plates 2 h after they are seeded (Subheading 3.3: Culturing on mwMEA Plates). Assessment of activity is made on DIV 2, 5, 7, 9, and 12. Cells experience a full media change and subsequent dosing on DIV 5 and 9. On DIV 12 two cytotoxicity assays are performed (Subheading 3.7: Viability). The CytoTox 96® Non-Radioactive Cytotoxicity Assay measures the amount of lactate dehydrogenase (LDH) released from lysed cells (remaining post-treatment) using absorbance. The CellTiter-Blue® Cell Viability Assay measures the fluorescence of an end product produced only by metabolically active cells. It should be noted that this procedure could easily be modified to a "chronic" or "repeated dosing" protocol by, for example, delaying the start of dosing until cultures have matured to DIV 12.

3.5.1 Preparation of 50× Dosing Plate

Test compounds are prepared at 1000× the final desired testing concentration in the appropriate solvent (e.g., dI H_2O, EtOH, DMSO). This results in a final vehicle concentration of 0.1% (vol/vol).

A dosing plate is prepared at 50× the final concentration (a 1:20 dilution of the 1000× stock in NB/B27). Compounds are tested at a range of concentrations (generally 0.03–30 μM). The 50× stock is used to dose the 48-well MEA plate. Addition of the 50× stock into the mwMEA plate is a 1:50 dilution resulting in the desired final concentration (*see* **Note 10**).

1. Aliquot ~5 mL of NB/B27 media in preparation for the 96-well dosing plate. Keep the aliquot in the water bath until use.
2. Rotate the 96-well dosing plate so that the letters are horizontal at the top of the plate (H to A from left to right) and the numbers are vertical (increasing from 1 to 12 on the right side of the plate. The first 6 rows (48 wells) are used for the initial dosing of the mwMEA plate (DIV 0) (*see* **Note 11**).
3. Pipette 5 μL of stock compound into the appropriate well as indicated by the 96-well Dosing Plate Map (Fig. 1) changing the pipette tip each time.
4. Add 95 μL of warmed NB/B27 media to each well using a multichannel pipette to achieve a 1:20 dilution.

3.5.2 Chemical Exposure (Cells Are Dosed on DIV 0, 5, and 9)

1. Dose MEA plate(s) using the 50× dilution prepared in advance. Using an adjustable multichannel pipette, add 10 μL of the 50× stock into the mwMEA plate to achieve the desired final concentration. Column H on the dosing plate corresponds to column 1 on the mwMEA plate, column G on the dosing plate corresponds to column 2 on the mwMEA plate, and so on (*see* **Note 12**). Repeat the transfer starting at the lowest concentration, and continue until all columns of the mwMEA plate have been dosed (*see* **Note 13**).
2. Upon completion, return the mwMEA plates to the incubator (*see* **Note 14**).
3. Repeat these steps for DIV 5 and 9.

3.5.3 MEA Recordings (MEA Plates Are Recorded on DIV 2, 5, 7, 9, and 12)

1. Using the Maestro MEA system, allow the 48-well MEA plate to sit in the Maestro for 10 min prior to recording (*see* **Note 15**).
2. Record a 15 min baseline (Subheading 3.4: MEA Recording). The following file types are collected for this assay: AxIS Raw (.raw), Spike Counts (.csv), and Spike List (.csv).

96 well Dosing Plate Map (μM)

	1	2	3	4	5	6	7	8	9	10	11	12
A	A 30	B 30	C 30	D 30	E 30	F 30						
B	A 10	B 10	C 10	D 10	E 10	F 10						
C	A 3	B 3	C 3	D 3	E 3	F 3						
D	A 1	B 1	C 1	D 1	E 1	F 1						
E	A 0.3	B 0.3	C 0.3	D 0.3	E 0.3	F 0.3						
F	A 0.1	B 0.1	C 0.1	D 0.1	E 0.1	F 0.1						
G	A 0.03	B 0.03	C 0.03	D 0.03	E 0.03	F 0.03						
H	A 0	B 0	C 0	D 0	E 0	F 0						

Fig. 1 Example 96-well dosing plate map for chemicals at 50×. Colored letters represent a particular test chemical, in μM

3.5.4 Refeed: Full Media Change with NB/B27 (Old Media is Replaced with Fresh Media on DIV 5 and 9)

1. Aliquot enough NB/B27media for the media change (25 mL of NB/B27 media is needed for one 48-well MEA plate) and place the aliquot in the water bath until it has warmed to 37 °C (approximately 30 min for 75 mL).
2. After media has reached 37 °C, remove it from the water bath. Inside the laminar flow hood, pipette 25 mL of media into a sterile reservoir.
3. Using the vacuum trap, gently aspirate most of the media from two columns of wells at a time, ensuring that a thin film is left covering the cells.
4. Gently add 500 μL of fresh media back into the wells using an adjustable multichannel pipette (*see* **Note 16**).
5. Repeat these steps for any subsequent plates.
6. After a full media change, place MEA plates back into the incubator until dosing.

3.5.5 Viability Measures

On DIV 12, two measures of viability are used to assess cell health: The CytoTox 96 Non-Radioactive Cytotoxicity Assay and the CellTiter Blue Cell Viability Assay. For a more detailed description of these assays, refer to Subheading 3.7: Viability.

3.6 Acute Screening

This procedure begins with the isolation of primary cortical neurons and their culture on 48-well MEA plates (Subheading 3.3: Culturing on mwMEA Plates). The cultures experience a full media change on DIV 5 and 9 (Subheading 3.5.4: Refeed: Full Media Change with NB/B27). When the cultures are exposed to chemicals on DIV 13, half of the media is changed on the day prior (DIV 12). On DIV 13, an assessment of baseline activity is made for 40 min. The cultures are then acutely exposed to chemicals and a second 40 min assessment of activity is recorded. After the second recording, two cytotoxicity assays are performed (Subheading 3.7: Viability).

3.6.1 MEA Recordings

1. Using the Maestro system, allow the 48-well MEA plate to equilibrate in the Maestro for 20 min prior to recording (*see* **Note 17**) (Subheading 3.4).
2. After 20 min, record a 40 min baseline.

3.6.2 Preparation of the Dosing Plate

1. During the baseline recording, prepare the dosing plate (similar to the procedure described in Subheading 3.5.1: Preparation of 50× Dosing Plate). Dosing solutions are prepared by making a 1:10 dilution of stock compound in NB/B27 (5 μL stock compound, and 45 μL NB/B27).
2. Prepare the control wells (Fig. 3) by preparing a 1:10 dilution of DMSO in NB/B27 in three wells of the dosing plate (solvent controls).
3. Add 70 μL of lysis solution provided by Promega's CytoTox 96 Non-Radioactive Cytotoxicity Assay kit to a single well on the dosing plate (total cellular LDH control) in preparation for the LDH release assay detailed in Subheading 3.7.2.
4. A 1:1000 dilution of 25 mM bicuculline is added directly to 2 wells of the 48-well MEA plate after the 40 min baseline.

3.6.3 Chemical Exposure

1. After recording the 40 min baseline, add 10 μL from the dosing plate into the corresponding wells on the mwMEA plate to achieve a 1:50 dilution.
2. Add 20 μL of lysis from the dosing plate to the appropriate well on the mwMEA plate.
3. Add a 1:1000 dilution of 25 mM bicuculline to 2 wells of the 48-well MEA plate.
4. Record the dosed activity for 40 min.
5. After 40 min, remove a 50 μL sample from each well of the mwMEA plate and transfer it to a new sterile 96-well clear plate (LDH release transfer plate) in preparation for measuring LDH release (Subheading 3.7.2).

3.7 Viability

3.7.1 CellTiter Blue Assay

This assay is a fluorescent method to determine the viability of cells following both acute and developmental exposure to chemicals. As written, this procedure has only been optimized for primary cortical cultures seeded in mwMEA plates at a density of 150 K cells/well.

(a) Solution Preparation

1. Determine the amount of solution needed by multiplying the number of 48-well plates by 10 mL and adding an additional 3–5 mL for loss during preparation.
2. Prepare a 1:6 dilution of the CellTiter Blue reagent by adding warm (37°C) NB/B27 media (*see* **Note 18**).
   a. (Ex) For three 48-well MEA plates: Add 5.5 mL of CellTiter Blue to 27.5 mL of media. Divide into 11 mL aliquots.

(b) Measurement of CellTiter Blue Fluorescence

1. Using the vacuum trap, gently aspirate most of the treatment media from two columns of wells at a time, ensuring that a thin film is left covering the cells.
2. Gently add 200 μL fresh CellTiter Blue solution to each well.
3. Add 150 μL of CellTiter Blue solution to three wells on a 96-well opaque plate (CellTiter Blue transfer plate) for blanks.
4. Place both the mwMEA plate as well as the transfer plate in the incubator for 45–60 min, ensuring to gently shake the mwMEA plate every 15 min (*see* **Notes 19** and **20**).
5. At the end of the incubation period, transfer 150 μL from each well of the MEA plate to the corresponding wells of the transfer plate (Fig. 2).
6. Allow the transfer plate to come to room temperature and lance any large bubbles using a syringe needle (*see* **Note 21**).
7. Measure the fluorescence using the BMG FLUOstar OPTIMA Fluorescence/Luminescence Microplate Reader (*see* **Note 22**).

3.7.2 LDH Release Assay

This procedure has been employed to measure LDH release from cortical cells exposed to chemicals using the CytoTox 96 Non-Radioactive Cytotoxicity Assay kit from Promega (Catalog # G1780), and is adapted from the product insert provided by Promega. This assay is an absorbance method to determine the health of cells following acute exposure to chemicals. As written, this operating procedure has only been optimized for primary cortical culture seeded in mwMEAs at a seeding density of 150 K cells/well. Use of this procedure with other culture types or conditions may require modification.

For both LDH release and total cellular LDH (Subheading 3.7.3: Total Cellular LDH Assay) measures, thaw assay buffer and substrate mix at room temperature and avoid excessive exposure to

48 Well MEA Plate Map

	1	2	3	4	5	6	7	8	Concentration (μM)
A	DMSO	30	10	3	1	0.3	0.1	0.03	Chemical 1
B	DMSO	30	10	3	1	0.3	0.1	0.03	Chemical 2
C	DMSO	30	10	3	1	0.3	0.1	0.03	Chemical 3
D	DMSO	30	10	3	1	0.3	0.1	0.03	Chemical 4
E	DMSO	30	10	3	1	0.3	0.1	0.03	Chemical 5
F	DMSO	30	10	3	1	0.3	0.1	0.03	Chemical 6

96 Well CellTiter Blue Transfer Plate

	1	2	3	4	5	6	7	8	9	10	11	12
A	DMSO	30	10	3	1	0.3	0.1	0.03	Chemical 1			
B	DMSO	30	10	3	1	0.3	0.1	0.03	Chemical 2			
C	DMSO	30	10	3	1	0.3	0.1	0.03	Chemical 3			
D	DMSO	30	10	3	1	0.3	0.1	0.03	Chemical 4			
E	DMSO	30	10	3	1	0.3	0.1	0.03	Chemical 5			
F	DMSO	30	10	3	1	0.3	0.1	0.03	Chemical 6			
G	Blank	Blank	Blank									
H												

Fig. 2 Example 48-well MEA plate map and 96-well CellTiter Blue transfer plate map. After the 45–1 h incubation period, spent CellTiter Blue is transferred from each well of the 48-well MEA plate to corresponding wells on a 96-well opaque plate (transfer plate)

light. Make up reconstituted substrate buffer by adding 12 mL thawed assay buffer at room temperature to a vial of substrate mix. Invert mixture a few times to ensure the substrate mix has dissolved completely. Prepare both the media blanks and positive controls prior to the beginning of the LDH release and total cellular LDH assays. In a typical experiment, three identically treated MEA plates are used.

(a) Preparation for LDH Release Measure

1. During the 40 min recording of treated activity (Subheading 3.6), prepare the positive control by making a 1:3000 dilution of purified LDH (provided by Promega's CytoTox 96® Non-Radioactive Cytotoxicity Assay kit) in NB/B27 on a 96-well clear plate (LDH release transfer plate). Transfer 50 μL to the appropriate well of the LDH transfer plates (*see* **Note 23**) (Fig. 3).

48 Well MEA Plate Map

	Controls	2	3	4	5	6	7	8	Concentration (μM)
A	DMSO	40	10	3	1	0.3	0.1	0.03	Chemical 1
B	DMSO	40	10	3	1	0.3	0.1	0.03	Chemical 2
C	DMSO	40	10	3	1	0.3	0.1	0.03	Chemical 3
D	BIC	40	10	3	1	0.3	0.1	0.03	Chemical 4
E	BIC	40	10	3	1	0.3	0.1	0.03	Chemical 5
F	LYSIS	40	10	3	1	0.3	0.1	0.03	Chemical 6

96 Well LDH Release Transfer Plate Map

	1	2	3	4	5	6	7	8	9	10	11	12
A	DMSO	40	10	3	1	0.3	0.1	0.03		Chemical 1		
B	DMSO	40	10	3	1	0.3	0.1	0.03		Chemical 2		
C	DMSO	40	10	3	1	0.3	0.1	0.03		Chemical 3		
D	BIC	40	10	3	1	0.3	0.1	0.03		Chemical 4		
E	BIC	40	10	3	1	0.3	0.1	0.03		Chemical 5		
F	LYSIS	40	10	3	1	0.3	0.1	0.03		Chemical 6		
G	Media Blank	Media Blank	Media Blank	Purified LDH								
H												

BIC: bicuculline

Fig. 3 Example 48-well MEA plate map and 96-well LDH release transfer plate map. In preparation for the LDH release assay, a 50 μL sample of treated media is transferred from each well of the 48-well MEA plate to the corresponding wells on the 96-well transfer plate

2. Prepare media blanks by adding 50 μL of NB/B27 media into three wells of the LDH release transfer plate (*see* **Note 24**).
3. The total cellular LDH control measure will be used in analysis (*see* **Note 25**) (Subheading 3.6).

(b) Measuring LDH Release

1. After the 40 min treatment recording, remove 50 μL of media from each well of the treated MEA plate and transfer to the corresponding wells on the LDH release transfer plate (Fig. 3).
2. Add 50 μL of reconstituted substrate mix to each well of the LDH release transfer plate.
3. Incubate at room temperature for 30 min in the dark (closed drawer).
4. After the 30 min incubation period, add 50 μL of stop solution to the wells.
5. Lance any large bubbles with a syringe needle and record the absorbance at a single wavelength of 490 nm within 1 h after the addition of stop solution.

(c) Calculations: Determining cytotoxicity (LDH Release)

OD: Optical Density

Released LDH OD = Raw Released LDH OD–Media Blank Mean OD

Cellular LDH OD (Lysis) = Raw Cellular LDH–Media Blank Mean OD

The percent LDH release in each well is calculated using the following formula:

$$\textbf{\% Release} = 100 \times \frac{(\text{Released LDH OD})}{(\text{Cellular LDH OD})}$$

3.7.3 Total Cellular LDH Assay

This procedure has been employed to measure Total Cellular LDH from cortical cells exposed to a variety of compounds over a period of 12 days in vitro using the CytoTox 96® Non-Radioactive Cytotoxicity Assay kit from Promega (Catalog # G1780), and is adapted from the product insert provided by Promega. This assay is an absorbance method to determine the health of cells after developmental exposure to chemicals in a culture. This procedure is performed after the DIV 12 recording for the developmental (ontogeny) screening assay. As written, this operating procedure has only been optimized for primary cortical culture seeded in mwMEAs at a seeding density of 150 K cells/well. Use of this procedure with other culture types or conditions may require modification.

(a) Preparation for Total Cellular LDH Measure

1. Prepare dilute lysis solution by adding 1.5 mL of lysis solution to 75 mL PBS.
2. Prepare media blanks by adding 25 μL of dilute lysis solution into 3 wells of a transparent 96-well LDH plate (total cellular LDH transfer plate) (*see* **Note 26**) (Fig. 4).
3. Prepare the positive control by making a 1:3000 dilution of purified LDH (provided by Promega's CytoTox 96® Non-Radioactive Cytotoxicity Assay kit) in dilute lysis solution. Transfer 50 μL to the appropriate well on the LDH plate(s).

(b) Measuring Total Cellular LDH

When measuring total cellular LDH, cells must be lysed. This procedure occurs after the CellTiter Blue Assay detailed in Subheading 3.7.1.

1. Remove the remaining CellTiter Blue from the 48-well MEA plate, and add 500 μL of dilute lysis solution to each well.
2. Incubate the plate at 37 °C for 30 min.
3. After the 30 min incubation period, thoroughly mix each column by pipetting up and down. Remove 25 μL from each well of the MEA plate and transfer to the corresponding well on the total cellular LDH transfer plate (Fig. 4).

48 Well MEA Plate Map

	1	2	3	4	5	6	7	8	Concentration (μM)
A	DMSO	30	10	3	1	0.3	0.1	0.03	Chemical 1
B	DMSO	30	10	3	1	0.3	0.1	0.03	Chemical 2
C	DMSO	30	10	3	1	0.3	0.1	0.03	Chemical 3
D	DMSO	30	10	3	1	0.3	0.1	0.03	Chemical 4
E	DMSO	30	10	3	1	0.3	0.1	0.03	Chemical 5
F	DMSO	30	10	3	1	0.3	0.1	0.03	Chemical 6

96 Well Total Cellular LDH Transfer Plate Map

	1	2	3	4	5	6	7	8	9	10	11	12
A	DMSO	30	10	3	1	0.3	0.1	0.03	Chemical 1			
B	DMSO	30	10	3	1	0.3	0.1	0.03	Chemical 2			
C	DMSO	30	10	3	1	0.3	0.1	0.03	Chemical 3			
D	DMSO	30	10	3	1	0.3	0.1	0.03	Chemical 4			
E	DMSO	30	10	3	1	0.3	0.1	0.03	Chemical 5			
F	DMSO	30	10	3	1	0.3	0.1	0.03	Chemical 6			
G	Blank	Blank	Blank	Purified LDH								
H												

Fig. 4 Example 48-well MEA plate map and 96-well total cellular LDH transfer plate map. After lysing the cells in the 48-well MEA plate, 25 μL is transferred from each well of the mwMEA plate to corresponding wells on the 96-well transfer plate

4. Add 25 μL of PBS to each well of the LDH plate, excluding the positive control well.
5. Add 50 μL of reconstituted substrate mix to each well of the LDH plate.
6. Incubate at room temperature for 30 min in the dark (closed drawer).
7. After the 30 min incubation period, add 50 μL of stop solution to each well (*see* **Note 27**).
8. Pop any large bubbles with a syringe needle and record the absorbance at a single wavelength of 490 nm within one hour after the addition of stop solution (*see* **Note 28**).

(c) Calculations

Total LDH is determined per well for each mwMEA plate used in a single experiment. The following calculations are used to determine cytotoxicity in each well.

OD: Optical Density

Blank-corrected cellular LDH OD = Raw Cellular LDH OD–Mean of Media Blank OD

The total LDH in each well is calculated using the following formula:

$$\%\text{Total Cellular LDH} = 100 \times \frac{(\text{Blank-corrected cellular LDH OD})}{(\text{Mean of Blank-corrected cellular LDH controls OD})}$$

4 Notes

1. PEI comes as 50% (w/v) in H_2O, when weighing only half of the weight is PEI. One must weigh twice as much.
2. PEI is toxic to cells, so PEI solution should not be allowed to dry on the plate prior to rinsing three times.
3. The drop should bead up on the surface and not spread.
4. **Steps 3–5** describe the default settings for neural recordings. The AxIS Software will open with the most recently used settings as a default. Thus, if the previous recording was from cardiomyocytes, the settings will be different. The experimenter should check these prior to recording to be sure they are correct. AxIS also allows saving "user profiles" which can be loaded upon starting an experiment. This will ensure that the correct settings are used and is recommended, but not required.
5. Burst detection is optional and is for viewing purposes. Collection of burst analysis should always be obtained by replaying the .raw file and not during collection of the .raw file, as this will cause AxIS to crash.
6. Always record the .raw file when conducting an experiment. This file can be replayed at a later date to produce other file types in AxIS. The .raw files are very large, and can be ~50 GB per hour of recording. Depending on how long you will be recording and the number of plates you will record from, check the hard drive of the computer to be sure that sufficient storage space is available. If sufficient space is not available, transfer files to an external hard drive.
7. The most commonly used files are as follows:

   Spike Counts: This file type can be used to determine Active Electrodes and Mean Firing Rate.

   Spike List: This file type can be used to derive a variety of bursting parameters.

8. Typical root mean squared (rms) noise levels should be around 2–4 μV, so experimenter can use their best judgment on grounding electrodes with rms noise between 4 and 6 μV.
9. 48-well MEA plates have 16 electrodes per well. If any given well has less than 10 active electrodes, it should not be used in analysis. Criteria may differ depending on culture quality and cell type. This quality criterion is specific to our laboratory.
10. The 50× dosing plate should be prepared while the cells are attaching (2 h; refer to Subheading 3.3: Culturing on mwMEA Plates). Time the preparation of the dosing plate so that the cells have attached at the same time or prior to the completion of the dosing plate.
11. In a typical experiment, one 96-well dosing plate can be used for two dosing sessions (i.e., DIV 0 and 5) if the dosing plate is prepared using half of the plate.
12. These steps are based on the assumption that control wells start in the first column of the MEA plate and that the concentration increases up to column 8.
13. Pipette tips only need to be changed when dosing a new plate as long as increasing from a low to high concentration.
14. After dosing on DIV 0, allow MEA plates to remain in the incubator of the laboratory in which they were plated overnight before moving them to another location. Cells dosed on DIV 5 and 9 can immediately be transferred to another incubator after dosing if necessary. This particular laboratory cultures cells as well as carries out the methods outlined in this procedure in a separate location of the recording system. We typically allow the MEA plates to sit in an incubator overnight in the same room as the recording system before taking the DIV 2,5,7,9, and 12 recordings.
15. It was previously determined [1] that there is a period of instability in mean firing rate that occurs when moving plates from the incubator into the Maestro.
16. Add media slowly to avoid washing the cells off the electrodes. When adding media, slowly dispense it against the side of the well for better control.
17. It was previously determined [1] that there is a period of instability in mean firing rate that occurs when moving plates from the incubator into the Maestro.
18. Adjustment of this dilution may be necessary depending upon the cell type and density used.
19. Plates are gently shaken every 15 min to ensure that all the reagents in a well has the opportunity to be reduced by the cells. Since cells are seeded in the center of a single well, the

reagent localized directly over the cells tends to be reduced more rapidly.

20. The rate that the CellTiter Blue is reduced is dependent on cell type and density.
21. Bubbles will interfere with the accuracy of the reading.
22. Acceptable values for negative control wells in this assay (using white opaque plates and the appropriate gain corresponding to this plate type) are between 26 and 40,000. Values outside of this range should be investigated by ensuring that the correct gain is used, and checking for air bubbles. Negative control well values below 26,000 could indicate that some cells may have been washed off during refeeding. Blank wells should be between 6 and 10,000, but can vary depending on the combination of plate type and gain setting.
23. The kit contains purified LDH, which can be used to verify that the kit itself is working within specifications and/or develop a standard curve from which to quantify LDH. Values should range between 2.5 and 3.5 OD for both the acute and developmental assays.
24. NB/B27 media blank absorbable values should range between 0.05 and 0.08 optical density (OD) units. If the OD value is greater than 0.08, ensure there are no air bubbles.
25. For the LDH release assay, a total cellular LDH control (lysis buffer) for inducing LDH release from cells is included on every plate. Acceptable values for the positive control (lysis buffer) range from 1.9 to 3.0 OD.
26. Dilute lysis solution media blank absorbable values should range between 0.03 and 0.06 OD units. If the OD value is greater than 0.06 OD, ensure there are no air bubbles.
27. For the total cellular LDH assay, after incubation and when stop solution is added, negative control wells should appear dark red. Wells that have little red indicate cell death in this assay. Negative control (DMSO only) wells should range between ~0.7 and 1.5 OD.
28. Following determination of the total LDH in each well, the data can be used to (1) determine viable cells in treated wells compared to control wells. (Total LDH in treatment ×/Total LDH in control ×100) or (2) Perform quality checks of the data (e.g., Total LDH in a well will be low if no cells were plated in that well). This can be determined by comparing LDH values in surrounding wells on the same plate in addition to the corresponding well (same compound and concentration) in the other plates from that culture.

Acknowledgments

Preparation of this document has been funded by the U.S. Environmental Protection Agency. This document has been subjected to review by the National Health and Environmental Effects Research Laboratory and approved for publication. Approval does not signify that the contents reflect the views of the Agency, nor does mention of trade names or commercial products constitute endorsement or recommendation for use.

Jasmine P Brown is supported by Student Services Contracts EP-14-D-000101. Brittany S Lynch is an Oak Ridge Institute for Science and Education Fellow.

References

1. Valdivia P, Martin MT, Houck K, Lefew WR, Ross J, Shafer TJ (2014) Multi-well microelectrode array recordings detect neuroactivity of ToxCast compounds. Neurotoxicology 44:204–217
2. Nicolas J, Hendriksen PJM, van Kleef RGDM, de Groot A, Bovee TFH, Rietjens IMCM, Westerink RHS (2014) Detection of marine neurotoxins in food safety testing using a multielectrode array. Mol Nutr Food Res 58: 2369–2378
3. Hogberg HT, Sobanski T, Novellino A, Whelan M, Weiss DG, Bal-Price AK (2011) Application of micro-electrode arrays (MEAs) as an emerging technology for developmental neurotoxicity: evaluation of domoic acid-induced effects in primary cultures of rat cortical neurons. Neurotoxicology 32:158–168
4. Robinette BL, Harrill JA, Mundy WR, Shafer TJ (2011) In vitro assessment of developmental neurotoxicity: use of microelectrode arrays to measure functional changes in neuronal network ontogeny. Front Neuroeng 4:1
5. Cotterill E, Hall D, Wallace K, Mundy WR, Eglen SJ, Shafer TJ (2016) Characterization of early cortical neural network development in multiwell microelectrode array plates. J Biomol Screen 21(5):510–519
6. Johnstone AF, Gross GW, Weiss DG, Schroeder OH, Gramowski A, Shafer TJ (2010) Microelectrode arrays: a physiologically based neurotoxicity testing platform for the 21st century. Neurotoxicology 31:331–350
7. Wallace K, Strickland JD, Valdivia P, Mundy WR, Shafer TJ (2015) A multiplexed assay for determination of neurotoxicant effects on spontaneous network activity and viability from microelectrode arrays. Neurotoxicology 49: 79–85

Chapter 14

Quantitative Ratiometric Ca^{2+} Imaging to Assess Cell Viability

Oliver Friedrich and Stewart I. Head

Abstract

Viability of cells is strongly related to their Ca^{2+} homeostasis. Ca^{2+} signal fluctuations can be on a slow time scale, e.g., in non-excitable cells, but also in the range of tens of milliseconds for excitable cells, such as nerve and muscle. Muscle fibers respond to electrical stimulation with Ca^{2+} transients that exceed their resting basal level about 100 times. Fluorescent Ca^{2+} dyes have become an indispensable means to monitor Ca^{2+} fluctuations in living cells online. Fluorescence intensity of such "environmental dyes" relies on a buffer-ligand interaction which is not only governed by laws of mass action but also by binding and unbinding kinetics that have to be considered for proper Ca^{2+} kinetics and amplitude validation. The concept of Ca^{2+} dyes including the different approaches using ratiometric and non-ratiometric dyes, the way to correctly choose dyes according to their low-/high-affinity properties and kinetics as well as staining techniques, and in situ calibration are reviewed and explained. We provide detailed protocols to apply ratiometric Fura-2 imaging of resting Ca^{2+} and Ca^{2+} fluctuations during field-stimulation in single isolated skeletal muscle cells and how to translate fluorescence intensities into absolute Ca^{2+} concentrations using appropriate calibration techniques.

Key words Ca^{2+} transients, Resting Ca^{2+}, In situ calibration, Dynamic range, Ratiometric Ca^{2+} indicators, Muscle cells, Buffer kinetics, Dissociation constant

1 Introduction

1.1 The Importance of Intracellular Ca^{2+} Levels for Cell Viability

Living cells require a tight and adjustable boundary towards the extracellular space to maintain their intracellular milieu against ion and compound gradients. Entry or exit of polar or charged molecules is governed by transport processes up- or down-hill electrochemical gradients, while lipophilic and unpolar substances are passively distributed via diffusion governed by sole chemical gradients. Entry or exit of signalling molecules gives rise to changes in their respective cytosolic concentrations but not necessarily to changes in cytoplasmic concentrations. This is because the cytoplasm of living cells is sputtered with organelles that can also take up or release molecules from the cytosol. Thus, in addition to signalling with the exterior, cells also regulate their internal

Daniel F. Gilbert and Oliver Friedrich (eds.), *Cell Viability Assays: Methods and Protocols*, Methods in Molecular Biology, vol. 1601, DOI 10.1007/978-1-4939-6960-9_14,

signalling through inner compartmentalization. Ca^{2+} ions, for which there may exist up to 10,000 times concentration gradients between the cytosol at rest (~100 nM) and the external space (~mM) or intracellular organelles (endoplasmic reticulum, ~several hundred μM), are one of the most important signalling molecules (Fig. 1a). Dynamic changes in intracellular cytosolic $[Ca^{2+}]_i$ need to be tightly regulated to a working range of normal homeostatic processes from tens to hundreds of times over resting levels, depending on cell types. However, if $[Ca^{2+}]_i$ exceeds critical levels (~1,000-fold increases), cell death via necrosis or apoptosis may occur [2]. Thus, our knowledge of mechanisms involved in cell signalling related to ion homeostasis, in particular Ca^{2+}, crucially depends on the availability of methods to detect and monitor Ca^{2+} ion distributions within cells, and even ion movements between organelles and the cytosol. Before the invention of optical technologies to monitor so-called "environmental dyes" (i.e., dyes that respond to changes in an environmental variable, such as Ca^{2+}, Na^{+} ions, pH, or membrane potential), ion fluxes across membranes were usually recorded using electrophysiological techniques (e.g., patch clamp, voltage clamp). Thus, recording chemo-electrical signals across a barrier membrane requires impalement of those membranes with electrodes against a reference electrode external to the compartment in question. While this works well in case of investigating ion fluxes across surface membranes, atraumatic functional assessment of such a configuration including intracellular organelles is, so far, impossible (Fig. 1b). Also, since ion current recordings may suffer from differentiation of the respective ion class of interest, because charges are superposed originating from various contributing ion species (unless sophisticated pharmacological blockade of respective ion channels is involved), techniques with an intrinsic high specificity and minimal invasiveness are required. With the invention of Ca^{2+}-sensitive chemical dyes of high quantum efficiency and photochemical stability [3], Ca^{2+} fluorescence microscopy has become a widely used cell technology in life sciences. Modular epifluorescence microscopes with fast CCD-based acquisition rates and high-quantum efficiency sensors have become very affordable for life science labs, and thus the capability to perform cellular Ca^{2+} imaging experiments is omnipresent today. Preconfigured systems are readily available for specific dye, light source, and detector combinations. However, although the use of Ca^{2+}-sensitive dyes may look straightforward, there are many pitfalls and constraints to reconcile that may give rise to false interpretations if not properly considered, in particular when it comes to quantitative Ca^{2+} signal analysis. The concept of non-ratiometric versus ratiometric Ca^{2+}-dyes needs to be weighed according to hardware availabilities and the type of biological preparation involved. More importantly, errors usually occur through neglecting to consider or understand the limitations of the buffer

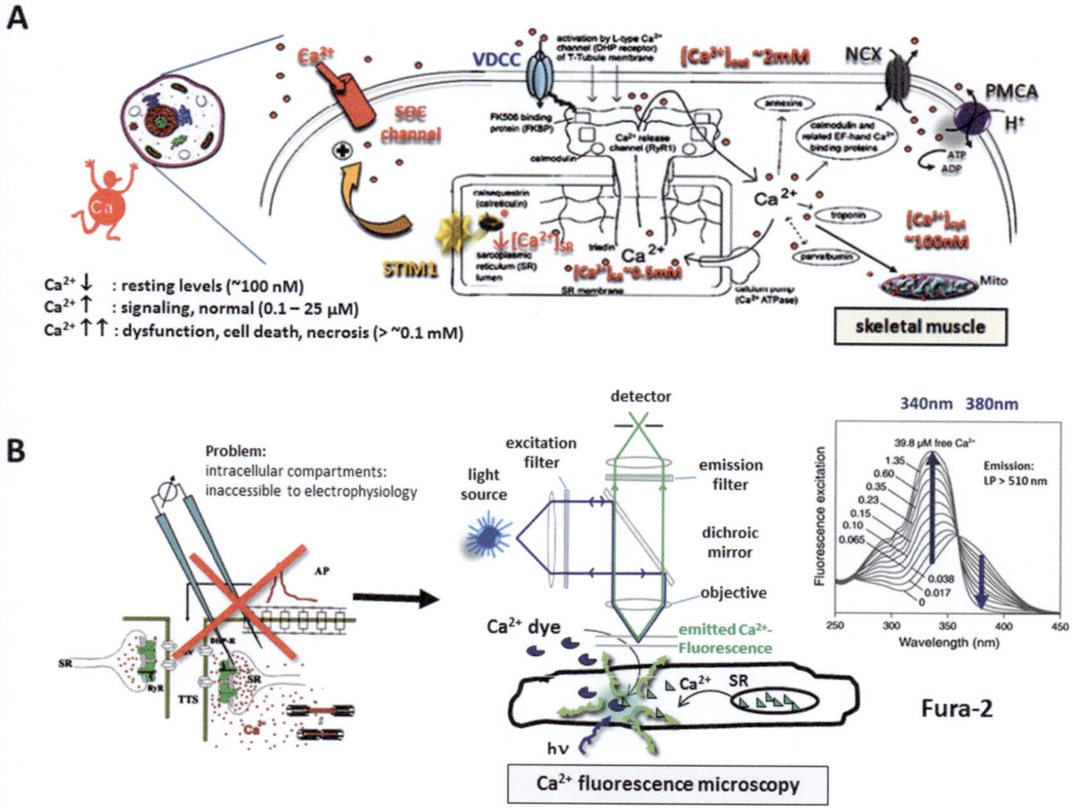

Fig. 1 Ca^{2+} homeostasis and optical metrologies to quantify cellular Ca^{2+} signalling by fluorescence microscopy. (**a**) Ca^{2+} entry and distribution into cells is tightly regulated by differently operating ion channels and transporters. Gradients across the cell membrane barrier are in the order of 10^4 external over cytosolic. Signalling during normal cell homeostasis happens within the range of 10^1 to some 10^2 of this gradient while from 10^3, cell dysfunction and irreversible damage may occur. Shown are some of the cell membrane (sarcolemma) and internal inter-compartment transport systems in skeletal muscle for Ca^{2+} ion movements (modified from [1]). (**b**) Ca^{2+} fluxes across membranes of intracellular organelles are not accessible to conventional electrophysiology techniques but Ca^{2+} fluorescence microscopy provides a minimally-invasive optical metrology to monitor free Ca^{2+} through specific binding to Ca^{2+}-selective fluorochromes. Induced fluorescence can be spectrally separated from the excitation light path by dichroic mirror optics exploiting the Stokes shift between excitation and emission light. A dye dispersed in the cytosol, e.g., the ratiometric dual-excitation wavelength Ca^{2+}-dye Fura-2, can be used to monitor cytosolic Ca^{2+} fluctuations by external Ca^{2+} entry or internal compartmental redistribution

chemistry behind fluorescent dyes, the corresponding dynamic range of dyes, calibration procedures, and correcting for dye kinetics. This chapter reviews the most important concepts of Ca^{2+}-buffer interaction that have to be kept in mind as well as gives step-by-step recipes for performing dynamic quantitative Ca^{2+} measurements in living cells. Although this chapter utilizes skeletal muscle as an example, the concepts are applicable to all cell types.

1.2 The Buffer-Ligand Concept Underlying Ca^{2+} Imaging, Dynamic Range of Dyes, Ratiometric and Non-ratiometric Approaches

A fluorescent Ca^{2+} indicator represents a chelating molecule that undergoes a reversible binding reaction with its ligand Ca^{2+} to form a complex with fluorescent properties. Thus, such molecules consist of a chelating Ca^{2+}-binding portion and a fluorescent residue portion, usually involving delocalized π-electrons. An example of a commonly used CA^{2+} chelator, BAPTA, and a BAPTA-derived Ca^{2+}-sensitive dye, Fluo-3, is shown in Fig. 2a. For Fluo-3, also its transition from a non-Ca^{2+}-binding acetoxy-methyl-ester (AM) form to its Ca^{2+}-binding de-esterified form is shown. Only the latter fluoresces when bound to Ca^{2+}, forming the Fluo-3-Ca complex. As detailed later, this AM form is heavily used in life science research to stain cells, because the AM form is cell permeant, unlike its de-esterified form. The buffer reaction for the indicator with one Ca^{2+} ion is governed by the equation

$$Ca^{2+} + \text{Buffer} \underset{k_{off}}{\overset{k_{on}}{\rightleftarrows}} \text{Buffer} - Ca \tag{1}$$

The concentration changes of either of the three constituents is governed by the rate constants k assuming first order kinetics (Buffer abbreviated as “B”):

$$\frac{\mathrm{d}[Ca^{2+}]}{\mathrm{d}t}(t) = k_{off}[BCa](t) - k_{on}\cdot[\mathrm{B}]\cdot[Ca^{2+}](t) = \frac{\mathrm{d}[\mathrm{B}]}{\mathrm{d}t}(t) \tag{2}$$

$$\frac{\mathrm{d}[BCa]}{\mathrm{d}t}(t) = -k_{off}\cdot[BCa](t) + k_{on}\cdot[\mathrm{B}]\cdot[Ca^{2+}](t) \tag{3}$$

The steady-state equilibrium ($t \longrightarrow \infty$) is described by the law of mass action:

$$\frac{[\mathrm{B}]\cdot[Ca^{2+}]}{[BCa]}\bigg|_{t\to\infty} = \frac{k_{off}}{k_{on}} = \frac{\tau_{on}}{\tau_{off}} = K_d \tag{4}$$

with τ as the time constant of the reaction (inverse of rate constant k). The ratio of the rate constants defines the dissociation constant K_d. The K_d value reflects the 50% binding strength of the buffer to its ligand: buffers with small K_d values are so-called “high-affinity” buffers while larger K_d values define “low-affinity” buffers. Importantly, one always has to keep in mind that the buffer equilibrium and thus the K_d values are vastly influenced by environmental variables such as ionic strength, temperature, pH, pressure, or presence of other binding partners, e.g., Ca^{2+}-binding proteins or divalent cations such as Mg^{2+} [4, 5]. This means when using tabulated K_d values provided by manufacturers, special focus has to be on the environmental conditions given. Moreover, uncritical re-use of such K_d values in own experiments at largely differing conditions may produce large errors in calculating $[Ca^{2+}]_i$ from fluorescence values unless an own conditional Ca^{2+}-calibration is used. We will provide a recipe for such a calibration in the later

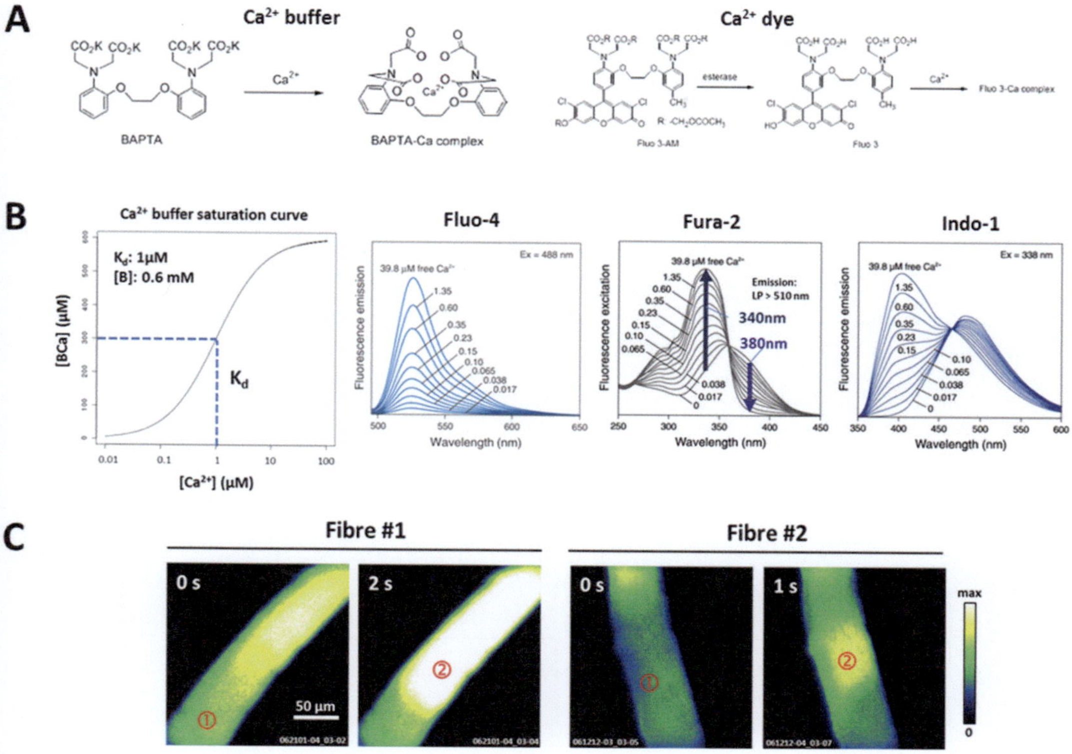

Fig. 2 Ca^{2+} dye properties of common dyes. (**a**) Structural similarities of Ca^{2+}-chelating buffers and dyes, emphasizing the buffer concept for fluorescent environmental dyes. (**b**) Steady-state sigmoidal "free Ca^{2+}"-"buffered complex" (BCa) relationship, defining binding strength by the dissociation constant K_d and the dynamic range through the slope of the curve. Shown are also excitation (Fluo-4, Indo-1) or emission (Fura-2) spectra of common ratiometric (Fura-2, Indo-1) and non-ratiometric (Fluo-4) dyes. (**c**) Example of a propagating Ca^{2+} wave within two intact muscle fibers stained with the non-ratiometric dye Fluo-4 (AM form) and imaged with confocal microscopy. Although suggestive from image comparison, without an exact knowledge of the total dye concentration within a cell, the conclusion of higher signal intensities in areas ① and ② in fiber #1 over fiber #2 reflecting higher Ca^{2+} concentrations, cannot be made and may in fact be wrong

sections. Table 1 provides physicochemical properties of some commonly used ratiometric and non-ratiometric Ca^{2+} dyes. As can be seen, although steady-state binding curves extracting K_d values are usually readily available, the kinetics data are sometimes very sparse even for very common dyes, because obtaining those usually requires special stopped-flow fluorescence experiments (e.g., [11]). The differentiation between high- vs. low-affinity dyes is usually determined by the buffer's unbinding kinetics which is fast for low-affinity dyes (high rate constant k_{off}) and slow for high-affinity dyes (small rate constant k_{off}) (Table 1). The binding kinetics (k_{on}) for all dyes is usually within the same order of magnitude set by diffusion kinetics.

Establishing fluorescence-[Ca^{2+}] curves is straightforward, in principle. One requires to have a solution or a compartment

Table 1
Fluorescence mode and chemical buffer properties of commonly used fluorescent Ca^{2+} dyes

Dye	Fura-2	Indo-1	Fura-Red	Fluo-4	Fluo-5N
Excitation	340/380 nm (UV)	346 nm (UV)	420/480 nm, 457/480 nm	~488 nm	~470 nm
Emission	> ~510 nm (LP)	405/475 nm (BP)	>600 nm (LP)	> 500 nm (LP)	>500 nm (LP)
Ratiometry mode	Dual excitation ratio	Dual emission ratio	Dual-excitation ratio	Non-ratiometric	Non-ratiometric
K_d (in vitro)	140–224 nM	250 nM	~150 nM	~350 nM	~90 μM
k_{on}	2.5–6.5 × 10^8 $M^{-1}s^{-1}$	5 × 10^8 $M^{-1}s^{-1}$	n.a.	1.5–3 × 10^8 $M^{-1}s^{-1}$	3 × 10^8 $M^{-1}s^{-1}$
k_{off}	84–96 s^{-1} (20 °C)	130 s^{-1} (20 °C, 0.1 M KCl)	n.a.	275–450 s^{-1}	1.2 10^5 s^{-1}
Affinity	High-affinity	High-affinity	High-affinity	High-affinity	Low-affinity

Source: The Molecular Probes® Handbook. LP: long-pass, BP: band-pass. Kinetics data: Fura-2 from [6, 7], Indo-1 from [8]; Fluo-4 from [9, 10]; Fluo-5N from [10]

containing a given concentration of total dye (buffer, [B]) and known free Ca^{2+} ($[Ca^{2+}]_{free}$) and measure the resulting fluorescence. This fluorescence value reflects the amount of bound [CaB]. Increasing $[Ca^{2+}]_{free}$ then, in turn, increases the amount of [CaB] by shifting the equilibrium towards the right (Eq. 1). Figure 2b shows an example of such a saturation curve where $[Ca^{2+}]_{free}$ is plotted in logarithmic scale against the concentration of CaB (or later, in fluorescence values). The crucial point here is that a known and fixed concentration of buffer [B] was used. In a cuvette situation, this usually does not represent an unknown condition; however, in living cells it usually does, in particular in cases where the cells were loaded with the membrane-permeant AM form of the dye. Figure 2c illustrates an example of two single intact muscle fibers that had been stained with 10 μM of the fast high-affinity dye Fluo-4 AM in the external bath and that were observed with a confocal fluorescence microscope. Staining was identical over 30 min at room temperature followed by 10 min at 37 °C to allow for de-esterification of the dye before imaging. In these two cells, a spontaneous Ca^{2+} wave traversing through the cells was observed. The fluorescence intensity is shown pseudo-color coded. Qualitatively, one would be tempted to make the relationship "*more brightness = more* Ca^{2+}", in fiber #1 over fiber #2, both for the regions of wave maximum ② as well as for the resting levels ①. However, the same result could be obtained simply by a larger dye concentration in fiber #1 over fiber #2 without any change in

$[Ca^{2+}]_{free}$. This can directly be seen from Eq. 1 by using the fact that the total amount of dye initially present in the system $[B]_{tot}$ splits into a Ca-bound and unbound portion:

$$[B]_{tot} = [B] + [BCa] \tag{5}$$

Replacing [B] in Eq. 4 yields

$$[BCa] = \text{observed Fluorescence} = \frac{[Ca^{2+}] \cdot [B]_{tot}}{K_d + [Ca^{2+}]} \tag{6}$$

where $[Ca^{2+}]$ represents the free Ca^{2+} concentration $[Ca^{2+}]_{free}$ present. The reason for different dye concentrations in different cells loaded with AM dyes lies in the fact that even if the total dye concentration of the AM form being applied externally to the cells was known, once the molecule is in the cells and the AM ester has been cleaved off by intracellular esterases, it represents a different molecule trapped inside the cytosol. By the conversion of the AM form in the cytosol to the charged dye molecule, the diffusion gradient for the AM molecules into the cells will not decrease noticeably and AM molecules will continue to diffuse into the cells as long as the AM dye is present and de-esterified dye molecules will accumulate to different and unknown contents among individual cells. Thus, one would either make sure to more or less exactly know the amount of dye inside the cell by means of intracellular injection of the charged form of the dye (using pressure injection of dye-salt or iontophoresis, see later) or by other means to circumvent the necessity to know $[B]_{tot}$ inside the cells. In particular, the latter has become available by use of so-called *ratiometric* dyes, i.e., dyes that show alternating peak behaviors within either their excitation or emission spectrum with $[Ca^{2+}]_{free}$. Figure 2b shows example spectra of commonly used Ca^{2+} dyes. Fluo-4 has one peak in its emission spectrum that scales with Ca^{2+} levels and does not readily fulfil the requirements for ratiometric quantitative Ca^{2+} imaging for viability. In principle, it could be used for quantitative Ca^{2+} imaging by establishing a Ca^{2+}-calibration curve at a given dye concentration and use this dye concentration in iontophoresis experiments exclusively, but this is very rarely approached. Alternatively, Fura-2 and Indo-1 are popular dyes that contain two peaks within their excitation (Fura-2) or emission (Indo-1) spectrum, and each peak behaves opposite to the other with respect to Ca^{2+}. Each fluorescence bin (e.g., the fluorescence recorded for Fura-2 at 340 nm excitation, F_{340}) can be regarded as an application of Lambert-Beer's law

$$F_{340} = [B] \cdot d \cdot K \cdot f([Ca^{2+}]) \tag{7}$$

with [B] as the dye concentration inside the cell, d the thickness of the optical path, K as an optical equipment constant and a function of free [Ca^{2+}]. It is not necessary to know *d* and *K*, because using the ratios yields:

$$R = \frac{F_{340}}{F_{380}} = \frac{[\mathrm{B}]\cdot d\cdot K\cdot f([Ca^{2+}])\,|_{340}}{[\mathrm{B}]\cdot d\cdot K\cdot f([Ca^{2+}])\,|_{380}} = f'([Ca^{2+}]) \qquad (8)$$

More importantly, also [B] is now eliminated from the ratio, rendering this approach independent of differences in dye concentration among cells that inevitably occur during staining of cells with AM dyes. The steady-state ratio *R* in Eq. 8 is then again a sigmoidal function $f'([Ca^{2+}])$ of Ca^{2+} as in Fig. 2b (with the ratio *R* on the ordinate).

1.3 In Situ calibrations for Fluorescent Ca^{2+} Dyes

Since the Ca^{2+}-saturation curve (Fig. 2b) is sigmoidal, and the steepness of the curve is determined by the stoichiometry of the Ca^{2+}-buffer-complex reaction (1:1 for many dyes, *see* Fig. 2a), this sigmoidal relationship is expected to be completely characterized by the minimum and maximum ordinate value (absolute fluorescence for non-ratiometric dyes, ratio for ratiometric dyes) and the buffer's K_d (e.g., [3, 12]):

$$[Ca^{2+}]_i = K_d \cdot \frac{F - F_{min}}{F_{max} - F'}, [Ca^{2+}]_i = \beta \cdot K_d \cdot \frac{R - R_{min}}{R_{max} - R} \qquad (9)$$

with $[Ca^{2+}]_i$ referring to intracellular cytosolic free Ca^{2+}. Note that for ratiometric dyes, the relationship also contains a factor *ß* that limits R_{max} at saturating Ca^{2+} conditions and is reflected by the ratio of the fluorescence intensity from the wavelength bin corresponding to the Ca^{2+}-bound buffer complex under Ca^{2+}-free and Ca^{2+}-saturating conditions [3]. Thus, if the K_d value is known, one could perform a two-point calibration to assess R_{min} (Ca^{2+}-free), R_{max} (excess Ca^{2+}), and F_{380} at both Ca^{2+} conditions to obtain *ß*, in case of the ratiometric dye Fura-2. The problem is however, that K_d values provided by manufacturers in their data sheets are usually obtained under cuvette conditions, containing only a Ca^{2+}-buffered dye solution without any other competing Ca^{2+} buffers (or maybe one known additional Ca^{2+} buffer, e.g., EGTA, to "clamp" free Ca^{2+} concentrations to desired values for calibration). The cellular cytosol, in contrast, contains a multitude of both soluble and insoluble Ca^{2+} buffering proteins, like parvalbumin, ATP, troponin C, SERCA pumps, and many more, in case of muscle cells. Other cell types may contain other Ca^{2+} binding proteins; thus, introducing a Ca^{2+} dye into the cytosol represents "adding another competitor to a field of competition" for Ca^{2+} ions. Since the dynamic range of any buffer is usually within roughly ±1.5 log steps around the K_d value (which corresponds to the approximate range of $0.03 \bullet K_d \leq [Ca^{2+}] \leq 30 \bullet K_d$), it is crucial to know how the natural cytosolic environment shifts the apparent K_d value. It is clear that the affinity of a given dye will be apparently reduced by entering a Ca^{2+} competition arena, as shown in Fig. 3, effectively increasing its apparent K_d value "in situ." Fig. 3a shows a scenario for skeletal muscle for which some of the most important Ca^{2+}-binding partners are known regarding their cytosolic concentrations and in vitro K_d values, as well as showing simulated Ca^{2+}-saturation curves superposed. As a rule of thumb, the buffer with highest affinity is served first while lower affinity buffers become served with Ca^{2+} ions only from larger concentrations on. For example, ATP with a very large K_d value but also at large cellular concentrations itself, becomes

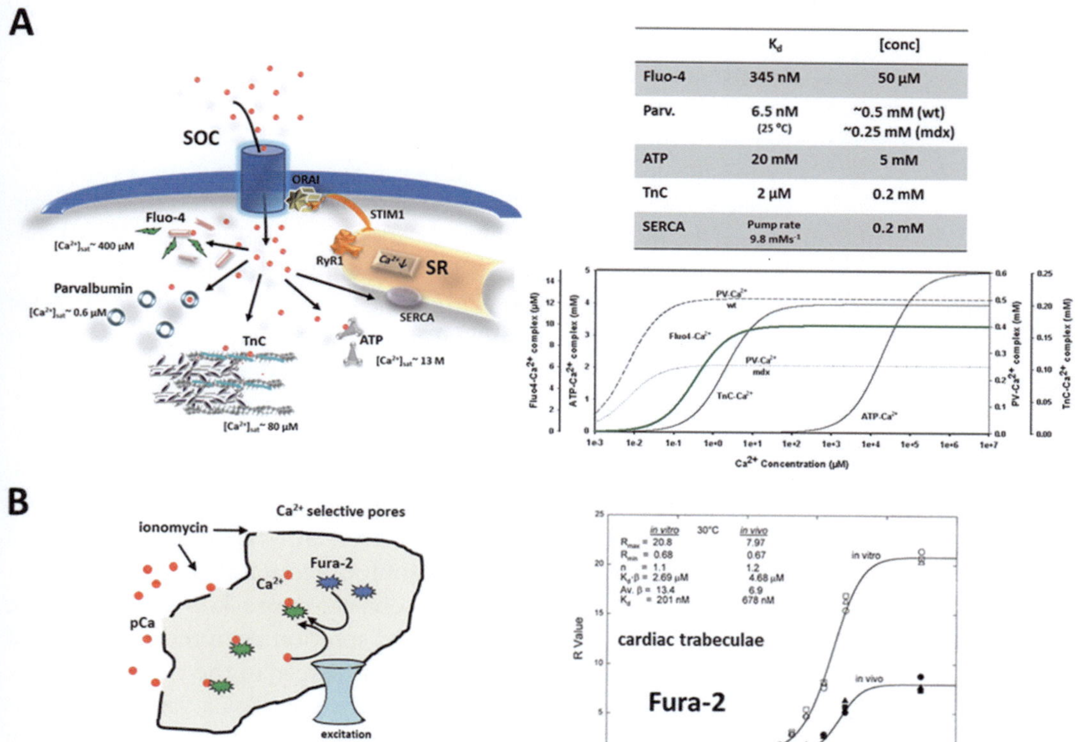

Fig. 3. Apparent K_d of a Ca^{2+} dye within the cellular environment increases by competing binding partners and requires in situ calibration. (**a**) Model of some of the main Ca^{2+}-binding partners in a muscle fiber stained with 50 μM Fluo-4, with their affinity constants and cellular concentrations given in the table (values taken from [13, 14]). From these values, Ca^{2+}-saturation curves for the respective buffer molecules were calculated and plotted (right panel, bottom) to estimate the approximate saturation concentrations of Ca^{2+} for the respective molecule plotted in the scheme on the left. (**b**) Strategy of following in situ calibration using a Ca^{2+} ionophore (in this case ionomycin) to allow Ca^{2+} equilibration over the sarcolemma for a cell stained with a Ca^{2+} dye (here: Fura-2). Calibration curves are reconstructed from fits to ratio-pCa pairs as explained in the text. An example of how different results from in situ and in vitro calibration can be due to cytosolic Ca^{2+} competition is shown in the right panel for a cardiac muscle preparation (from [15]); see [16] as an example for a skeletal muscle study

fully saturated from about 13 M of Ca^{2+} while parvalbumin saturates at only 0.6 μM of Ca^{2+}. Thus, if all binding partners were known, one could simulate the fluorochrome's apparent K_d. However, as this is usually not the case, there is still the possibility of performing an "in situ" calibration which is more elaborate than the simple two-point cuvette calibration. After explaining the concept here, the next sections will include a step-by-step example. To perform in situ calibrations, first cells need to be stained with the respective dye, either using AM ester form staining or iontophoresis or pressure injection of the ionized form. While the former does usually only make sense for ratiometric dyes, for reasons given above, the latter can also be used for non-ratiometric dyes, since the dye concentration after iontophoresis is more readily known to the experimenter. With the dye trapped in the cytosol, the cell membrane is then Ca^{2+} permeabilized using a Ca^{2+} ionophore (e.g., ionomycin) that introduces Ca^{2+}-selective pores into the membrane (Fig. 3b). The cells are then sequentially incubated with external solutions that only differ in their respective pCa (free Ca^{2+} concentration) and are allowed to equilibrate. After equilibration, extracellular and intracellular $[Ca^{2+}]_{free}$ can be assumed equal and fluorescence can be assessed by imaging at this given pCa before proceeding to the next Ca^{2+} concentration. It is clear that for reconstruction of a sigmoidal curve with unknown flexion point (the apparent K_d to determine!), one requires at least three recordings, at Ca^{2+}-free, Ca^{2+}-saturating conditions, and at the K_d value. Since the latter however, because unknown, is the fit parameter to extract from the data, it is clear that the fit improves with increasing number of pCa steps included in the calibration. We usually use at least six points that can be conveniently adapted by mixing a Ca^{2+}-free and a Ca^{2+} saturating solution containing the Ca^{2+} chelator EGTA, for which binding constants are known, to set the required total Ca^{2+} in the buffered solution required for a desired $[Ca^{2+}]_{free}$ in the external bath and ultimately, in the cytosol. A table with solution recipes and mixtures for given pCa values we use is given in the later sections (*see* **Note 2**). Figure 3b also shows an example from a study documenting the marked differences obtained from in vitro vs. in situ calibration [15]. Once plotting all steady-state ratio values over pCa, a sigmoidal fit can be obtained by applying a Hill fit with the constraint of setting the Hill coefficient to 1 (acknowledging the 1:1 binding ratio). We usually use the following equation which can be implemented in SigmaPlot, Origin, or GraphPrism:

$$R(pCa) = R_{\min} + \frac{R_{\max} \cdot (10^{-pCa})^b}{(K_d)^b + (10^{-pCa})^b} \tag{10}$$

where R(pCa) represents the column of the measured ratios, and pCa the column of the pCa values to the set $[Ca^{2+}]_{free}$ in the analysis program spread sheet. *b* is the Hill coefficient and is set to a constraint of 1. After fitting, the K_d value can be extracted. We also recommend having the extrapolated fitting curve plotted into the data plot (such as in Fig. 3b) which then provides the in situ calibration curve allowing to relate steady-state (this is important, e.g., resting values!) ratios to corresponding cytosolic Ca^{2+} levels. One could follow the following concept:

1. Record resting ratios first in all intact single cells within a recording dish.
2. Then Ca^{2+}-permeabilize the whole dish with ionomycin.
3. Successively bath the dish with different pCa solutions, allow for equilibration, and record ratios in all cells in the same order as in (1).
4. Plot ratio over pCa for each cell.
5. Perform sigmoidal fit for each cell individually, extract individual K_d value, and reconstruct individual fit curve.
6. Read resting $[Ca^{2+}]_{free}$ to corresponding resting ratio of each individual cell.
7. Perform statistics on resting $[Ca^{2+}]_{free}$ and K_d.

Some cellular systems may not survive a complete set of pCa incubations or some Ca^{2+} levels may trigger cellular reactions that may affect subsequent readings. In such cases, several studies also use group analyses where ratio-pCa pairs are lumped from a substantial number of cells and mean calibration curves are reconstructed (e.g., [17]). However, this also yields only one single group K_d value from the fitting procedure and is, of course, less accurate.

1.4 Assessing Fast Ca^{2+} Signals with Fluorescent Dyes Requires Corrections for Buffer Binding/Unbinding Kinetics

The above-mentioned concept provides accurate quantitative assessment of resting cytosolic Ca^{2+} levels regardless of the kinetics properties of the involved dye because recordings and calibrations refer to steady-state levels at long equilibration times. Elevated $[Ca^{2+}]_{free}$ levels can be a good indicator of cellular vitality however, in particular for excitable cells like muscle and nerve, Ca^{2+} fluctuations in response to an external stimulus (e.g., an electrical field pulse) can be a more sensitive parameter to assess cellular viability, i.e., reflecting the intactness of the Ca^{2+} regulation machinery. For excitable cells, such Ca^{2+} responses can be extremely fast; in case of skeletal muscle single twitches, Ca^{2+} transients only last a few hundreds of ms with an even faster activation kinetics (tens of ms), followed by an exponential decay [18]. In order to faithfully map such fast signals, the dynamic range of the fluorescent dye not only must cover the expected Ca^{2+} level range of signal fluctuations but also must possess a relatively fast binding and unbinding kinetics. Those kinetics parameters, in particular k_{on} and k_{off}, limit the temporal resolution of the underlying process. For example, imaging the twitch Ca^{2+} transient in a muscle fiber is reflected by a time constant of <10 ms for activation [19]. If the τ_{on} of the used dye was 100 ms, even a ten times faster process like the transient activation could not be tracked any faster than the dye binding kinetics. Fura-2 has a k_{on} in the range of 100 s^{-1} equivalent to a τ_{on} of ~10 ms, thus suitable to faithfully track fast temporal signals in skeletal muscle cells. The measured fluorescence in time lapse recordings still contain the superposition of temporal change in $[Ca^{2+}]_{free}$ "per se" and the binding kinetics to the indicator dye (e.g., Fura-2). However, the latter can be mathematically corrected for from the time course of the raw fluorescence recording $F(t)$ or the ratio $R(t)$ by using the relationship [19]:

$$[Ca^{2+}]_{corr}(t) = R(t) + \frac{dR(t)}{dt} \cdot \frac{1}{k_{off} \cdot \left(1 + \frac{R(t)}{K_d}\right)} \tag{11}$$

For fast Fura-2 dual-excitation recordings, the acquisition rate is primarily set by the way the hardware switches between wavelengths. In order to even increase temporal resolution, one could also use an approach recording only the time course represented by the Ca^{2+}-bound complex excitation bin F_{380}, followed by an isosbestic wavelength (358 nm) recording F_{358} before and after the transient recording to generate ratios [19, 20]. The approach is visualized step by step in Fig. 4, where the first panel in Fig. 4b reflects a transient recording from a single FDB fiber field-stimulated (details see below), achieving an acquisition rate of 20 kHz recording F_{380} during the transient. The middle panel shows the transient signal converted to $[Ca^{2+}]_{free}$ using Eq. 9 which still contains the buffer kinetics in the signal. The right panel shows the superposition of this uncorrected $[Ca^{2+}]$-trace (black) alongside with the corrected signal (red) after applying Eq. 10. Details will be provided in the step-by-step recipes below. As can be seen from Fig. 4b, the apparent signal amplitude is much higher following correction; the dye kinetics underestimates the apparent maximum peak $[Ca^{2+}]_i$ by the slowing-down effect of the k_{on} of Fura-2 being somewhat slower than the "per se" faster cytosolic Ca^{2+} rise. Since the decaying phase of such a Ca^{2+}-transient is at least 10–20 times slower than the activation phase, its kinetics is still reliably mapped also by the uncorrected data which also is less noisy. The reason for the larger noise in the corrected data is because of including the first time derivative of the signal in the correction procedure which significantly amplifies the noise in mathematical product operations of Eq. 10. One has to bear in mind, however, that this approach is only as good as is the knowledge of the kinetics constants under "as close as possible to experimental condition." For instance, the association kinetics of Fura-2 in frog myoplasm at 16 °C was reported to be roughly four times slower as compared to in vitro at 20 °C [21].

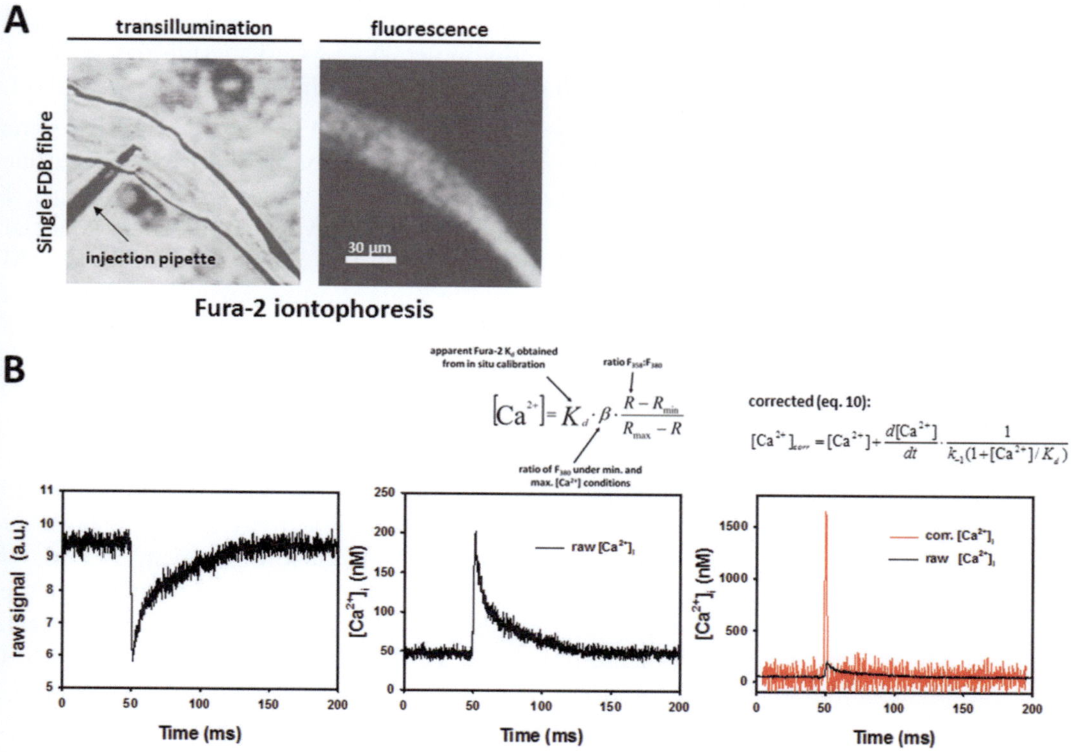

Fig. 4 Correcting fast fluctuations in recorded Ca^{2+} fluorescence for dye binding kinetics. (**a**) Experimental setting of staining a single FDB fiber with 50 μM Fura-2 by iontophoresis. (**b**) Stepwise correction of the F_{380} transient into buffer-kinetics corrected $[Ca^{2+}]_{corr}$ using Eqs. 9 and 10

2 Materials

2.1 Epifluorescence Microscopy Hardware

For quantitative Ca^{2+} fluorescence imaging, a fluorescence microscope is essential. For sake of speed and due to the fact that for global cytosolic recordings, no overly high resolution imaging capabilities are required, commercial epifluorescence systems (inverted or upright) will be sufficient. Apart from the specified hardware we use (*see* **Note 1**), any other imaging hardware suitable for dual-excitation of dual-emission ratiometric imaging can be used.

2.2 Resting Cytosolic Ca^{2+} Level Determination and In Situ Calibration

1. As preparation, we use intact single muscle fibers isolated from toe muscles *flexor digitorum brevis* (FDB) or *Mm. interossei* (*see* Subheading 3.1). For enzymatic dissociation, collagenase type IA (from *Clostridium histolyticum*) is recommended.
2. A stereomicroscope for dissection and single fiber transfer is required.
3. Extracellular normal physiological solutions used are either Krebs or Ringer's solution. Krebs solution: 4.75 mM KCl, 118 mM NaCl, 1.18 mM KH_2PO_4, 1.18 mM $MgSO_4$, 24.8 mM $NaHCO_3$, 2.5 mM $CaCl_2$, 10 mM glucose. Ringer's solution:

145 mM NaCl, 5 mM KCl, 2.5 mM $CaCl_2$, 10 mM hepes (2-[4-(2-hydroxyethyl)piperazin-1-yl]ethanesulfonic acid), 1 mM $MgCl_2$, 10 mM glucose. pH adjusted to 7.4 with NaOH.

4. Fura-2 staining solution: For AM staining, add the same amount of mL of either DMSO (dimethylsulfoxide) or acetone as the powder weight in mg to one vial of Fura-2 AM (e.g., 1 mL to 1 mg) to dissolve the powder. With the molecular weight of 1001.85 mg/mmol, this approach yields a ~1 mM stock solution of dye. For iontophoresis: for direct injection of the membrane impermeant Fura-2 pentapotassium salt form of the dye, dissolve in *aqua dest.* to a stock solution of 1 mM.
5. Pluronic® F-127.
6. 1 M hepes (in *aqua dest*).
7. Ca^{2+}-ionophoresis: Use the Ca^{2+} ionophore ionomycin from *Streptomyces conglobatus.* Its ion selectivity is $Ca^{2+} > Mg^{2+} >> Sr^{2+} = Ba^{2+}$. Dissolve each 1 mg of ionomycin in 141 μL of DMSO to yield a 10 mM stock solution. Alternatively, the 4-bromo-Ca^{2+} ionophore A23187 can be used. Dissolve each 1 mg of the powder in 166 μm of DMSO to yield a 10 mM stock solution.
8. Mechanical paralysis of single fibers: To mechanically block contractile activation of single fibers, in particular for the in situ calibration, the myosin-ATPase II inhibitor BTS (*N*-benzyl-*p*-toluene sulphonamide) shall be used. During in situ calibration with high Ca^{2+} solution, not blocking activation of the contractile apparatus would ultimately result in an irreversible contracture that can be prevented by 50 μM BTS in all calibration solutions (see below). With a molecular weight of 261.3 g/mol and solubility in DMSO up to 100 mM, dissolve each 1 mg of BTS in 76.5 μL of DMSO to yield a 50 mM stock solution.
9. Calibration base solutions HR (high relaxing) and HA (high activation): To produce internal solutions with "clamped" free Ca^{2+}-levels, one requires a buffer-ligand chemical chelator program. An example is React II (developed by Godfrey Smith, University of Glasgow) that uses tabulated critical stability constants for Ca^{2+} chelating buffers with divalent and monovalent cations at given environmental conditions to calculate free Ca^{2+} within a solution. Alternatively, the software MaxChelator (developed by Chris Patton, University of Stanford, http://maxchelator.stanford.edu/) can be used. The concept is to produce defined free Ca^{2+} concentrations from a Ca^{2+}-free (HR) and a high-Ca^{2+} containing EGTA-buffered solution (HA) from the resulting total Ca^{2+} concentration of mixing. Both solutions can be kept as stock solutions in vials in the freezer, and are thawed and mixed on the day of experimentation. To produce physiological internal solutions of defined $[Ca^{2+}]_{free}$ or pCa (negative decadic logarithm of $[Ca^{2+}]_{free}$),

Table 2
Pipette scheme for calibration solutions of defined pCa values

HA:HR	Volume HA (μL)	Volume HR (μL)	$[Ca^{2+}]_{free}$ (μM)	pCa
0:100	0	1,000	~0.001	9.00
30:70	300	700	0.182	6.74
50:50	500	500	0.416	6.38
55:45	550	450	0.505	6.30
60:40	600	400	0.613	6.21
65:35	650	350	0.751	6.12
70:30	700	300	0.929	6.03
80:20	800	200	1.51	5.82
90:10	900	100	2.97	5.53
95:5	950	50	4.94	5.31
98:2	980	20	7.77	5.11
100:0	1,000	0	11.90	4.92

prepare appropriate mixtures of HA and HR to a total volume of 1 mL as given in Table 2.

High relaxing solution (HR): 30 mM hepes (4-(2-hydroxyethyl)-1-piperazineethanesulfonic acid), 6.25 mM Mg(OH), 30 mM EGTA (ethylene glycol tetraacetic acid), 0 mM $CaCO_3$, 8 mM Na_2ATP (adenosine triphosphate), 10 mM Na_2CP (creatine phosphate). High-activating solution (HA): 30 mM hepes, 6.05 mM $Mg(OH)_2$, 30 mM EGTA, 29 mM CaCO, 8 mM Na_2ATP, 10 mM Na_2CP. Both HA and HR solutions are adjusted to pH 7.2 using KOH. Usually, about 40–50 mM of KOH are required for titration, thus adding equimolar K^+ to the solution. Use a commercial pH electrode to adjust.

10. Alternatively, use a Fura-2 Calcium *Imaging Calibration Kit* available from a local supplier. This is a quick and easy way of calibrating your system for calcium if you do not have the capability of producing the complex solutions described above. Simply follow the instructions provided with the kit.
11. Plasticware: Use any Perspex® or glass chambers you are familiar with that have an open bottom to attach a glass coverslip (0.17 mm thickness) to.
12. Polydimethyl-siloxane (PDMS) can be used to seal the coverslip against the chamber body.
13. Pasteur glass pipettes for trituration and single fiber transfer.

2.3 Fura-2 Iontophoresis and Electrical Field-Stimulation of Intact Single Fibers

1. Micromanipulator: for Fura-2 iontophoresis, any fine movement micromanipulator with three axes which can be attached to the microscope and viewed under high magnification >40 is suitable.
2. Microelectrode amplifier: Similarly, for iontophoresis of the free acid form of Fura-2, any good microelectrode amplifier with a current injection facility and the bridge balance capability will suffice.
3. For extracellular electrical stimulation of the isolated muscle fiber, any good isolated stimulator will do the job.
4. Microelectrode puller: Any good micropipette collar can be used to produce the fine glass microelectrodes needed to penetrate the isolated skeletal muscle fiber. We use the microelectrode puller P 1000.
5. Glass capillaries should be pulled from borosilicate and may contain an internal filament to ease backfilling. For example, heavy polished thick/standard and thin-walled borosilicate tubing with filament from *Sutter Instruments®* is a good choice.
6. 3 M KCl or 1 M K-acetate to backfill the electrode. Filled electrodes should have an electrical resistance in the range of 6–12 MΩ.
7. Bipolar platinum electrodes made from fine platinum wire held in any suitable electrode holder attached to the micromanipulator (the micromanipulator companies provide a range of electrode holders for this purpose).

2.4 Analysis Tools

1. Any scientific analysis software package to visualize scientific data and apply fitting and mathematical operation procedures can be used. We use Origin, SigmaPlot, or GraphPrism.

3 Methods

3.1 Single Fiber Isolation

1. After dissection of the FDB or *interossei* muscles, transfer muscles into a fresh dish containing physiological saline (Krebs or Ringer's solution, *see* **2.2.3**).
2. For enzymatic digestion, add 2–3 mg/mL collagenase I and incubate for 30 min at 37 °C (or 45 min at 30 °C) under aeration with 95% O_2/5% CO_2.
3. Wash muscles several times with fresh solution to remove collagenase.
4. For single fiber isolation, triturate the muscle chunk using a tip-broken Pasteur glass pipette with the enlarged orifice fire polished. Using a pipette ball at the back, the muscle is vigorously sucked in and squeezed out of the pipette several times.

Single fibers detach from the muscle and sink to the dish bottom.

5. Manually transfer single fibers to a recording chamber using an intact Pasteur pipette(see 4.3.-4.7.).

3.2 Resting Fura-2 Fluorescence Determination in Single Cells, In Situ Calibration and Quantitative $[Ca^{2+}]_{free}$ Determination

1. Before single fiber transfer, clean the recording chamber with 70% ethanol.
2. After transfer of single fibers, incubate the fibers with Fura-2 staining solution. For 1 mL of staining solution, add 10 μL from the Fura-2AM stock solution to a mixture of 975 μL of Ringer (or Krebs), 5 μL of the non-ionic surfactant Pluronic® F-127, and 10 μL of 1 M hepes (for additional pH buffering in the presence of the Fura-2). Exchange the normal saline in the dish for the staining solution, cover the chamber in aluminum foil or inside a box and keep light-protected at room temperature for 30 min.
3. Then place the dish inside a culture incubator for another 15 min to allow for de-esterification of the dye.
4. After that, exchange the staining solution for fresh Ringer's or Krebs solution.
5. Place the dish on the stage of your fluorescence microscope and set the appropriate filters for dual excitation 340/380 nm ratiometric fluorescence recordings (refer to your microscope administrator for exact settings).
6. Within the dish, record a resting fluorescence ratio image from each single fiber. Note the order of the fibers to revisit for calibration. Alternatively, some microscopes provide automated stages where initially, recording positions can be assigned which are then repetitively revisited during the recording protocol. For each single fiber, a F_{340}/F_{380} ratio image (R_{rest}) and both F_{340} and F_{380} raw images should be now available.
7. For performing the in situ calibration, add 1 μL from the 50 mM stock of BTS to the 1 mL of each of the calibration solution caps (*see* **item 9** of Subheading 2.2) to yield a final concentration of 50 μM.
8. To the initial Ca^{2+}-free solution (1,000 μL HR, pCa 9), also add 10 μL from either the ionomycin or A23187 stock to yield a ~10 μM concentration.
9. For the Ca^{2+} ionophoresis and equilibration with "zero" Ca^{2+}, slowly exchange the Ringer's (or Krebs) solution for the first calibration solution ("Ca^{2+}-free" HR, 50 μM BTS, 10 μM ionomycin or A23187).
10. After complete solution exchange, incubate for at least 20 min. The incubation in Ca^{2+}-free calibration solution takes longest

in order to withdraw all Ca^{2+} from the cell through the Ca^{2+} permeable pores by diffusion.

11. Revisit each fiber in the dish with the microscope and take another ratio image which now refers to R_{min}. See 4.8.-4.10.
12. Successively exchange the solution in the dish for the other calibration solutions and repeat the imaging procedure. With the last solution (Ca^{2+}-saturated, HA, 50 μM BTS), obtain the maximum ratio image R_{max}.
13. For quantitation of resting Fura-2 ratio R_{rest} to absolute $[Ca^{2+}]_{rest}$ levels, the pCa-R calibration curves need to be reconstructed for each individual cell. In your analysis software, collect the pCa-R values in two columns and plot the data for each cell.
14. For fitting a curve to the data, use Eq. 10 with x as the pCa column and y as the R column. Set the slope factor b to unity and let the software calculate the fit. Check the r^2 value for optimum output (we require $r^2 > 0.9$ to keep a data set in the analysis).
15. From each fit to the individual data, extract the fitted R_{min}, R_{max}, and K_d value for computation of mean, SD, and SEM. This yields the in situ K_d value.
16. To express R_{rest} of each single fiber initially recorded before the calibration in terms of absolute Ca^{2+} concentration, the pCa value and thus the $[Ca^{2+}]$ value corresponding to R_{rest} can be computed either from the inverse function of Eq. 10 using the obtained fit parameters or just be looked up from the fitted curve x–y pair columns.

3.3 Fura-2 Iontophoresis and Electrical Field-Stimulation of Intact Single Fibers to Extract Action Potential-Induced Global Ca^{2+} Transients Corrected for Dye Kinetics

1. In order to fill the electrode with a free acid form Fura-2 for iontophoresis, dissolve the entire 1 mg free acid Fura-2 powder in filtered (one micron filter) distilled water. Use a vortex mixer to dissolve.
2. Use a 1 μL Hamilton syringe to inject 1 μL of the Fura-2 solution into the microelectrode. Use a fine needle on the Hamilton syringe inserted inside the glass to inject 1 μL of dye as close to tip of the electrode as possible. After a few minutes, you will see the Fura-2 solution move towards the tip of the electrode by capillary forces down the internal filament.
3. Now use 1 M K-acetate to backfill the remaining portion of the electrode leaving an air gap between the Fura-2 solution and the *K*-acetate filling solution. The filling solution can be introduced into the body of the microelectrode by attaching a fine hollow filament to a syringe and inserting this filament into the body of microelectrode.

4. Connect the microelectrode to the amplifier via the head stage (normally provided with the amplifier).
5. After transfer of single fibers to the recording chamber, choose one single fiber for iontophoretic staining.
6. Place the microelectrode in the bathing solution surrounding the isolated muscle fiber and use the micromanipulator to press the tip of the electrode against the muscle fiber.
7. Once inside the fiber, if there is a current injection facility on board the amplifier, use this to inject –9 nA of current down the electrode; do this for between 30 s and 1 min. Using the 380 nm excitation of the light source, view the fiber down the microscope during filling to ensure an optimal fill (Fig. 4). If you overfill the fiber, the Fura-2 will potentially over-buffer the internal calcium transient, and there will be no contraction visible.
8. After filling, quickly remove the microelectrode from the field of view of the ×40 fluorescence objective so as not to interfere with calcium recording.
9. Allow about 20 min for the dye to completely distribute within the cytosol. Estimates about the internal dye concentration of Fura-2 from this method are in the range of 5–50 μM [22].
10. Ensure your photomultiplier tube has been activated.
11. In order to stimulate the fiber, use the bipolar platinum electrode and approach the fiber with the stimulating electrode until you can see both the fiber stimulating electrode in the same field of view at ×40. Use a single 1 ms square pulse from the stimulator to make sure the isolated muscle fiber contracts and relaxes.
12. Increase the amplitude of stimulus intensity somewhere between 10 and 100 V until you get a maximum twitch response, as measured visually, or which produces the largest calcium transient. Take this as a maximal stimulation, increase the voltage slightly beyond this point to ensure supra-maximum stimulation. See 4.11.-4.13.
13. Use the stimulator module to deliver a range of pulses to the muscle fiber, for instance, to generate a calcium-frequency curve between 1 and 200 Hz. Alternatively, apply repeated (one second off) 500 ms bursts of the hundred Hertz 1 ms stimuli to produce muscle fatigue. During the twitch tetanic contractions, you should be able to observe the signals produced by the 340 and 380 nm excitation signals moving in response to the calcium changes in the muscle fiber.
14. For assessment of calibration, use either the same steps for in situ calibration as detailed above in **3.1**, or restrict your calibration to R_{min}, R_{max}, and *kß* determination as in [22].

15. At the end of the experiment, use a microelectrode with a broken tip to suck up the muscle fiber. Transfer this single fiber to an appropriate denaturation buffer for eventual later Western blot protein analysis. This way is it possible to directly correlate the calcium transient data, for example, with fiber type and sarcoplasmic reticulum proteins. The ability to correlate the protein fingerprint of the fiber with its calcium transient measures makes this an extremely powerful technique for probing mechanisms of muscle contraction in processes of muscle disease.
16. Plot the time course of $R(t)$ for each fiber and identify the regions of global transients in response to pulse stimulation. From the calibration parameters extracted from each fiber, convert the ratio into $[Ca^{2+}]$ by using the inverse of Eq. 10. With R_{min}, R_{max}, and K_d available, $[Ca^{2+}]$ can be calculated from the R-value column by applying the following transform to it:

$$[Ca^{2+}](t) = K_d \cdot \frac{R(t) - R_{min}}{R_{max} - (R(t) - R_{min})} \tag{12}$$

It is advised to place the transform column next to the R column and then use this one to produce a time derivative transform to be placed in the subsequent column. Using the apparent K_d value from the in situ calibration and the k_{off} rate constant for Fura-2 as 40 s^{-1} in muscle [19], one can apply Eq. 11 as a transform to the data using the $[Ca^{2+}]$ column and the $d[Ca^{2+}]/dt$ column of the spreadsheet. The result can be placed in the subsequent column labelled as $[Ca^{2+}]_{corr}$.

17. Plot $[Ca^{2+}]_{corr}$ as function of time to yield the graphical output of the corrected Ca^{2+}-transient. On this graph, now determination of peak amplitudes, basal levels, time-to-peak, and exponential transient decay kinetics can be analyzed.

4 Notes

1. Imaging systems used: We use an inverted microscope with a Perfect Focus System® and a Lambda DG4 ultra-high speed wavelength switcher illumination system for fast dual excitation of Fura-2 (OF), and an epifluorescence microscope to which an OptoLED light source for ultra-high stability and "instantaneous" (sub-microsecond) vibration-free wavelength switching (SIH) is connected. On the excitation side, the appropriate excitation filters (340, 380, 358 nm) are required which can be purchased through the microscope system supplier or by optical suppliers. Emission filters for Fura-2 are usually LP >510 nm filters and only need to be modified for other wavelength bins if multispectral imaging is pursued.

2. For processing the Fura-2 AM, we prefer the special packages of 20 × 50 μg because stock solutions can be freshly prepared and will last for a couple of stains.
3. After collagenase treatment of muscle, in our experience, we have better yields of intact single fibers when continuing with fiber isolation after a period of 1–2 h of storage of the digested muscles at 4 °C in a fridge.
4. To have better manual control for single fiber transfer using a Pasteur pipette, we prefer more rigid rubber balls that can be more finely handled with two finger tips.
5. We usually transfer 20–30 single fibers from the dish with the triturated muscle chunk and single fiber pool to the recording chamber. By that time, the fibers have already equilibrated to room temperature, restoring membrane potential to polarized values (important for field stimulation).
6. We use an inverted microscope for visualization and recording, thus a 170 μm glass coverslip is attached to the bottom of the recording chamber. We use poly-dimethyl-siloxane (PDMS) coating of cover slips to the bottom of the recording chamber. This allows to firmly seal the cover slip to the Perspex® chamber bottom for multiple re-use. Cleaning with ethanol makes the single fibers stick more tightly to the glass bottom. Alternatively, a laminin coating can be used at 5–10 mg/mL.
7. It is best to use a custom-made, small Perspex® chamber with a total volume of ~500 μL to save staining solution (the more expensive solution!).
8. For the Ca^{2+} ionophoresis and equilibration with "zero" Ca^{2+}, the solution exchange must be performed slowly and in several exchange steps (and this is really important!) to the first calibration solution ("Ca^{2+}-free" HR, 50 μM BTS, 10 μM ionomycin or A23187). We perform this solution exchange with two 200 μL pipettes, sucking off solution from one side of the ~500 μL chamber and adding the same amount of calibration solution from the other side. We do this five times to use up all the solution in the cap.
9. During the successive Ca^{2+} calibration steps with increasing Ca^{2+} concentrations, direct Ca^{2+}-activation of the contractile apparatus should effectively be blocked by BTS. If not so, increase the BTS concentration on subsequent experiments.
10. Fitting of pCa-R relationships: sometimes, allowing the b value to float may also improve the fit r^2. However, we only allow variations between 0.95 and 1.05 (5% variability to account for experimental variability of data from ideal values).
11. After back-injection of dye solution into the glass microelectrode, electrical continuity of the tip of the electrode with the

amplifier is ensured by the capillary action of the internal glass filament within the electrode, even though it may appear that there is a large bubble within the pipette. It is important to maintain the separation of the potassium acetate and the Fura-2 to dye by air bubble otherwise one will get significant dilution of the Fura-2 dye. It is very costly to fill the whole electrode with the Fura-2 dye, and it will have an extremely high resistance due to the fact it is present in a pure dye solution in water (few conducting ions).

12. For electrical stimulation, remember to place a suitable silver chloride pellet in the bathing solution surrounding the isolated muscle fiber in order to make a circuit via earth (a good source of hands-on information using electrophysiology techniques is provided in The Axon Guide, [23]).
13. Before impalement of the single fiber with the microelectrode, the bridge circuit on the amplifier should be adjusted so the output reading is 0 mV (balanced). Once touching with the electrode tip outside the muscle cell, tap lightly on the microscope or table and the electrode will penetrate the muscle fiber. When this happens, one will see an immediate negative deflection in the voltage recording to around –60 mV to –80 mV. The negative internal resting membrane potential is a sign that the fiber is healthy. During penetration, the fiber may twitch briefly due to the depolarizing effect of the penetration.

Acknowledgments

Supported by mobility funds through the German Academic Exchange Service (DAAD) to OF and the Universities Australia scheme to SIH.

References

1. Berchtold MW, Brinkmeier H, Müntener M (2000) Calcium ion in skeletal muscle: its crucial role for muscle function, plasticity, and disease. Physiol Rev 80(3):1215–1265
2. Barros LF, Castro J, Bittner CX (2002) Ion movements in cell death: from protection to execution. Biol Res 35(2):209–214
3. Grynkiewicz G, Poenie M, Tsien RY (1985) A new generation of Ca2+ indicators with greatly improved fluorescence properties. J Biol Chem 260(6):3440–3450
4. Paredes RM, Etzler JC, Watts LT, Zheng W, Lechleiter JD (2008) Chemical calcium indicators. Methods 46:143–151
5. Takahashi A, Camacho P, Lechleiter JD, Herman B (1999) Measurement of intracellular calcium. Physiol Rev 79(4):1089–1125
6. Pawley J (ed) (2006) Handbook of biological confocal microscopy, 3rd edn. New York, Springer
7. Kao JP, Tsien RY (1988) Ca^{2+} binding kinetics of Fura-2 and azo-1 from temperature relaxation measurements. Biophys J 53(4): 635–639
8. Jackson AP, Timmermann MP, Bagshaw CR, Ashley CC (1987) The kinetics of calcium binding to Fura-2 and indo-1. FEBS Lett 216(1):35–39

9. Rüdiger S (2014) Stochastic models of intracellular calcium signals. Phys Rep 534:39–87
10. Kong CHT, Laver DR, Cannell MB (2013) Extraction of sub-microscopic Ca fluxes from blurred and noisy fluorescent indicator images with a detailed model fitting approach. PLoS Comput Biol 9(2):e1002931
11. Eberhard M, Erne P (1989) Kinetics of calcium binding to Fluo-3 determined by stopped-flow fluorescence. Biochem Biophys Res Comm 163(1):309–314
12. Bruton JD, Cheng AJ, Westerblad H (2012) Methods to detect Ca^{2+} in living cells. In: Islam MS (ed) Calcium signaling, advances in experimental medicine and biology 740. Springer, Dordrecht, pp 27–43
13. Rios E, Stern MD, Gonzalez A, Pizarro G, Shirokova N (1999) Calcium release flux underlying Ca2+ sparks of frog skeletal muscle. J Gen Physiol 114:31–48
14. Sano M, Yokota T, Endo T, Tsukagoshi H (1990) A developmental change in the content of parvalbumin in normal and dystrophic mouse (mdx) muscle. J Neruol Sci 97:261–272
15. Jiang Y, Julian FJ (1997) Pacing rate, halothane, and BDM affect Fura 2 reporting of [Ca2+] in intact rat trabeculae. Am J Physiol 273:C2046–C2056
16. Bakker AJ, Head SI, Williams DA, Stephenson DG (1993) Ca^{2+} levels in myotubes grown from the skeletal muscle dystrophic (*mdx*) and normal mice. J Physiol 460:1–13
17. Friedrich O, Reiling SJ, Wunderlich J, Rohrbach P (2014) Assessment of Plasmodium falciparum PfMDR1 transport rates using Fluo-4. J Cell Mol Med 18(9):1851–1862
18. Launikonis BS, Stephenson DG, Friedrich O (2009) Rapid Ca2+ flux through the transverse tubular membrane, activated by individual action potentials in mammalian skeletal muscle. J Physiol 587(10):2299–2312
19. Bakker A, Head SI, Stephenson DG (1997) Time course of calcium transients derived from Fura-2 fluorescence measurements in single fast twitch fibres of adult mice and rat myotubes developing in primary culture. Cell Calcium 21(5):359–364
20. Head SI, Chan S, Houwelling PJ, Quinlan KGR, Murphy R, Wagner S, Friedrich O, North KN (2015) Altered Ca^{2+} kinetics associated with α-actinin-3 deficiency may explain positive selection for ACTN3 null allele in human evolution. PLoS Genet 11(2):e1004862
21. Baylor SM, Hollingworth S (1988) Fura-2 calcium transients in frog skeletal muscle fibres. J Physiol 403:151–192
22. Head SI (1990) Membrane potential, resting calcium and calcium transients in isolated muscle fibres from normal and dystrophic mice. J Physiol 469:11–19
23. Sherman-Gold R (Ed.) (1993) The axon guide for electrophysiology and biophysics laboratory techniques. Axon Instruments, Foster City (http://www.imbb.forth.gr/images/facilities/Cells_Animals/Electrophysiology%20Unit/Axon_Guide.pdf)

Chapter 15

Functional Viability: Measurement of Synaptic Vesicle Pool Sizes

Jana K. Wrosch and Teja W. Groemer

Abstract

Neurons and their function of conveying information across a chemical synapse are highly regulated systems. Impacts on their functional viability can occur independently from changes in morphology. Here we describe a method to assess the size of synaptic vesicle pools using live cell fluorescence imaging and a genetically encoded probe (pHluorin). Assessing functional parameters such as the size of synaptic vesicle pools can be a valuable addition to common assays of neuronal cell viability as they demonstrate that key cellular functions are intact.

Key words Electrical stimulation, Live cell fluorescence imaging, pHluorin, Synaptic vesicle pools, Synaptic vesicle recycling

1 Introduction

The measurement of synaptic vesicle pools sizes is a well-established tool to verify neurons' functional viability.

The presynaptic bouton of a synapse contains synaptic vesicles. Their function is to concentrate and transport neurotransmitter. When an action potential reaches the synapse, synaptic vesicles are exocytosed and their content is released into the synaptic cleft [1]. The released neurotransmitter will then activate receptors on the postsynaptic side. To compensate for the lost vesicles and the disembogued membrane, new vesicles are formed through endocytosis and reacidification [2].

The synaptic vesicle recycling cycle is a carefully regulated mechanism and is playing a role in synaptic plasticity [3–5]. According to the activation intensity—action potential frequency and duration—different vesicle populations are released from the synapse: Vesicles that are docked to the membrane are released upon low intensity

The original version of this chapter was revised. The erratum to this chapter is available at: DOI 10.1007/978-1-4939-6960-9_23

Daniel F. Gilbert and Oliver Friedrich (eds.), *Cell Viability Assays: Methods and Protocols*, Methods in Molecular Biology, vol. 1601, DOI 10.1007/978-1-4939-6960-9_15, © Springer Science+Business Media LLC 2017

activation and form the readily releasable pool (RRP) [6]. Spontaneously fusing vesicles originate from this same pool of readily releasable vesicles [7]. Other vesicles are only released upon stronger activation. Both of these vesicle populations undergo synaptic vesicle recycling and form the recycling pool (RECP) [8]. A third population of vesicles, the resting pool (RESTP) cannot be exocytosed [3]. The total pool (TOTALP) of vesicles in a synapse is made up by the sum of these three vesicle pools.

The effect of pharmacological drugs on synaptic vesicle pool sizes can be a valuable indication of presynaptic drug targets and has been researched in context with, e.g., antidepressants [9], G-protein signaling [10], and synaptic plasticity [3–5].

To quantify synaptic vesicle pool sizes, the vesicles are labeled with an exocytosis-dependent fluorescent tag. Upon synaptic stimulation of controlled intensity the different vesicle pools are released. The intensities of the fluorescence responses can be recorded and represent the number of vesicles released from the different pools.

There are two common methods for exocytosis-dependent staining of synaptic vesicles: FM-dyes and pH-dependent fluorophores.

When using FM-dyes, vesicles are first loaded with the fluorophores; upon exocytosis the fluorophores are released and fluorescence intensity decreases with every vesicle exocytosed [1, 11–13]. On the one hand, this method is very robust and doesn't require genetic modifications. On the other hand, it can only provide the absolute value of synapses' vesicle pool sizes at a given time and cannot be normalized to account for varying synapse sizes across the recorded cells [3, 5, 14].

In the second approach, neurons express vesicular proteins, tagged with a pH-dependent GFP, called pHluorin [15]. Fluorescence is quenched in the acidic lumen of the vesicle and upon exocytosis and the exposure to the neural pH in the synaptic cleft the fluorophores light up [2, 16]. The total pool of all vesicles can be made visible by dequenching all fluorescence with freely diffusing ammonium ions. With that, the fluorescence response upon stimulation can be normalized to the total size of the synapse, which yields more reliable results than the first method [9]. Whenever it is feasible to use cells, expressing proteins tagged with pHluorins, this second method thus is the better choice.

Synapto-pHluorin (SpH) is the pH-dependent GFP pHluorin, fused to the vesicular protein Synaptobrevin2, also known as VAMP2 [17]. It is known that SpH overexpression does not perturb presynaptic function and is therefore a suitable staining method to visualize synaptic transmission [18]. If the cells are expressing SpH, the total pool of vesicles can be made visible by adjusting extracellular pH levels [3]. This allows the measured vesicle pool sizes to be normalized to each synapse's indivudal size and is a great advantage of using SpH over other methods. [3].

For the vesicle pool size measurement using pHluorin dyes, cells expressing the tagged proteins are imaged alive with a suitable fluorescence microscope. During imaging, the cells will be electrically stimulated. As a response to the stimulation, synaptic vesicles will be released and the resulting fluorescence increase will be recorded. Stimulation with specific intensities can release specifically the readily releasable pool or the recycling pool. To normalize the recorded fluorescence increase to the total number of vesicles in a specific synapse, fluorescence of all vesicles—also those of the resting pool—will be dequenched by perfusing the cells with ammonium ions.

For accurate results the vesicular ATPase inhibitor concanamycin A will be present in the imaging buffer throughout the recording time. When released membrane is recycled into new vesicles, the vesicular ATPase reacidifies these newly formed vesicles and quenches the fluorescence of pHluorin tagged membrane proteins taken up into the vesicle together with the recycled membrane. This quenching of recycled proteins blurs the amplitude of fluorescence increase upon an electrical stimulation and is prevented by inhibiting the vesicular ATPase from quenching the fluorophores.

2 Materials

2.1 Neurons

Use newborn rat primary hippocampal neurons or any other established neuronal culture method. The culture can be pure neurons or mixed; it can be primary culture, stem cell derived, or any other type of cell line (*see* **Note 1**).

2.2 Culture Medium

Cultivate the cells in Minimum Essential Medium (Thermo Fisher Scientific) supplemented with 1% B-27 (Thermo Fisher Scientific) or use a culture medium suitable for your specific cell type.

2.3 12-Well Culture Plates and Glass Cover Slips

Cultivate the cells on glass coverslips of 18 mm diameter (or a diameter suitable for your imaging chamber) in 12-well cell culture plates (e.g., TPP).

2.4 Matrigel

Prior to cell seeding coat the glass coverslips with 2% Matrigel (Corning Inc.) or another coating suitable for your specific cell type (*see* **Note 2**).

2.5 Synapto-pHluorin

Use the calcium-phosphate method to transfect the neurons with the Synapto-pHluorin-Plasmid (Plasmid factory) (*see* **Note 3**).

2.6 Fluorescence Microscope

These experiments require no specialized imaging equipment. The setup should resolve synapses (60× objective or higher) and should be able to record GFP wavelength (dichroic mirror with a cutoff wave length of 488 nm). The probe should be kept at 37°C and an imaging frame rate of 4–10 frames/s is recommended (*see* **Note 4**).

2.7 Imaging Chamber

Use a Series 20 heated imaging chamber (Warner Instruments) or another type that fits with your specific microscope.

2.8 Perfusion Equipment

During imaging, the synapses need to be perfused with imaging buffer. Also here there are no special requirements except delivering the solution to the field of view at a given time (*see* **Note 5**).

2.9 Stimulation Equipment

During the recording the neurons need to be excited electrically. The switching needs to be fast enough to deliver millisecond pulses (*see* **Note 6**).

2.10 Normal Imaging Buffer and Ammonium Chloride Buffer

1. For the two solutions, prepare half of the final volume deionized water in a graduated cylinder each.
2. Weigh in the salts and dilute (*see* **Note 7**):

   Normal imaging buffer: 140 mM NaCl, 2.4 mM KCl, 10 mM HEPES, 10 mM Glucose, 2 mM MgCl, 2 mM $CaCl_2$; adjust pH to 7.5 using NaOH and HCl.

   Ammonium chloride buffer: 140 mM NaCl, 2.4 mM KCl, 10 mM HEPES, 10 mM Glucose, 2 mM MgCl, 2 mM $CaCl_2$, and 50 mM NH_4Cl.
3. Adjust each solution's pH to 7.5 using NaOH and HCl solutions.
4. Fill up each solution with deionized water to the final volume.
5. Use Mannitol to adjust both solutions' osmolarity to 320 mOsm or to match the osmolarity of your specific cell culture medium.
6. Both imaging buffer and ammonium chloride buffer can be stored at 4 °C for years (*see* **Note 8**).

2.11 Concanamycin A

Concanamycin A is in the literature also sometimes referred to as Folimycin (*see* **Note 9**).

2.12 Image Processing Software

Using MATLAB (Mathworks Inc.) is recommended due to the complexity of the necessary programming steps.

3 Methods

3.1 Neuronal Culture

The cultivated neurons can be used for vesicle pool size measurements as soon as they are electrically excitable (15 days in culture for primary hippocampal neurons). Transfect the cells with Synapto-pHluorin at least 24 h prior to the recordings.

3.2 Preparation of Imaging Buffer

Before starting the recordings, freshly add 80 nM Concanamycin A to the normal imaging buffer.

3.3 Preparation of Imaging Chamber

Move the glass coverslip with the cultivated neurons into the imaging chamber and cover with 1 ml of normal imaging buffer (*see* **Note 10**).

3.4 Imaging

3.4.1 General Outline

For the vesicle pool size measurement, the neuronal cultures are imaged with a fluorescence microscope during the perfusion with imaging buffers and electrical stimulation. For an overview *see* Fig. 1. The fluorescence response to these manipulations is recorded.

During the first recording phase, cells are perfused with normal imaging buffer, containing concanamycin A, and are electrically stimulated at different strengths to release synaptic vesicles of the different vesicle pools. Then perfusion is switched to ammonium chloride solution to dequench and record the total fluorescence of the total vesicle pool.

Configure the microscope recording so that you know the time points of electrical stimulations and of perfusion change for the later image processing.

3.4.2 Detailed Description of the Imaging Phases

1. *Stable baseline*: In the beginning of the recording start perfusion with normal imaging buffer and allow 2 min time for equilibration (*see* **Note 11**).
2. *Readily releasable poolstimulation*: Electrically stimulate with 40 1-ms-pulses of alternating polarity (*see* **Note 12**) at 20 Hz to release and dequench the readily releasable pool [6, 8] (*see* **Note 13**).
3. *Pause for equilibration*: Allow fluorescence to stabilize for 1 min.

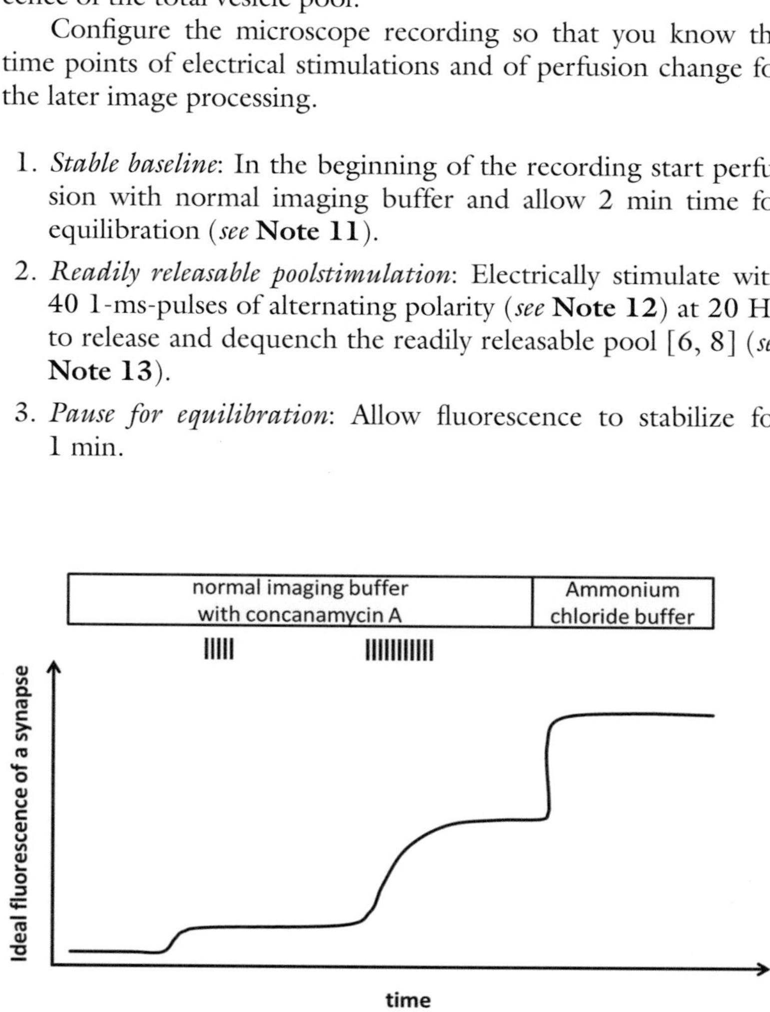

Fig. 1 Recording scheme of vesicle pool size measurement. Cells are perfused with normal imaging buffer supplemented with concanamycin A during the recording. Electrical stimulation of 40 pulses at 20 Hz and 900 pulses at 10 Hz is delivered after an equilibration phase of 1 min each, to release vesicles of the RRP and RECP. Then perfusion solution is switched to ammonium chloride buffer to record the total fluorescence from all—also resting pool—vesicles

4. *Recycling poolstimulation*: Electrically stimulate with 900 1-ms-pulses of alternating polarity at 10 Hz to release and dequench the rest of the recycling pool [19, 20].
5. *Pause for equilibration*: Allow fluorescence to stabilize for 1 min.
6. *Ammonium chloride buffer perfusion*: Switch perfusion from normal imaging buffer to ammonium chloride buffer. The freely diffusing Ammonium ions dequench fluorescence in all vesicles, also the resting pool that cannot be released [9].

3.5 Data Analysis

1. ROI detection

   Before automatic synapses detection fluorescent artifacts are filtered out by using an image showing only those regions whose fluorescence intensity increases during the stimulation. For that, pixel wise average 2–5 image frames just before and just after the second electrical stimulation (*see* **Note 14**). Then subtract the two images (*see* **Note 15**).

   Now run an automatic ROI detection algorithm such as feature point detection [21].
2. Calculate fluorescence trace

   Calculate fluorescence curves of each region of interest by averaging over all pixels of the region in each time frame respectively (*see* **Note 16**).
3. Subtract baseline fluorescence

   Average the baseline fluorescence for each ROI and subtract it from the trace (*see* **Note 17**).
4. Normalize fluorescence traces to the total pool of vesicles

   Average the total pool fluorescence; then divide the trace in each time point by that average to normalize the fluorescence level to 100% for the total of all vesicles (*see* **Note 18**).
5. Calculate vesicle pool sizes

   During electrical stimulation the targeted vesicle pool becomes fluorescent (additionally to the already fluorescent vesicles). With that, the relative size of the vesicles pools can be read from the increase of fluorescence during stimulation.

   To calculate, average the fluorescence in each ROI over 2–5 image frames before the onset and after the completion of the electrical stimulations; then subtract the two averages to get the increase in fluorescence triggered by the stimulation. This calculated difference is the size of the targeted vesicle pool relative to the total pool of that synapse (*see* **Note 19**).
6. Averaging vesicle pool sizes

   The calculated relative vesicle pool sizes can now be averaged over all synapses in a recording (*see* **Note 20**).

4 Notes

1. An exemplary protocol for the cultivation of primary hippocampal neurons can be found in [22]. An exemplary protocol for the cultivation of stem cell derived neurons can be found in [23].
2. Alternatives to using Matrigel would be 0.1 mg/ml Poly-L-Lysin, 0.1 mg/ml Poly-L-Ornithine, 50 μg/ml Laminin, or other extra cellular matrix proteins. Depending on the glass surface neurons might grow well enough even without coating.
3. An exemplary protocol for calcium phosphate transfection can be found in [24]. Other transfection methods such as electroporation [25] or lipofectamine transfection [26] are also possible.

   PHluorins have also been fused to other synaptic and vesicular proteins, such as for example Synaptophysin (SypHy) [10, 27]. These variants are also available and can equally well be used for these measurements.
4. Make sure to set up the imaging with the excitation intensity so low that synapses during baseline are visible, but so that the fluorescence increase upon stimulations can still be recorded and does not get lost in high-intensity clipping.
5. A possible setup consists of reservoirs for the solutions with outlets (50 ml syringes) on a raised mount, suitable tubes, and a small glass pipette outlet. Automatic valves for the tubes are necessary to guarantee an immediate and accurate switch between the solutions during the recording. The flow rate can be controlled by attaching a variable clip to the tube. A flow rate of 0.5–1 ml/min is a good trade-of between immediate delivery and sheer stress. Be sure to use non-autofluorescent glass for the outlet.
6. A common method to do this is to deliver electrical pulses through parallel field electrodes. Other setups such as MEAs can be used if the cells can be sufficiently excited.

   When using parallel field electrodes, the polarity should alternate to prevent hydrolysis. In the imaging buffer described here and a distance of 10 mm between the electrodes, a current of 51 mA is sufficient to excite almost all cells in the culture. This may differ in your setup. Tune the excitation intensity to get a reliable response. Be aware that too high intensities would damage the cells.
7. Both media can be prepared as a 10× stock (ten times the end concentrations of the salts) and the diluted in portions as working stock.
8. If the imaging buffer will be stored for a long time, we recommend to filter it and to store it under sterile conditions.

After the imaging buffer was stored for a long time, be sure to double-check osmolarity before use.

9. Concanamycin A inhibits the vesicular ATPase which is acidifying the lumen of newly formed vesicles. With this mechanism blocked, SpH once released and dequenched will not be requenched even if recycled into a new vesicle. This way, vesicular membrane recycling can't deceive the calculation of vesicle pool sizes. During ammonium chloride, concanamycin A is not required anymore, since all fluorophores are dequenched anyways and the substance is quite expensive.

   Concanamycin A should be dissolved in the working solution fresh each day and not stored.

10. The time between the first application of concanamycin A and the beginning of the recording should be held as short as possible to avoid inaccuracies through spontaneous vesicle fusion.

11. Equilibration may need more time. Try it out and make it so long that fluorescence is stable; if it does not stabilize but reach a constant slope, it can be corrected during data analysis.

12. Alternate polarity to prevent hydrolysis.

13. Different vesicle pools can alternatively be targeted by working with different calcium concentrations [4].

14. Averaging several images is reducing camera noise. For exemplary MATLAB code please refer to the supplementary files.

15. Subtracting the images before and after perfusion switch clears the image from most autofluorescent artifacts or unresponsive cells and synapses and from surface expression of SpH [17]. For exemplary MATLAB code please refer to the supplementary files.

16. A detailed description of this step can be found in Jia et al., **Box 1** [28]. Use only **step 1**; the further steps of that protocol are not recommended in the current application. For exemplary MATLAB code please refer to the supplementary files.

17. To get the baseline fluorescence level for each ROI, average 5–10 fluorescence values. Then go through the entire recording and subtract this baseline level from the recorded values.

    For exemplary MATLAB code please refer to the supplementary files.

    If the baseline didn't stabilize to a constant level, but to a constant slope—even after giving it a few more minutes' time—you can fit a linear fit to the baseline and subtract it from the trace to correct for the constant increase in fluorescence.

18. This normalization to the fluorescence level of the total pool of each region of interest is a big strength of using pHluorin over FM-dyes. It compensates for different synapse sizes and makes the relative vesicle pool sizes comparable and reliable across

different synapses, cells, and cultures. For exemplary MATLAB code please refer to the supplementary files.

19. To get the relative vesicle pool size of the readily releasable and the recycling pool, calculate the average fluorescence of a ROI before and after the corresponding electrical stimulation. Then subtract the two, to get the final size of the respective vesicle pool in percent of this ROI's total vesicle pool. For exemplary MATLAB code please refer to the supplementary files.
20. Do not repeat the recording in a different field of view in the same culture even after a washout of concanamycin A and a new dequenching of fluorescence in the recycled vesicles, since electrical stimulation and ammonium chloride treatment during the first recoding can affect the results of a flowing one.

To account for systematic errors such as variance in cultures, choice of field of view, pipetting errors, and temperature, the results should be trusted only if based on a sufficient sample size of at least ten different coverslips per experimental condition. Be sure to double-check statistical power of your results [29].

Acknowledgments

This work was supported by Else-Kröner-Fresenius Stiftung grant 2012_A35.

J.K.W. performed this work in fulfillment of the requirements for obtaining the degree "Dr. rer. biol. hum." at the University of Erlangen-Nuremberg.

Supplementary Files

Exemplary MATLAB code for processing a vesicle pool size recording can be found in the GitHub repository available at https://github.com/janawrosch/VesiclePoolSizes.

References

1. Ryan TA, Smith SJ (1995) Vesicle pool mobilization during action potential firing at hippocampal synapses. Neuron 14(5):983–989
2. Wienisch M, Klingauf J (2006) Vesicular proteins exocytosed and subsequently retrieved by compensatory endocytosis are nonidentical. Nat Neurosci 9(8):1019–1027
3. Atwood HL, Karunanithi S (2002) Diversification of synaptic strength: presynaptic elements. Nat Rev Neurosci 3(7):497–516
4. Wang X, Pinter MJ, Rich MM (2016) Reversible recruitment of a homeostatic reserve pool of synaptic vesicles underlies rapid homeostatic plasticity of quantal content. J Neurosci 36(3):828–836
5. Waters J, Smith SJ (2002) Vesicle pool partitioning influences presynaptic diversity and weighting in rat hippocampal synapses. J Physiol 541(Pt 3):811–823
6. Rosenmund C, Stevens CF (1996) Definition of the readily releasable pool of vesicles at hippocampal synapses. Neuron 16(6):1197–1207
7. Wilhelm BG, Groemer TW, Rizzoli SO (2010) The same synaptic vesicles drive active and spontaneous release. Nat Neurosci 13(12):1454–1456

8. Schikorski T, Stevens CF (2001) Morphological correlates of functionally defined synaptic vesicle populations. Nat Neurosci 4(4):391–395
9. Jung J, Loy K, Schilling EM et al (2014) The antidepressant fluoxetine mobilizes vesicles to the recycling pool of rat hippocampal synapses during high activity. Mol Neurobiol 49(2):916–930
10. Tagliatti E, Fadda M, Falace A et al (2016) Arf6 regulates the cycling and the readily releasable pool of synaptic vesicles at hippocampal synapse. Elife 5:e10116
11. Welzel O, Tischbirek CH, Jung J et al (2010) Synapse clusters are preferentially formed by synapses with large recycling pool sizes. PLoS One 5(10):e13514
12. Ryan TA, Li L, Chin LS et al (1996) Synaptic vesicle recycling in synapsin I knock-out mice. J Cell Biol 134(5):1219–1227
13. Marra V, Burden JJ, Crawford F et al (2014) Ultrastructural readout of functional synaptic vesicle pools in hippocampal slices based on FM dye labeling and photoconversion. Nat Protoc 9(6):1337–1347
14. Welzel O, Henkel AW, Stroebel AM et al (2011) Systematic heterogeneity of fractional vesicle pool sizes and release rates of hippocampal synapses. Biophys J 100(3):593–601
15. Miesenbock G, De Angelis DA, Rothman JE (1998) Visualizing secretion and synaptic transmission with pH-sensitive green fluorescent proteins. Nature 394(6689):192–195
16. Hua Y, Sinha R, Thiel CS et al (2011) A readily retrievable pool of synaptic vesicles. Nat Neurosci 14(7):833–839
17. Sankaranarayanan S, De Angelis D, Rothman JE et al (2000) The use of pHluorins for optical measurements of presynaptic activity. Biophys J 79(4):2199–2208
18. Rother M, Brauner JM, Ebert K et al (2014) Dynamic properties of the alkaline vesicle population at hippocampal synapses. PLoS One 9(7):e102723
19. Li Z, Burrone J, Tyler WJ et al (2005) Synaptic vesicle recycling studied in transgenic mice expressing synaptopHluorin. Proc Natl Acad Sci U S A 102(17):6131–6136
20. Groemer TW, Klingauf J (2007) Synaptic vesicles recycling spontaneously and during activity belong to the same vesicle pool. Nat Neurosci 10(2):145–147
21. Sbalzarini IF, Koumoutsakos P (2005) Feature point tracking and trajectory analysis for video imaging in cell biology. J Struct Biol 151(2):182–195
22. Kaech S, Banker G (2006) Culturing hippocampal neurons. Nat Protoc 1(5):2406–2415
23. Takahashi K, Tanabe K, Ohnuki M et al (2007) Induction of pluripotent stem cells from adult human fibroblasts by defined factors. Cell 131(5):861–872
24. Threadgill R, Bobb K, Ghosh A (1997) Regulation of dendritic growth and remodeling by Rho, Rac, and Cdc42. Neuron 19(3):625–634
25. Transfection of mammalian cells by electroporation (2006) Nat Method. 3(1):67–68
26. Zhang XS, Huang J, Zhan CQ et al (2016) Different influences of lipofection and electrotransfection on in vitro gene delivery to primary cultured cortex neurons. Pain Physician 19(3):189–196
27. Royle SJ, Granseth B, Odermatt B et al (2008) Imaging phluorin-based probes at hippocampal synapses. Methods Mol Biol 457:293–303
28. Jia H, Rochefort NL, Chen X et al (2011) In vivo two-photon imaging of sensory-evoked dendritic calcium signals in cortical neurons. Nat Protoc 6(1):28–35
29. Button KS, Ioannidis JP, Mokrysz C et al (2013) Power failure: why small sample size undermines the reliability of neuroscience. Nat Rev Neurosci 14(5):365–376

Chapter 16

Phenotyping Cellular Viability by Functional Analysis of Ion Channels: GlyR-Targeted Screening in NT2-N Cells

Katharina Kuenzel*, Sepideh Abolpour Mofrad*, and Daniel F. Gilbert

Abstract

Glycine receptor chloride channels (GlyRs) are attractive drug targets for therapeutic intervention and are also more and more recognized in the context of in vitro neurotoxicity and developmental neurotoxicity testing. Assaying the functional properties of GlyR can serve as an indicator of cellular viability and the integrity of the developing and mature central nervous system. Human pluripotent NTERA-2 (NT2) stem cells undergo neuronal differentiation upon stimulation with retinoic acid and express a large variety of neuronal proteins—including GlyR. YFP-I152L, a halide-sensitive variant of yellow fluorescent protein, allows high-throughput fluorescence-based functional analysis of GlyRs in NT2 cells. Here we describe a protocol for phenotyping of cellular viability by functional analysis of GlyR in neuronally differentiated NT2 (NT2-N) cells using YFP-I152L as a reporter of functional integrity of GlyRs. The protocol describes neuronal differentiation of NT2 stem cells, transient transfection of NT2-N cells with YFP-I152L as well as functional imaging and analysis of data from high-content imaging.

Key words Human pluripotent embryonal teratocarcinoma stem cells, NT2 cells, NT2-N cells, Glycine receptor chloride channel (GlyR), YFP-I152L, Cell viability, Toxicological screening

1 Introduction

Glycine receptors (GlyRs) are ligand-gated ion channels which mediate inhibitory neurotransmission in the central nervous system (CNS). In mature neurons, upon activation by the amino acid and neurotransmitter glycine, GlyRs conduct a hyperpolarizing inward-directed anion current into the cells [1–3]. Impaired channel function, by, e.g., genetic or molecular perturbation, can lead to severe neurological disorders including epilepsy [4–6], neuropathic pain [7], chronic pain sensitization [2, 8, 9], and hyperekplexia [10, 11] and has also been associated with neurotoxicity [12–15]. Due to

The original version of this chapter was revised. The erratum to this chapter is available at: DOI 10.1007/978-1-4939-6960-9_24

*Katharina Kuenzel and Sepideh Abolpour Mofrad contributed equally to this work.

Daniel F. Gilbert and Oliver Friedrich (eds.), *Cell Viability Assays: Methods and Protocols*, Methods in Molecular Biology, vol. 1601, DOI 10.1007/978-1-4939-6960-9_16, © Springer Science+Business Media LLC 2017

their important role in inhibitory neurotransmission, these channels are considered attractive drug targets for therapeutic intervention [16] and are also increasingly recognized in the context of in vitro neurotoxicity (NT) [12, 13, 17–19] and developmental neurotoxicity (DNT) testing [14, 15]. Assaying the functional properties of GlyR can serve as an indicator of cellular viability and the integrity of the developing and mature central nervous system.

Human pluripotent NTERA-2 (NT2 or TERA2.cl.SP12) stem cells undergo neuronal differentiation upon exposure to retinoic acid and mimic the process of differentiation in the developing brain including developmental stages ranging from nondifferentiated stem cells, committed neural progenitors to differentiated neuronal—so-called NT2-N cells—and glial cells [20–25]. Electrophysiological studies with NT2-N cells have demonstrated voltage-activated calcium, TTX-sensitive sodium and potassium currents, spontaneous synaptic currents as well as glutamate, *N*-methyl-D-aspartate (NMDA), GABA and strychnine-sensitive glycine-induced currents [14, 21, 25–29], indicating that these cells exhibit properties similar to those described in native human neurons.

YFP-I152L, a variant of yellow fluorescent protein (YFP) with strongly enhanced anion sensitivity, is quenched by small anions and is thus suitable to reporting anionic influx into cells [30]. Figure 1 shows the principle of cellular viability phenotyping by functional analysis of GlyRs using the fluorescence reporter YFP-I152L. The fluorescent protein has been successfully and repeatedly applied for compound screening and structure-function analysis with many different chloride channel types [14, 15, 31–39].

Here we describe a protocol for phenotyping of cellular viability by functional analysis of GlyRs in NT2 cells using recombi-

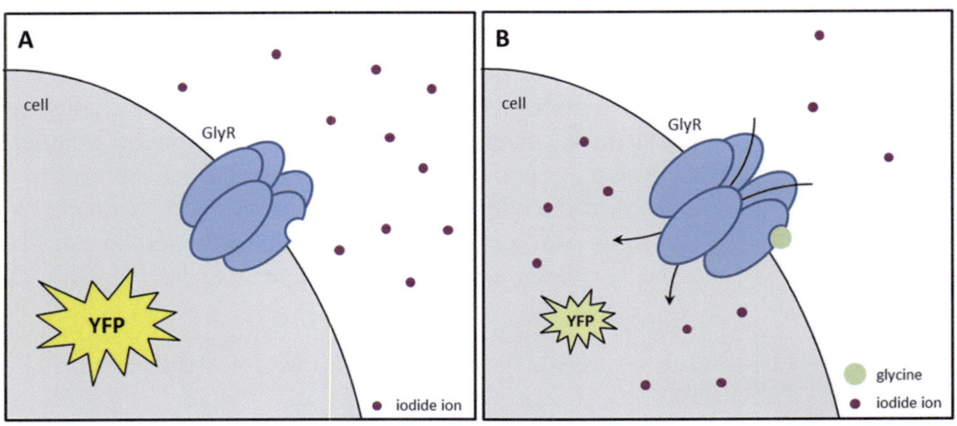

Fig. 1 Principle of cell viability phenotyping by functional analysis of GlyRs using the fluorescence reporter YFP-I152L. (**a**) Fluorescence intensity of YFP-I152L is strong in resting cells. (**b**) Upon activation by its ligand, GlyRs conduct an inward-directed anion current, leading to fluorescence quenching of YFP-I152L

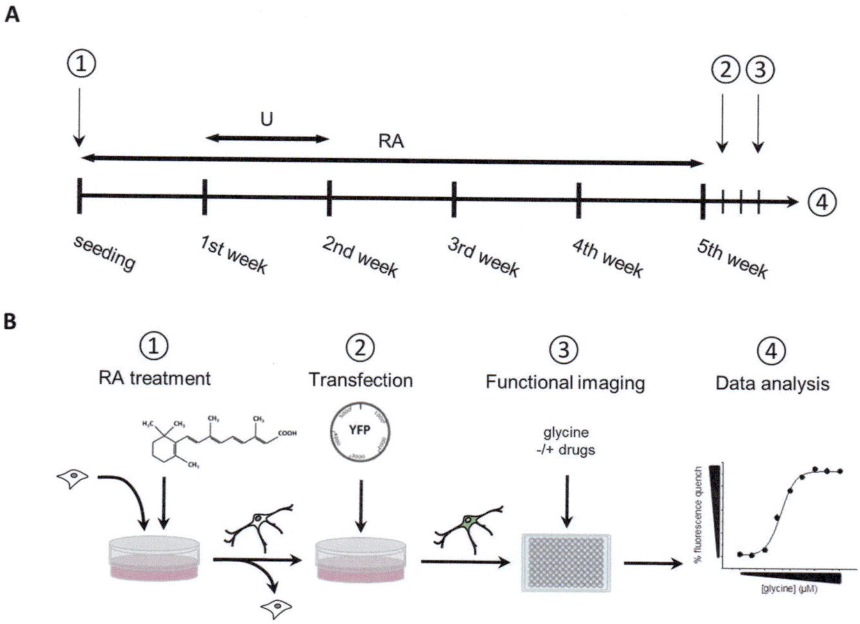

Fig. 2 Timeline of neuronal differentiation of NT2 cells in monolayer cultures (**a**) and experimental workflow (**b**). NT2 cells are differentiated into NT2-N cells by treatment with retinoic acid (RA) and exposure to uridine (U) [1]. Differentiated cells are replated and transiently transfected with YFP-I152L using the calcium-phosphate precipitation method [2]. After cell seeding into 96-well plates [3], fluorescence imaging experiments are conducted and cellular viability is assessed by analyzing GlyR function based on, e.g., agonist concentration-response and half-maximal activation concentration

nantly expressed YFP-I152L as a reporter of GlyR activation. YFP-I152L is expressed under the control of the human ubiquitin promoter C. The promoter has been reported to drive selective protein expression in principal neurons in the mammalian brain [40]. NT2 cells have previously been reported to provide a suitable system for expressing exogenous proteins in terminally differentiated neurons [21]. A timeline of neuronal differentiation of NT2 cells in monolayer cultures previously developed in our laboratory [41] and the experimental workflow are shown in Fig. 2.

2 Materials

1. Human pluripotent teratocarcinoma cells (NTERA-2.cl.D1, CRL-1973™, ATCC).
2. Cell culture medium: 10% fetal bovine serum (FBS), 100 U/mL penicillin, 0.1 mg/mL streptomycin in Dulbecco's Modified Eagle Medium (DMEM).
3. Phosphate buffered saline (PBS).
4. Trypsin-EDTA solution: 0.05% trypsin containing 0.025% EDTA.

5. Poly-D-lysine (PDL): 0.01 mg/mL in water.
6. Matrigel solution: 25% Matrigel® Basement Membrane Matrix in DMEM (Phenol Red Free).
7. Retinoic acid (RA): 100 mM stock in dimethyl sulfoxide (DMSO).
8. Differentiation medium: Cell culture medium supplemented with 10 μM retinoic acid.
9. Uridine: 100 mM stock in DMSO.
10. 2× Hepes buffered saline (HBS): 280 mM NaCl, 50 mM Hepes, 1.5 mM Na_2HPO_4 in water, pH 7.1 with NaOH. Sterile filtered.
11. $CaCl_2$ solution: 1 M $CaCl_2$ in water.
12. YFP-I152L-encoding vector.
13. Ultrapure water.
14. NaCl control solution: 140 mM NaCl, 5 mM KCl, 2 mM CaCl2, 1 mM $MgCl_2$, 10 mM HEPES, and 10 mM glucose in water, pH 7.4 (NaOH).
15. NaI test solution: 140 mM NaI, 5 mM KCl, 2 mM CaCl2, 1 mM $MgCl_2$, 10 mM HEPES, and 10 mM glucose in water, pH 7.4 (NaOH).
16. Glycine stock: 1 M in water.
17. T75 tissue culture flasks.
18. 60-mm cell culture dish.
19. 96-well plates with transparent bottom.
20. 15 mL conical centrifuge tube.
21. 1.5 mL reaction tube.
22. Hemocytometer.

3 Methods

Perform all cell culture work inside a laminar flow hood. Functional imaging experiments can be conducted under nonsterile conditions.

3.1 Cell Culture, Harvesting and Counting

1. Retrieve frozen NT2 cells from liquid nitrogen and thaw quickly in a 37 °C water bath.
2. Dilute the cells in 15 mL cell culture medium in a T75 tissue culture flask and incubate under standard cell culture conditions at 37 °C, 5% CO_2 and 95% relative humidity.
3. Exchange the medium after 24 h.
4. Passage the cells when approximately 80–90% confluent. Remove the old medium and wash the cells carefully with

5 mL PBS. Remove PBS, add 2 mL of Trypsin-EDTA solution and incubate at standard culture conditions for about 3 min. Detach cells from the bottom of the flask by carefully tapping the flask a few times and add 8 mL cell culture medium to stop the enzymatic reaction. After dislodging cells, pipette suspension up and down to singularize the cells.

5. Mix 100 μL cell suspension with 900 μL cell culture medium in a 1.5 mL reaction tube for counting the cells with a hemocytometer (e.g., Neubauer chamber). Pipette 10 μL cell suspension into each compartment of the counting chamber. Count the number of cells in a total of eight grids using a microscope. Calculate the average number of cells per milliliter, using the following formula:

$$\frac{\sum_{\text{cells}}}{8} \times 10^5 = \frac{\text{cells}}{mL}$$

6. Seed the cells in suitable dilution, e.g., 1 × 10⁶ cells/T75 flask and incubate under standard cell culture conditions until approximately 80–90% confluent.

3.2 Cell Differentiation

1. Coat 60-mm cell culture dishes with PDL. Add 2–4 mL PDL solution into each dish and make sure to cover the complete surface. Store dishes overnight or for at least 2 h at 4 °C. Remove PDL solution and let the dishes dry under sterile conditions (*see* **Note 1**).
2. Harvest cells as described in Subheading 3.1, **step 4**.
3. Seed 200,000 cells/cm² in standard cell culture medium into PDL-coated 60-mm dishes.
4. Remove medium after 24 h, wash the cells with 1 mL PBS, and add 4 mL cell culture medium supplemented with 10 μM retinoic acid. Culture the cells in differentiation medium for 4–5 weeks at standard conditions and change medium every 3–4 days. In the second week of differentiation, add 10 μM of the mitotic inhibitor uridine to the medium in order to prevent proliferation of nondifferentiating cells (*see* **Notes 2** and **3**).

3.3 Calcium Phosphate Transfection

1. Coat 60-mm cell culture dishes with Matrigel solution. Cover the surface of the dishes with Matrigel-DMEM mixture and remove excessive solution. Incubate the dishes for about 15 min at 37 °C followed by room temperature (RT) until seeding of cells (*see* **Note 4**).
2. Separate neuronal clusters from nondifferentiated cells. To this end, remove the medium from dishes prepared as described in Subheading 3.2, wash with 2 mL PBS, add 1 mL Trypsin-EDTA solution, and incubate at RT for 2–3 min. Tap dish carefully to detach cell clusters from monolayer of nondifferen-

tiated cells (*see* **Note 5**). Add slowly 4 mL cell culture medium, transfer medium with neuronal clusters to a 15 mL conical centrifuge tube, and pipette up and down to separate the neurons from clusters. Seed the cells into a Matrigel-coated 60-mm cell culture dish and incubate for 24 h under standard culture conditions (*see* **Note 6**).

3. Approximately 1 h prior to transfection, remove medium of the 60-mm cell culture dish and add 3 mL fresh cell culture medium.
4. Prepare 150 μL DNA solution containing 0.25 M $CaCl_2$ and 9 μg YFP-I152L-encoding vector in water.
5. Add dropwise 150 μL of 2× HBS (final 1×) to DNA solution and incubate for 30 min at RT.
6. Add mixture dropwise to cells and incubate at standard culture conditions for about 12 h (*see* **Notes 7** and **8**).
7. Remove transfection medium from the cells and wash three times with 2 mL PBS (*see* **Note 9**). Add 3 mL culture medium and incubate for further 12 h at standard culture conditions.

3.4 Functional Imaging

1. Coat 96-well plate with Matrigel solution according to Subheading 3.3, **step 1**.
2. Prepare NT2-N-YFP-I152L cells for functional imaging. Harvest cells as described above using adapted volumes for 60-mm dishes (2 mL PBS, 1 mL Trypsin-EDTA solution, 4 mL cell culture medium). Determine cell number as described above (*see* **Note 10**), seed cells in a Matrigel-coated 96-well plate in a density of 20,000 cells per well and incubate for 24 h at standard culture conditions (*see* **Notes 11–13**).
3. Approximately 30 min prior to functional imaging experiments, remove medium, wash cells two times with 50 μL NaCl solution, add 50 μL control solution, and incubate under standard conditions (*see* **Note 14**).
4. During incubation, supplement NaI test solution with 10 mM glycine and serially dilute with test solution to obtain agonist solutions containing 0.01, 0.03, 0.1, 0.3, 1, 3, 10, 30, 100, 300, 1,000, 3,000 and 10,000 μM final agonist concentration.
5. Place the 96-well plate onto a motorized stage of a fluorescence imaging system.
6. Record initial fluorescence intensity of cells in control situation, add 100 μL of test solution with desired agonist concentration to the cells, and image each well for further 30 s after receptor activation (*see* **Note 15**).
7. For analyzing cell-based responses, select YFP-I152L-expressing cells and measure the fluorescent signal as the mean of all pixel values within the area of a cell before and 30 s after the addition of test solution.

8. Calculate fluorescence quench for each cell by using the following equation:

$$\%\text{Fluorescence quench} = \left(F_{\text{init}} - F_{\text{final}}\right) \times \frac{100}{F_{\text{init}}}$$

where F_{init} and F_{final} are the initial and final values of fluorescence, respectively. Determine the mean value of fluorescence quench as well as standard deviation for all agonist concentrations.

9. For calculating the half-maximal activation concentration (EC_{50} value), fit concentration-response relationships using a Hill equation.
10. If desired, repeat experiment with test solution supplemented with EC_{50} agonist concentration and different concentrations of test compounds (*see* **Note 16**).
11. Calculate half-maximal inhibition concentration (IC_{50} value) as described in **steps 7–9**.

4 Notes

1. For reusing PDL solution, store PDL at 4 °C during incubation of dishes and after removing from 60-mm dish.
2. For long-term storage we recommend storing RA in aliquots at −80 °C protected from light.
3. If the medium is yellowish after a short time during the first 4 days of differentiation, change medium every day.
4. Prepare Matrigel-coated cell culture dishes always fresh prior to seeding cells. For reusing Matrigel solution we recommend using ice-cold tips for pipetting and cooled dishes as well as 96-well plates; otherwise Matrigel may form a gel at room temperature.
5. Don't incubate too long with Trypsin-EDTA solution to avoid detaching of undifferentiated cells and dissociation of neuronal clusters.
6. For transient transfection, it is easier to combine the neuronal cells of four differentiation dishes in a single Matrigel-coated 60-mm dish.
7. Avoid areas with high concentration of transfection solution because pure transfection solution is toxic for cells.
8. It is not necessary to incubate exactly 12 h, incubation overnight is also suitable.
9. Wash carefully in order to prevent flushing the neurons away.

10. It is not necessary to dilute cells 1:10 prior to counting as described in Subheading 3.1 when cell density is low but recommended when the cell density is very high.
11. We recommend functional imaging 48 h upon transfection as YFP-I152L-expression of YFP is higher compared to shorter incubation times.
12. Functional imaging could also be conducted in 384-well plates in order to increase experimental throughput.
13. In our experience, 1/3 to 1/2 of a 96-well plate can be equipped with cells at the given number from four individual differentiation dishes.
14. If necessary, e.g., in the case of many dead cells, remove control solution immediately prior to imaging and replace by fresh control solution.
15. We recommend using 1,000 μL tips for pipetting as the solution can be added at a higher speed and with less pressure preventing detachment of cells from the bottom of the plate.
16. We recommend applying a broad range of concentrations for test compounds with unknown toxicity profile.

Acknowledgments

The authors gratefully acknowledge funding of the Staedtler Stiftung and Bavarian Equal Opportunities Sponsorship—Förderung von Frauen in Forschung und Lehre (FFL)—Promoting Equal Opportunities for Women in Research and Teaching. Mofrad SA performed the present work in fulfillment of the requirements for obtaining the degree Dr. rer. biol. hum. at the Friedrich-Alexander University of Erlangen-Nürnberg (FAU).

References

1. Lynch JW (2004) Molecular structure and function of the glycine receptor chloride channel. Physiol Rev 84:1051–1095
2. Lynch JW, Callister RJ (2006) Glycine receptors: a new therapeutic target in pain pathways. Curr Opin Invest Drug 7:48–53
3. Webb TI, Lynch JW (2007) Molecular pharmacology of the glycine receptor chloride channel. Curr Pharm Des 13:2350–2367
4. Meier JC, Henneberger C, Melnick I et al (2005) RNA editing produces glycine receptor alpha 3(P185L), resulting in high agonist potency. Nat Neurosci 8:736–744
5. Eichler SA, Kirischuk S, Juttner R et al (2008) Glycinergic tonic inhibition of hippocampal neurons with depolarizing GABAergic transmission elicits histopathological signs of temporal lobe epilepsy. J Cell Mol Med 12:2848–2866
6. Eichler SA, Forstera B, Smolinsky B et al (2009) Splice-specific roles of glycine receptor alpha 3 in the hippocampus. Euro J Neurosci 30:1077–1091
7. Xiong W, Cui T, Cheng K et al (2012) Cannabinoids suppress inflammatory and neuropathic pain by targeting α3 glycine receptors. J Exp Med 209:1121–1134
8. Harvey RJ, Depner UB, Wassle H et al (2004) GlyR alpha 3: an essential target for spinal PGE2-mediated inflammatory pain sensitization. Science 304:884–887
9. Zeilhofer HU (2005) The glycinergic control of spinal pain processing. Cell Mol Life Sci 62:2027–2035

10. Bode A, Lynch JW (2014) The impact of human hyperekplexia mutations on glycine receptor structure and function. Molec Brain 7:2–2
11. Chung SK, Vanbellinghen JF, Mullins JG et al (2010) Pathophysiological mechanisms of dominant and recessive GLRA1 mutations in hyperekplexia. J Neurosci 30:9612–9620
12. Hall RCW, Hall RCW (1999) Long-term psychological and neurological complications of lindane poisoning. Psychosomatics 40:513–517
13. Islam R, Lynch JW (2012) Mechanism of action of the insecticides, lindane and fipronil, on glycine receptor chloride channels. Br J Pharmacol 165:2707–2720
14. Kuenzel K, Friedrich O, Gilbert DF (2016) A recombinant human pluripotent stem cell line stably expressing halide-sensitive YFP-I152L for GABAAR and GlyR-targeted high-throughput drug screening and toxicity testing. Front Mol Neurosci 9:51
15. Talwar S, Lynch JW, Gilbert DF (2013) Fluorescence-based high-throughput functional profiling of ligand-gated ion channels at the level of single cells. PLoS One 8:e58479
16. Alexander SP, Peters JA, Kelly E et al (2015) The concise guide to pharmacology 2015/16: ligand-gated ion channels. Br J Pharmacol 172:5870–5903
17. Vale C, Fonfra E, Bujons J et al (2003) The organochlorine pesticides γ-hexachlorocyclohexane (lindane), α-endosulfan and dieldrin differentially interact with GABAA and glycine-gated chloride channels in primary cultures of cerebellar granule cells. Neuroscience 117:397–403
18. Narahashi T (2002) Nerve membrane ion channels as the target site of insecticides. Mini-Rev Med Chem 2:419–432
19. Mohamed F, Senarathna L, Percy A et al (2004) Acute human self-poisoning with the N-phenylpyrazole insecticide fipronil—a GABAA-gated chloride channel blocker. J Toxicol, Clin Toxicol 42:955–963
20. Lee VM, Andrews PW (1986) Differentiation of NTERA-2 clonal human embryonal carcinoma cells into neurons involves the induction of all three neurofilament proteins. J Neurosci 6:514–521
21. Pleasure SJ, Page C, Lee VM (1992) Pure, postmitotic, polarized human neurons derived from NTera 2 cells provide a system for expressing exogenous proteins in terminally differentiated neurons. J Neurosci 12:1802–1815
22. Sandhu JK, Sikorska M, Walker PR (2002) Characterization of astrocytes derived from human NTera-2/D1 embryonal carcinoma cells. J Neurosci Res 68:604–614
23. Stewart R, Christie VB, Przyborski SA (2003) Manipulation of human pluripotent embryonal carcinoma stem cells and the development of neural subtypes. Stem cells 21:248–256
24. Ozdener H (2007) Inducible functional expression of Bcl-2 in human astrocytes derived from NTera-2 cells. J Neurosci Method 159:8–18
25. Coyne L, Shan M, Przyborski SA et al (2011) Neuropharmacological properties of neurons derived from human stem cells. Neurochem Int 59:404–412
26. Laurenza I, Pallocca G, Mennecozzi M et al (2013) A human pluripotent carcinoma stem cell-based model for in vitro developmental neurotoxicity testing: effects of methylmercury, lead and aluminum evaluated by gene expression studies. International journal of developmental neuroscience: the official journal of the International Society for. Dev Neurosci 31:679–691
27. Gao L, Lyons AR, Greenfield LJ Jr (2004) Hypoxia alters GABAA receptor function and subunit expression in NT2-N neurons. Neuropharmacology 46:318–330
28. Neelands TR, Greenfield LJ Jr, Zhang J et al (1998) GABAA receptor pharmacology and subtype mRNA expression in human neuronal NT2-N cells. J Neurosci 18:4993–5007
29. Munir M, Lu L, Wang YH et al (1996) Pharmacological and immunological characterization of N-methyl-D-aspartate receptors in human NT2-N neurons. J Pharmacol Exp Therap 276:819–828
30. Galietta LJ, Haggie PM, Verkman AS (2001) Green fluorescent protein-based halide indicators with improved chloride and iodide affinities. FEBS Lett 499:220–224
31. Balansa W, Islam R, Fontaine F et al (2010) Ircinialactams: subunit-selective glycine receptor modulators from Australian sponges of the family Irciniidae. Bioorg Med Chem 18:2912–2919
32. Balansa W, Islam R, Fontaine F et al (2013) Sesterterpene glycinyl-lactams: a new class of glycine receptor modulator from Australian marine sponges of the genus Psammocinia. Org Biomol Chem 11:4695–4701
33. Balansa W, Islam R, Gilbert DF et al (2013) Australian marine sponge alkaloids as a new class of glycine-gated chloride channel receptor modulator. Bioorg Med Chem 21:4420–4425
34. Gebhardt FM, Mitrovic AD, Gilbert DF et al (2010) Exon-skipping splice variants of excitatory amino acid transporter-2 (EAAT2) form heteromeric complexes with full-length EAAT2. J Biol Chem 285:31313–31324
35. Gilbert D, Esmaeili A, Lynch JW (2009) Optimizing the expression of recombinant alphabetagamma GABAA receptors in HEK293 cells for high-throughput screening. J Biomolec Screen 14:86–91

36. Gilbert DF, Islam R, Lynagh T et al (2009) High throughput techniques for discovering new glycine receptor modulators and their binding sites. Front Molec Neurosci 2:17
37. Gilbert DF, Wilson JC, Nink V et al (2009) Multiplexed labeling of viable cells for high-throughput analysis of glycine receptor function using flow cytometry. Cytomet Part A 75:440–449
38. Kruger W, Gilbert D, Hawthorne R et al (2005) A yellow fluorescent protein-based assay for high-throughput screening of glycine and GABAA receptor chloride channels. Neurosci Lett 380:340–345
39. Walzik MP, Vollmar V, Lachnit T et al (2015) A portable low-cost long-term live-cell imaging platform for biomedical research and education. Biosens Bioelectron 64:639–649
40. Wilhelm F, Winkler U, Morawski M et al (2011) The human ubiquitin C promoter drives selective expression in principal neurons in the brain of a transgenic mouse line. Neurochem Int 59:976–980
41. Abolpour Mofrad S, Kuenzel K, Friedrich O et al (2016) Optimizing neuronal differentiation of human pluripotent NT2 stem cells in monolayer cultures. Developm Grow Differen 58:664–676

Chapter 17

Systematic Cell-Based Phenotyping of Missense Alleles

Aenne S. Thormählen and Heiko Runz

Abstract

Sequencing of the protein-coding genome, the exome, has proven powerful to unravel links between genetic variation and disease for both Mendelian and complex conditions. Importantly, however, the increasing number of sequenced human exomes and mapping of disease-associated alleles is accompanied by a simultaneous, yet exponential increase in the overall number of rare and low frequency alleles identified. For most of these novel alleles, biological consequences remain unknown since reliable experimental approaches to better characterize their impact on protein function are only slowly emerging.

Here we review a scalable, cell-based strategy that we have recently established to systematically profile the biological impact of rare and low frequency missense variants in vitro. By applying this approach to missense alleles identified through cohort-level exome sequencing in the low-density lipoprotein receptor (*LDLR*) we are able to distinguish rare alleles that predispose to familial hypercholesterolemia and myocardial infarction from alleles without obvious impact on LDLR levels or functions. We propose that systematic implementation of such and similar strategies will significantly advance our understanding of the protein-coding human genome and how rare and low frequency genetic variation impacts on health and disease.

Key words High-throughput functional profiling, Missense alleles, RNAi, Exome sequencing, *LDLR*, Rare-variant association studies, RVAS

1 Introduction

Since it has first been shown to reliably map protein-coding variation across the entire human genome, sequencing of the exome, the ~1.5% of the human genome that encode for proteins, has inevitably transformed human genetics. Whole-exome sequencing (WES) has now become a routine tool to unravel and better understand the molecular basis of monogenetic diseases and it is increasingly being applied in clinical diagnostic settings [1]. Moreover, WES has further demonstrated its value when being applied to large clinical cohorts and extreme families where it has helped to associate multiple protein-coding low frequency and rare variants to complex human traits and diseases, thereby massively increasing the resolution of genome-wide association studies (GWAS) [2].

Daniel F. Gilbert and Oliver Friedrich (eds.), *Cell Viability Assays: Methods and Protocols*, Methods in Molecular Biology, vol. 1601, DOI 10.1007/978-1-4939-6960-9_17, © Springer Science+Business Media LLC 2017

However, with WES being conducted in an ever-increasing number of individuals and across all ethnicities, the overall number of human alleles identified through sequencing by far exceeds the number of variants that can reliably be linked to a distinct physiological function or human disease [3]. To a large degree this owes to the fact that methods that would allow to thoroughly distinguish alleles that disrupt protein functions from the overwhelming number of alleles without obvious functional consequences are still in their infancies. This is particularly true for the missense variants as the largest and clinically most relevant group of human alleles [4]. Our inability to distinguish relevant from less relevant genetic variation poses a considerable bottleneck for contemporary human genetics.

We here review an experimental strategy that may help to partially address this problem and that is based on the systematic parallelized testing of missense alleles for a biological consequence in cultured cells. We have applied this strategy to thoroughly characterize the impact of WES-identified rare and low frequency missense alleles in the low-density lipoprotein receptor gene (*LDLR*) on LDLR functions in cells [5]. We further validated our in vitro findings through comparison with clinical parameters from >3000 individuals, specifically with plasma lipid levels and the incidence of myocardial infarction. Through merging our experimental with clinical data, we could massively increase the power of rare-variant association testing that had previously linked rare variation in *LDLR* to the risk of myocardial infarction [6].

Our experimental strategy includes two separate, but complementary workflows that are high throughput, unbiased, and quantitative. First, we use systematic overexpression of GFP-tagged cDNAs encoding human *LDLR* to compare the uptake of fluorescent-labeled LDL into cells expressing mutated *LDLR*-GFP relative to wild-type-*LDLR*. Second, the same experiments are conducted under a "complementation" setting where endogenous *LDLR* is knocked down with siRNA prior to overexpression. Then, it is tested by how much cells expressing the siRNA-resistant wild-type-*LDLR* or mutated *LDLR*-GFP are able to reconstitute LDLR function. Automated high-content automated microscopy joint with customized multiparametric computational image analysis routines allows to selectively quantify *LDLR* expression and LDL uptake from large numbers of GFP-positive and GFP-negative cells. Definition of distinct cellular phenotypes makes it possible to relate the tested genetic variants to the mechanisms by which LDLR function is impaired. When applying this strategy to large numbers of exome-identified rare missense alleles we could objectively distinguish disruptive missense alleles with a high likelihood to be relevant for disease from variants without obvious consequence on LDLR functions. This allowed us to independently validate Mendelian missense mutations that had previously been described as causing Familial Hypercholesterolemia (FH), but also to identify variants previously described as of unclear significance as very probably causing FH. Importantly, carriers of *LDLR* missense alleles classified by our experimental strategy as disruptive on protein functions showed significantly higher plasma LDL levels as well as a higher risk for early-onset myocardial infarction than carriers of alleles classified as neutral. Our systematic in vitro variant profiling strategy has proven powerful to functionally characterize rare and low frequency missense alleles in *LDLR* and can be applied in a high-throughput manner to any other gene for which gene function can be monitored in cells.

2 Materials

1. Bovine Serum Albumin (BSA): stock solution: 20%, sterile filter after dissolving and store at 4 °C.
2. Cellprofiler 2.0 software.
3. Dapi: Bisbenzimide trihydrochloride (1 mg/ml).
4. DiI-LDL: Low-Density Lipoprotein from human plasma, DiI complex (1 mg/ml).
5. Dimethylsulfoxid (DMSO).
6. Draq5: (5 mM).
7. Dulbecco's Modified Eagle Medium (DMEM): Low Glucose (1 g/l), with L-Glutamine, High Glucose (4 g/l), with L-Glutamine.
8. Fetal Bovine Serum: FBS.
9. Filipin III: dissolved in 1 mg/ml DMF.
10. Hela Kyoto: strongly adherent epithelioid cervix carcinoma cells (gift from S. Narumiya, Kyoto University, Japan).
11. HPCD: 2-Hydroxy-propyl-beta-cyclodextrin.
12. Image J 1.46r.
13. Imaging Medium: Minimum Essential Medium Eagle Modified with Non-Essential Amino Acids and Earl's Salts, without L-Glutamine, Phenol Red, Sodium Carbonate: Dissolve 1 vial of powder with about 700–800 ml Millipore water, adjust pH to 7.2, add 30 mM Hepes (30 ml of 1 M Stock Solution), and fill up to 1 l with H_2O for pH 3.5, add HCl.
14. LDLR siRNA: (s224006), primer sequences: 5′-cagcgaagatgcgaagata-3′ (forward), 5′-tatcttcgcatcttcgctg-3′ (reverse).
15. Lipofectamine 2000.
16. Oligofectamine.
17. Olympus IX81 automated microscope using an UPlanApo 2 × 0.7 NA objective.

18. Opti-MEM R I + GlutaMAX TM I.
19. Paraformaldehyde (PFA).
20. Phosphate-buffered saline (PBS): 80 g/l NaCl, 2 g/l KCl, 2 g/l KH2PO4, 17.7 g/l Na2HPO4 × 2 H_2O, Medium was adjusted to ph 7.4 and autoclaved.
21. Plasmid DNA Purification NucleoBond Xtra Midi Plus.
22. QuikChange Lightning Site-Directed Mutagenesis Kit.
23. ScanR software vs. 2.1.0.15.
24. Scramble siRNA: (s229174), 5′-ttctccgaacgtgtcacgt-3′ (forward), 5′-acgtgacacgttcggagaa-3′ (reverse).

3 Methods

Reviewed are the workflow for cell-based experiments, high-content microscopy, and multiparametric image analysis. For description of genetic data and statistical approaches used to empower rare-variant association testing, please refer to our publication [5].

3.1 cDNA Cloning, siRNAs, and Site-Directed Mutagenesis

Any protein-coding *LDLR* missense variant to be tested is introduced into a wild-type *LDLR*-cDNA (transcript variant: NM_000527.4; ENST00000558518, Ensembl73) carboxy-terminally linked to EGFP using site-directed mutagenesis. The construct was adapted from previous studies within the laboratory and has been verified to show similar properties as endogenous *LDLR* [7]. Since *LDLR*-siRNAs used for complementation experiments target not only expression of endogenous *LDLR* but also overexpressed wild-type constructs, cDNAs have to be modified to confer siRNA resistance. For this, three mutations at wobble bases not affecting protein formation (c. A1053G, c. C1056T, and c. A1059G) are inserted into the *LDLR* template. This is performed within the 19-nucleotide consensus sequence (5′-cagcgaagatgcgaagata-3′) of *LDLR*-siRNA s224006 (Applied Biosciences) using site-directed mutagenesis: primer sequences: 5′-ctggtggcccagcgaaggtgtgaggatatcgatgagtgtca-3′ (forward) and 5′-tgacactcatcgatatcctcacaccttcgctgggccaccag-3′ (reverse). The cDNA construct which corresponds to wild-type *LDLR* but is resistant against *LDLR*-siRNA is named *LDLR*' in the following. Comparability of *LDLR*' and *LDLR* was systematically assessed using Western blots (for details *see* [5, 7]).

For mutagenesis of cDNA constructs QuikChange Lightning Site-Directed Mutagenesis Kit is used following manufacturer's instructions. Primers are designed using a web-based QuikChange Primer Design Program and ordered from Metabion. Every sample of cDNA is individually sequenced to verify successful mutagenesis before they are included into the following experiments.

3.2 Overexpression, Complementation, and Biological Assays

Figure 1 illustrates the two workflows used for systematic missense variant profiling under either an overexpression or complementation setting. Individual steps are numbered and described in the following.

1. Cells are seeded on glass coverslips in 12-well plates at a density of 4×10^4 cells per well and cultured in DMEM, 2 mM L-glutamine, 10% FBS overnight at 37 °C, and 5% CO_2 (*see* **Note 1**).
2. For overexpression 2 mg cDNA per well are fluid-phase transfected using Lipofectamine 2000 following manufacturer's instructions. Each 12-well plate includes three controls: one negative control without cDNA only containing Lipofectamine ("Mock"), one positive control transfected with the *LDLR*'-wild-type, and one well transfected with a cDNA only containing GFP.
3. Transfection reagents are replaced by medium after 4–6 h and cells are incubated using DMEM and 0.2% BSA overnight (*see* **Note 2**).

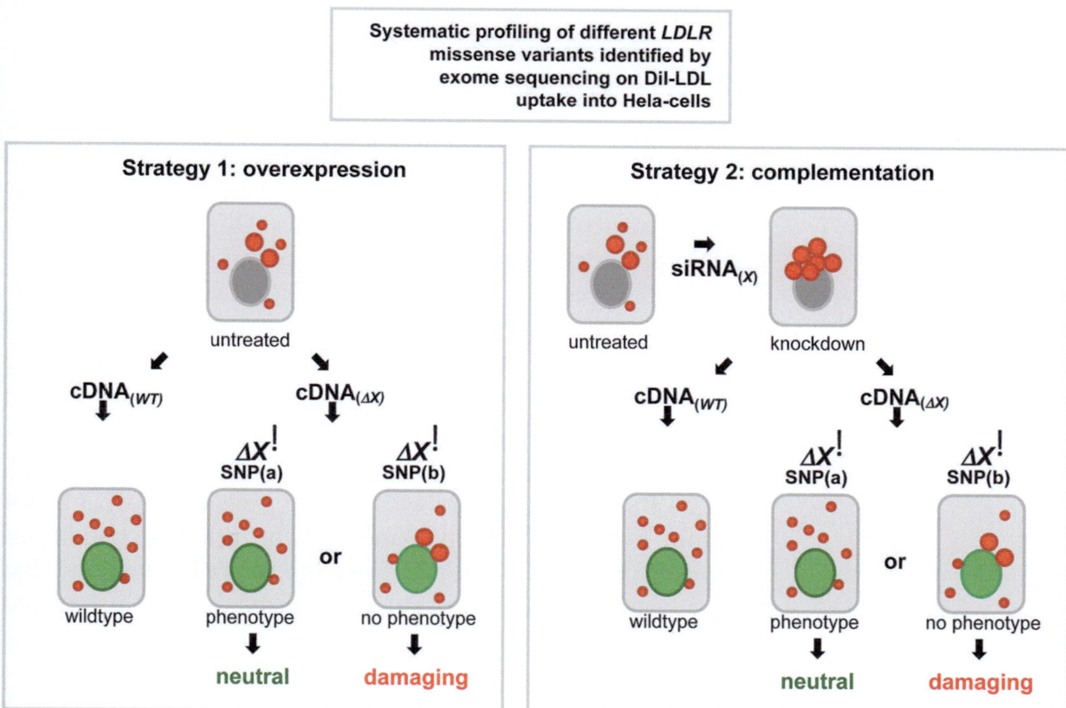

Fig. 1 Experimental strategies: cDNA overexpression and siRNA interference followed by cDNA complementation. Overexpression: Cells are treated with mutated cDNAs containing respective rare missense alleles of *LDLR* (ΔX). Phenotypes including the amount of DiI-LDL internalized as well as the intracellular accumulation of the dye are then compared to cells expressing the *LDLR* wild-type (WT). Complementation: Prior to cDNA treatment cellular *LDLR* expression is knocked down using a silencing siRNA, which induces a LDLR deficient phenotype. Overexpression of cDNAs containing *LDLR* missense alleles is then tested on successful reconstitution of LDL uptake when compared to *LDLR* wild-type. Abbreviation: SNP: Single nucleotide polymorphism

4. For complementation an additional step includes siRNA transfection 1 day before cDNA transfection is performed as described. For this, cells seeded on coverslips on the same density are fluid-phase transfected with 0.5 μl per well of 30 μM *LDLR*-siRNA (s224006) using Oligofectamine following manufacturer's instructions. In addition to all three controls used for overexpression, a fourth well is treated with a different nonsilencing siRNA (s229174, "scramble") and no cDNA serving as control to monitor successful knockdown when compared to "Mock." Cells are incubated for 4 h before DMEM and 30% FBS is added and incubated overnight. Transfection of cDNAs is performed the following day as described for overexpression.
5. All cells are then equally tested on uptake of fluorescently labeled DiI-LDL. After incubation overnight without FBS, 10 mg/ml HPCD (3-Hydroxypropyl-cyclodextrin) is added to each well and incubated for another 45 min (*see* **Note 3**). During this time the oligosaccharides bind excess cholesterol within the medium. This sensitizes cells for the subsequent DiI-LDL uptake step, as it exposes cells to cholesterol-depleting conditions and stimulates cell homeostatic responses aiming for enhanced cholesterol internalization.
6. Following HPCD exposure, cells are washed for 2 min with Imaging medium containing 0.2% BSA and incubated for 30 min with DiI-LDL on ice (*see* **Note 4**). DiI-LDL is diluted at a ratio of 1:800 in Imaging medium using 1.25 μg per coverslip. Incubating cells on ice induces DiI-LDL binding to LDL receptors at the plasma membrane but prohibits its internalization. For a 15 mm diameter coverslip 80 μl of DiI-solution is dropped on a plate covered by parafilm. Coverslips are put on top of these droplets with cells facing down. Cells are incubated for another 20 min at 37 °C in order for internalization of DiI-LDL to occur (*see* **Note 5**).
7. Following up, cells are washed once with Imaging medium +0.2% BSA for 2 min and with Imaging medium (pH 3.5) for 1 min both on ice (*see* **Note 6**). The acidified conditions remove excess DiI-LDL particles from the cell membrane that have been bound, but not yet internalized due to the lowered temperature that blocks clathrin-mediated endocytosis of the LDL-LDLR complex.
8. After washing again with Imaging medium +0.2% BSA and Imaging medium both for 2 min on ice, cells are fixed using paraformaldehyde 3% (PFA) for 20 min.
9. Nuclei and cell outlines are stained using Draq5 (diluted 1:750 in PBS) for 5 min and Dapi (diluted 1:1000 in PBS) for 10 min at room temperature (*see* **Note 7**).
10. Finally, coverslips are mounted in 6 μl Mowiol on microscope slides (*see* **Note 8**).

11. All cells are systematically analyzed using automated microscopy. Images are acquired with a ScanR-System (version 2.1.0.15) using an Olympus IX 81 wide-field microscope. This includes a MT20 Xenon lamp as well as an UPlanApo 20 × 0.7 objective. Samples of images of all different wavelengths analyzed can be seen in Figure 2 ("original"). The autofocus uses the nuclear stain by Hoechst "Dapi" (excitation: 350 nm, emission 447 nm), to first focus broadly (± 4 μm, 6 steps) and then finely (±0.83 μm, 6 steps). For 12-well experiments pictures are taken at 30 spots of each coverslip. At each position the microscope takes pictures at four different wavelengths. Besides a first image of nuclear staining (Dapi), Cy5 (excitation: 620 nm, emission: 700 nm) is used to visualize cytoplasms. Additionally, the Cy3 channel (excitation: 545 nm, emission: 610 nm) is used to show DiI-LDL and GFP2 (excitation: 470 nm, emission: 525 nm) to distinguish transfected from untransfected cells (*see* **Note 9**).

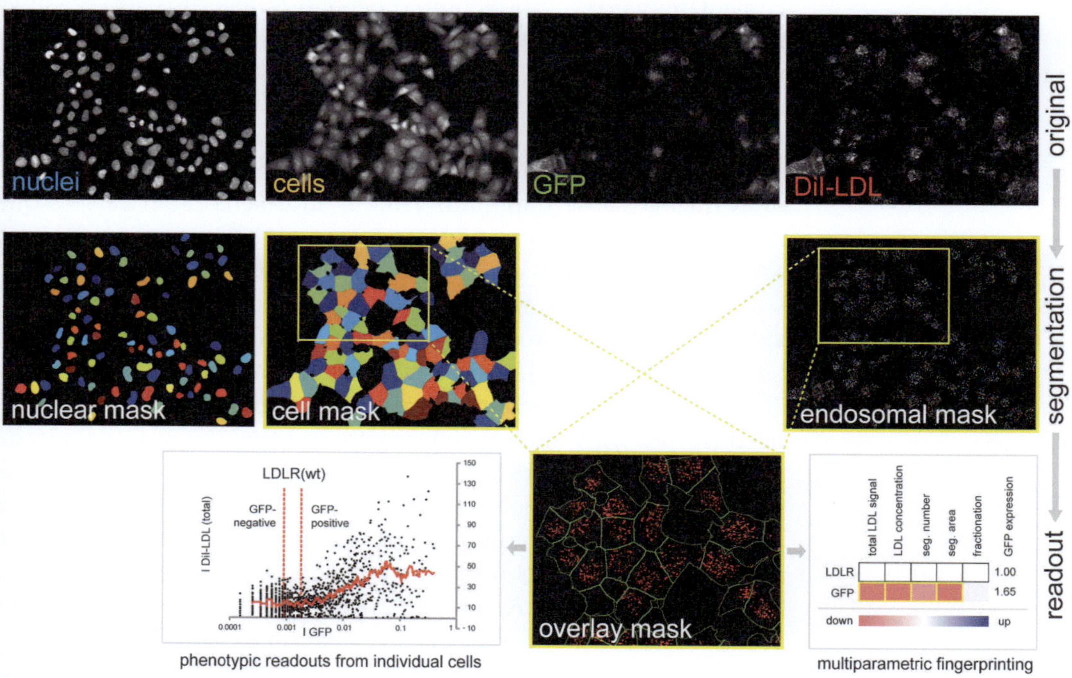

Fig. 2 Systematic image analysis and automated quantification of phenotypes. Cellprofiler 2.0 is used to automatically identify nuclei (nuclear mask), cytoplasms of cells (cellular mask) as well as spots of fluorescently labeled DiI-LDL (endosomal mask). Overlay of images containing cytoplasms and endosomes makes it possible to relate all LDL endosomes to respective cells. Number of endosomes and intensity of DiI-LDL signal among others can then be individually quantified for each cell. GFP expression of cells is used to determine which ones have been successfully transfected by cDNAs. Comparison of DiI-LDL uptake of GFP-positive versus GFP-negative cells shows impact of *LDLR*-cDNA constructs. A multiparametric fingerprint illustrates effects of respective alleles within five parameters using colors (*red*: downregulating when compared to *LDLR* wild-type, *blue*: upregulating when compared to *LDLR* wild-type)

3.3 Image Data Analysis

To systematically analyze the large amount of images a specific Cellprofiler 2.0 pipeline with the following modules was developed based on previous work within the laboratory [7]. Figure 2 illustrates the concept behind systematic image analysis.

After rescaling the intensity of all images, nuclei are detected using Dapi images (Identify Primary Objects). Each nucleus is then systematically extended with the help of the corresponding Cy5 image (Identify Secondary Objects) to recognize corresponding cytoplasms. Taking detected cytoplasm of cells as a reference, Cellprofiler only quantifies those DiI spots and GFP signals that refer to corresponding cell masks (Mask Image, Measure Image Intensity). DiI Spots are enhanced (Enhance or Suppress Features) in order to produce images largely composed of the objects of interest. From these intensified images, another module (Identify Primary Objects) can identify DiI spots within cells separately. Once DiI spots and cytoplasms are identified by Cellprofiler several parameters such as signal intensity (Measure Object Intensity), size or shape (Measure Object Size Shape), number and area of objects (Calculate math) can be acquired beyond others. It is also possible to quantify intensity of GFP, total DiI as well as enhanced DiI spots within cytoplasms. All data is related to corresponding cytoplasms of cells (Relate Objects) and finally exported into a spreadsheet, which can be further processed using R-Studio or Excel.

Cellprofiler measures GFP-background signal of each well individually whose minimum value is subtracted afterwards from all pictures of the respective treatment. Cells are then filtered according to cell size and area of DiI spots compared to total cell size. Boundaries for filtering are adjusted for each data set individually, excluding too small apoptotic as well as too large cell clusters. To distinguish between transfected and untransfected cells GFP signal of transfection-control cells "Mock," which have only been treated with transfection reagent but no cDNA, is measured. The value that includes 97% of those cells is set as lower threshold value. Cells below this value are classified as being untransfected and therefore not including cDNAs. Cells with a GFP signal that is larger than twice the threshold are classified as being cDNA transfected. Because *LDLR*-cDNAs within complementation experiments exceed endogenous *LDLR* function already at about 20% GFP expression levels, a supplemental upper threshold is introduced to exclude those cells expressing DiI-LDL more than endogenously normal. This reassures that complementation only takes restored LDLR functionality into account and does not exceed endogenous *LDLR* levels. For this additional threshold, a mean value of three components is determined. This includes (1) 5 times the DiI-LDL uptake of cells treated with *LDLR*'-siRNA and transfection mix only, (2) 5 times the DiI-LDL uptake of cells transfected with GFP alone "GFP-control" and *LDLR*'-siRNA, and (3) 1.25 times the DiI-LDL uptake of cells treated with "scramble" and transfection mix only. This serves as upper boundary for DiI-LDL signal of the set of transfected cells with all cells exceeding this threshold being excluded from the following analysis. From all data obtained mean values for cell area, cell number, DiI total signal, DiI mean signal, number of DiI spots, area of DiI spots, intensity of DiI spots compared to DiI signal of the whole cell, GFP total signal and GFP mean signal are calculated. These measurements are compiled for both transfected and untransfected cells of experimental replicas individually and consequently related to all specific treatments.

3.4 Statistical Analysis of Imaging Data

For all experiments five parameters are quantified (1) total signal intensity of segments above background within cell masks ("total LDL signal"), (2) mean intensity of segments per cell ("LDL concentration"), (3) number of segments within cell masks ("seg. Number"), (4) area of segments within cell masks ("seg. Area"), and (5) mean GFP expression. To quantify the impact of respective variants on LDL uptake, ratios of all parameters of transfected versus untransfected cells are determined. An experiment is regarded as successful if at least data of 25 transfected cells fulfilling all quality criteria can be acquired. All mean values of parameters for each variant and all experiments are compared to all values of wild-type LDL receptor and regarded as significant when a paired, two-tailed Student's *t*-test results in *p*-values below 0.05. A variant is categorized as disruptive missense or likely impairing protein functionality if the ratio of transfected to untransfected cells within overexpression differs significantly from the LDL receptor in at least three parameters including total signal as well as the corresponding z-score deviation of total DiI-LDL is below one. Alternatively, variants are regarded as likely benign if none of all ten parameters of both overexpression and complementation reach significance at 5% level. Variants not belonging to either group are labeled as being of unclear significance.

Figure 3 shows sample images of variants classified into different functional categories. First, cells transfected with wild-type *LDLR*'-constructs show increased DiI-LDL uptake as illustrated by bright DiI-LDL spots within GFP-positive cells (positive control). In contrast, cells transfected with GFP-control only show slight DiI-LDL uptake from endogenous *LDLR* expression not resulting in bright intracellular DiI spots (negative control). Most importantly, variants previously described as causing FH show significantly decreased DiI-LDL uptake of GFP-positive cells in our screen (FH-mutant G549D) almost resembling GFP-negative control. Several more variants show similar functional characteristics and are subsequently classified as "disruptive missense" likely being of disease relevance. On the contrary, many other variants show similar properties as wild-type *LDLR*' and are hence classified as "nondisruptive" by our analyses. Using the cell-based

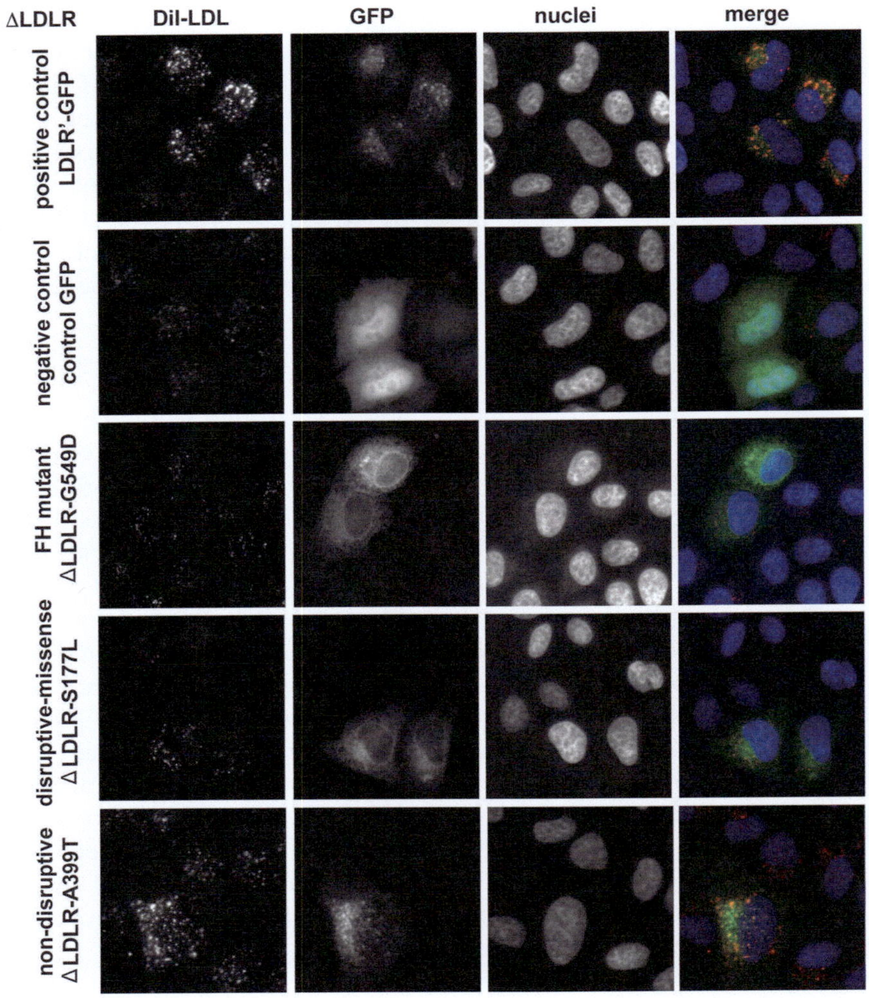

Fig. 3 Sample images from automated microscopy. For all treatments pictures at different wavelengths are obtained illustrating DiI-LDL spots, GFP-transfected cells as well as cytoplasms and nuclei of cells. Cells transfected with wild-type-LDLR'-constructs show successful DiI-LDL uptake as illustrated by intracellular fluorescently labeled LDL spots (positive control). Untransfected cells (negative control) only show slight endogenous LDL uptake. Cells transfected with known FH mutants or alleles with previously unknown functional characteristics can then be classified as either "disruptive missense" or "nondisruptive" according to individual LDL uptake in vitro

approach described it is therefore not only possible to validate alleles previously known to cause FH, but also to classify previously unknown rare missense alleles on likely functional relevance. A full list of variants screened can be found in our publication [5].

3.5 Conclusions

The cell-based missense variant functionalization strategy reviewed here enabled us to systematically profile the consequence of rare and low frequency *LDLR* missense alleles on LDL receptor functions. Importantly, it allowed us to distinguish missense alleles with a high probability to cause Familial Hypercholesterolemia from alleles without a measurable impact on protein and activities. Following this strategy, novel alleles that are being identified through clinical testing for FH or large-scale genome sequencing can be tested to conclude on disease relevance and validate hypotheses from human genetics.

While we could demonstrate that this approach sufficiently discriminates between disruptive and nondisruptive alleles in *LDLR*, the question arises how easily it can be transferred to other genes, for instance genes linked to alterations in plasma lipids or predisposition to myocardial infarction through family-based studies or GWAS [7]. In principle, given a human-derived cell model and an in vitro reporter assay that well translates to changes in a trait or disease predisposition in vivo, any protein-coding gene can be subjected to cDNA overexpression and siRNA interference. As opposed to CRISPR or knockout mice RNA interference may prove particularly powerful where knockdown needs to be fine-tuned, for instance when analyzing essential genes, or to reflect phenotypes that result from moderate inhibition, as opposed to complete absence of a gene or its function. When compared to other genes and conditions of high medical relevance, major advantages of studying lipid metabolism are that the underlying biology has been very thoroughly studied, that LDLR function can be directly measured through efficiency of LDL uptake, and that plasma LDL levels directly correlate with disease phenotypes (ranging from familial hypercholesterolemia to myocardial infarction). It can be assumed that it will be more challenging to assess significance for human conditions from in vitro experimentation if the correlation between genotype, biomarker, and clinical endpoint is less direct, or mechanisms are not as well understood.

Overall, we successfully applied the described strategy to functionally profile 70 *LDLR* missense alleles identified through exome sequencing of 3235 case and control subjects from an early-onset myocardial infarction cohort with known plasma lipid levels. We could identify 14 alleles as disruptive on LDLR functions, likely causing Familial Hypercholesterolemia. Importantly, a majority of 46 alleles did not show disruptive characteristics and were subsequently regarded as likely benign or of no functional significance. The remaining 10 variants ranked in between were classified as of unclear significance. Most importantly, carriers of alleles classified as likely disease causing showed significantly higher plasma LDL as well as early-onset MI when compared to likely benign variation. Results therefore directly correlate with clinical parameters commonly used to estimate cardiovascular risk [5].

As for any novel approach, it is of interest how well our strategy compares to existing methods to functionally test *LDLR* missense alleles, particularly protocols to individually test novel *LDLR* alleles as potentially causative for FH in diagnostic settings.

Noteworthy, our systematic profiling confirmed all previously known FH mutants present in the studied cohort, and also more than 90% of alleles for which impairment of receptor function had been observed in fibroblasts [5]. However, while it classifies a rate of 20% of alleles as disruptive missense and is therefore consistent with previous predictions [8] other sources routinely used in clinical practice annotate up to 50% of *LDLR* missense alleles as being of likely disease relevance, suggesting high false-positive rates. This includes the Human Genome Mutation Database (HGMD) as well as several of four frequently applied bioinformatic prediction tools. Conversely, the threshold we chose to assign *LDLR* missense alleles as disruptive may have been too conservative, and more alleles than classified by us as such may result in classical FH. Refined clinical follow-up, particularly segregation studies in families, should help to further substantiate our findings, as any finding that is based on in vitro experiments.

It can be expected that cell-based approaches like ours will become an integral feature of the genetics of the future that will necessitate to make sense of the multiple private mutations that every sequenced individual carries. At present, genetics is lacking a versatile strategy to unambiguously assess a variant's relevance in order to find the few high-risk alleles that cause disease or predispose to common traits. Similar to our strategy focusing on hypercholesterolemia/myocardial infarction and *LDLR*, few other studies have used cell-based assays to systematically extract evidence on the biological consequences of missense variants from genetic data, among these for Cystic Fibrosis and *CFTR* [9], as well as for type-2-diabetes and PPARG [10] or MTNR1B [11]. This reflects an increasing recognition of unbiased high-throughput strategies to systematically characterize variants in the face of a flood of variants from an increasing numbers of human genomes. It can be expected that these types of studies will contribute significantly to our understanding of genetic variation underlying both Mendelian and complex human diseases.

4 Notes

1. It is extremely important to seed cells evenly all over the coverslip in order to ensure successful automated microscopy in the following.
2. Cells often detach from the coverslip when being cDNA transfected for 6 h. This is why the medium was preferably replaced after 4 h.
3. Incubation with HPCD should not exceed 45 min in order to ensure comparability in between experimental replicas.
4. Due to light sensitivity of the dye all following steps should be performed under light protected conditions.

5. It is very important to adhere to the exact timing as one cycle of LDL internalization usually occurs within about 20 min.
6. The most important step to keep experiments precise is the exact timing of each part of the DiI-LDL uptake. Especially, cells should not be exposed to the acidified Imaging Medium for more than a minute.
7. Dilutions of Draq5 and Dapi staining may have to be adjusted according to individual samples from different suppliers.
8. One major challenge using this method is to apply it at a high-throughput level. For this, it is necessary to start several experiments in parallel using more than one 12-well plate at a time. However, by using several 12-well plates in parallel for all previous steps, only the uptake assay has to be performed with little time difference between plates making it possible to screen many variants in parallel. Overexpression experiments are performed in at least three different experimental replicates. Additional complementation experiments lead to at least four tests per variant.
9. The time of light exposure is adjusted to each image set and channel individually. A systematic quality control of pictures ensures discarding those with inadequate cell densities or of minor technical quality.

References

1. Green RC, Rehm HL, Kohane IS (2013) Chapter 9—Clinical genome sequencing. In: Genomic and personalized medicine, 2nd edn. Academic, San Diego, pp 102–122
2. Lange LA, Hu Y, Zhang H, Xue C, Schmidt EM, Tang ZZ, Bizon C, Lange EM, Smith JD, Turner EH, Jun G, Kang HM, Peloso G, Auer P, Li KP, Flannick J, Zhang J, Fuchsberger C, Gaulton K, Lindgren C, Locke A, Manning A, Sim X, Rivas MA, Holmen OL, Gottesman O, Lu Y, Ruderfer D, Stahl EA, Duan Q, Li Y, Durda P, Jiao S, Isaacs A, Hofman A, Bis JC, Correa A, Griswold ME, Jakobsdottir J, Smith AV, Schreiner PJ, Feitosa MF, Zhang Q, Huffman JE, Crosby J, Wassel CL, Do R, Franceschini N, Martin LW, Robinson JG, Assimes TL, Crosslin DR, Rosenthal EA, Tsai M, Rieder MJ, Farlow DN, Folsom AR, Lumley T, Fox ER, Carlson CS, Peters U, Jackson RD, van Duijn CM, Uitterlinden AG, Levy D, Rotter JI, Taylor HA, Gudnason V Jr, Siscovick DS, Fornage M, Borecki IB, Hayward C, Rudan I, Chen YE, Bottinger EP, Loos RJ, Saetrom P, Hveem K, Boehnke M, Groop L, McCarthy M, Meitinger T, Ballantyne CM, Gabriel SB, O'Donnell CJ, Post WS, North KE, Reiner AP, Boerwinkle E, Psaty BM, Altshuler D, Kathiresan S, Lin DY, Jarvik GP, Cupples LA, Kooperberg C, Wilson JG, Nickerson DA, Abecasis GR, Rich SS, Tracy RP, Willer CJ (2014) Whole-exome sequencing identifies rare and low-frequency coding variants associated with LDL cholesterol. Am J Hum Genet 94(2):233–245
3. Kiezun A, Garimella K, Do R, Stitziel NO, Neale BM, McLaren PJ, Gupta N, Sklar P, Sullivan PF, Moran JL, Hultman CM, Lichtenstein P, Magnusson P, Lehner T, Shugart YY, Price AL, de Bakker PI, Purcell SM, Sunyaev SR (2012) Exome sequencing and the genetic basis of complex traits. Nat Genet 44(6):623–630
4. Richards CS, Bale S, Bellissimo DB, Das S, Grody WW, Hegde MR, Lyon E, Ward BE (2008) ACMG recommendations for standards for interpretation and reporting of sequence variations: Revisions 2007. Genet Med 10(4):294–300
5. Thormaehlen AS, Schuberth C, Won HH, Blattmann P, Joggerst-Thomalla B, Theiss S, Asselta R, Duga S, Merlini PA, Ardissino D, Lander ES, Gabriel S, Rader DJ, Peloso GM, Pepperkok R, Kathiresan S, Runz H (2015)

Systematic cell-based phenotyping of missense alleles empowers rare variant association studies: a case for LDLR and myocardial infarction. PLoS Genet 11(2):e1004855

6. Do R, Stitziel NO, Won HH, Jorgensen AB, Duga S, Angelica Merlini P, Kiezun A, Farrall M, Goel A, Zuk O, Guella I, Asselta R, Lange LA, Peloso GM, Auer PL, Girelli D, Martinelli N, Farlow DN, DePristo MA, Roberts R, Stewart AF, Saleheen D, Danesh J, Epstein SE, Sivapalaratnam S, Hovingh GK, Kastelein JJ, Samani NJ, Schunkert H, Erdmann J, Shah SH, Kraus WE, Davies R, Nikpay M, Johansen CT, Wang J, Hegele RA, Hechter E, Marz W, Kleber ME, Huang J, Johnson AD, Li M, Burke GL, Gross M, Liu Y, Assimes TL, Heiss G, Lange EM, Folsom AR, Taylor HA, Olivieri O, Hamsten A, Clarke R, Reilly DF, Yin W, Rivas MA, Donnelly P, Rossouw JE, Psaty BM, Herrington DM, Wilson JG, Rich SS, Bamshad MJ, Tracy RP, Cupples LA, Rader DJ, Reilly MP, Spertus JA, Cresci S, Hartiala J, Tang WH, Hazen SL, Allayee H, Reiner AP, Carlson CS, Kooperberg C, Jackson RD, Boerwinkle E, Lander ES, Schwartz SM, Siscovick DS, McPherson R, Tybjaerg-Hansen A, Abecasis GR, Watkins H, Nickerson DA, Ardissino D, Sunyaev SR, O'Donnell CJ, Altshuler D, Gabriel S, Kathiresan S (2015) Exome sequencing identifies rare LDLR and APOA5 alleles conferring risk for myocardial infarction. Nature 518(7537):102–106
7. Blattmann P, Schuberth C, Pepperkok R, Runz H (2013) RNAi–based functional profiling of loci from blood lipid genome-wide association studies identifies genes with cholesterol-regulatory function. PLoS Genet 9(2):e1003338
8. Zuk O, Schaffner SF, Samocha K, Do R, Hechter E, Kathiresan S, Daly MJ, Neale BM, Sunyaev SR, Lander ES (2014) Searching for missing heritability: designing rare variant association studies. Proc Natl Acad Sci U S A 111(4):E455–E464
9. Sosnay PR, Siklosi KR, Van Goor F, Kaniecki K, Yu H, Sharma N, Ramalho AS, Amaral MD, Dorfman R, Zielenski J, Masica DL, Karchin R, Millen L, Thomas PJ, Patrinos GP, Corey M, Lewis MH, Rommens JM, Castellani C, Penland CM, Cutting GR (2013) Defining the disease liability of variants in the cystic fibrosis transmembrane conductance regulator gene. Nat Genet 45(10):1160–1167
10. Majithia AR, Flannick J, Shahinian P, Guo M, Bray MA, Fontanillas P, Gabriel SB, Rosen ED, Altshuler D (2014) Rare variants in PPARG with decreased activity in adipocyte differentiation are associated with increased risk of type 2 diabetes. Proc Natl Acad Sci U S A 111(36):13127–13132
11. Bonnefond A, Clement N, Fawcett K, Yengo L, Vaillant E, Guillaume JL, Dechaume A, Payne F, Roussel R, Czernichow S, Hercberg S, Hadjadj S, Balkau B, Marre M, Lantieri O, Langenberg C, Bouatia-Naji N, Charpentier G, Vaxillaire M, Rocheleau G, Wareham NJ, Sladek R, McCarthy MI, Dina C, Barroso I, Jockers R, Froguel P (2012) Rare MTNR1B variants impairing melatonin receptor 1B function contribute to type 2 diabetes. Nat Genet 44(3):297–301

Chapter 18

Second Harmonic Generation Microscopy of Muscle Cell Morphology and Dynamics

Andreas Buttgereit

Abstract

Microscopy in combination with contrast-increasing dyes allows the visualization and analysis of organs, tissues, and various cells. Because of their better resolution, the development of confocal and laser microscopes enables the investigations of cell components, which are labeled with fluorescent dyes. The imaging of living cells on subcellular level (also in vivo) needs a labeling by gene transfection of GFP or similar labeled proteins. We present a method for visualization of cell structure in skeletal and heart muscle by label-free Second Harmonic Generation (SHG) microscopy and describe analytic methods for quantitative measurements of morphology and dynamics in skeletal muscle fibers.

Key words Second harmonic generation, Skeletal muscle, Multiphoton microscopy, Image processing

1 Introduction

The exploration of cell biology began with the development of light microscopy in the early sixteenth century. Since the microscope has become one of the most important tools in biology and medicine, there are many different microscopy techniques. One of the microscopy techniques is the laser scanning microscope which scans thick biological samples point by point [1, 2]. The use of pulsed near infrared lasers (NIR) increases the penetration depth (optical window of biological tissues) of the scanning laser and enabled new physical effects for visualization of cell components (nonlinear optical microscopy, NLOM).

The first effect is multiphoton fluorescence. Two or more photons are absorbed simultaneously by a fluorescent dye. These effects can only be detected in volumes with a high density of photons, i.e., in the focus of an objective lens. Therefore, the fluorescent dye is excited only in this point and practically, nowhere else (optical pinhole effect and less bleaching of fluorescence dye) [3].

The other effect is higher harmonic generation. The strong electrical field of a high density volume of photons interacts with

Daniel F. Gilbert and Oliver Friedrich (eds.), *Cell Viability Assays: Methods and Protocols*, Methods in Molecular Biology, vol. 1601, DOI 10.1007/978-1-4939-6960-9_18, © Springer Science+Business Media LLC 2017

electrons in a nonlinear material to generate new photons with double (second harmonic generation, SHG) or triple (third harmonic generation, THG) the frequency of the excitation laser beam (half or third of wavelength). Similar to the multiphoton fluorescence, we obtain only a signal from the focus, so it is possible to compose a three-dimensional image. Nonlinear materials showing such behavior in biological tissue are collagen [4], myosin, and tubulin [5].

The visualization of myosin (type II) enables the analysis of ultrastructure in skeletal muscle fibers [6] and cardiomyocytes. Myosin, the motor protein in striated muscle tissue, is located in sarcomeres, the smallest functional unit responsible for force generation. Therefore, knowledge about the ultrastructure of myosin related to the whole muscle fiber (morphology) allows predictions on the dynamics of muscle fibers [6, 7].

This chapter describes the important components of a multiphoton microscope for optimized SHG recordings, the preparation of muscle tissue or single fibers, and some tools for quantitative analysis of ultrastructure in muscle fibers.

2 Materials

2.1 Tools and Solutions for Muscle Preparation for SHG Microscopy

1. Ringer's Solution (physiological saline, pH = 7.4): 140 mM NaCl, 2 mM $CaCl_2$, 1 mM $MgCl_2$, 10 mM HEPES, 4 mM KCl, 5 mM Glucose.
2. High-potassium Solution (HPS, pH = 7.0): 140 mM monopotassium glutamate, 10 mM $MgCl_2$, 10 mM HEPES, 1 mM EGTA, 10 mM Glucose.
3. Two Dumont tweezers.
4. 35 mm petri dish, silicone coated 50 mm petri dish, needles.
5. Sylgard 184 Silicone Elastomer Kit.
6. Micro (eye) scissor (spring type).
7. Collagenase Type IA or Type II.
8. Fire polished Pasteur pipette.
9. Suitable recording chamber with cover slips.
10. Formaldehyde or glutaraldehyde for fixation.
11. Ice, Vaseline.
12. Water bath, lab shaker.
13. Binocular preparation microscope.

2.2 Microscope

A multiphoton microscope is similar to a confocal laser scanning microscope without pinhole aperture. The point spread function (PSF) and by association the spatial resolution of multiphoton microscopy is primarily dependent on laser wavelength and the type of objective lens according to

$$d = \frac{\lambda}{2NA},$$

where d is the resolution limit, λ the laser wavelength, and NA = $n \sin \alpha$ the numerical aperture of the objective lens.

2.2.1 Laser Sources

The Second Harmonic Generation (SHG) effect requires a high density of photons and therefore, a high power laser source. But the interaction between a laser and a biological sample can lead to unwanted reactions like photothermal, photodisruptive, or photochemical effects, which may change or even destroy the biological sample. The solution for this problem is using a short pulse laser (ps or fs). This kind of laser emits enough energy for the SHG effect in small pulses and the time between the pulses is needed for thermal relaxation. Dependent on the optical path and the different parts like mirrors or filters in it as well as the quantum efficiency of the detector, for a sufficient contrast, I recommend a mode-locked laser which generates a minimal power of 50 mW after the objective lens (*see* **Note 1**).

2.2.2 Objective Lens

The choice of the right objective lens depends on various conditions. Therefore, it is important to review the properties of lenses (*see* **Note 2**).

1. Numerical aperture (NA):

   The NA describes the range of angles α in which the light is focused and collected. Resolution d and NA are inversely proportional to each other and therefore a high NA leads to a high resolution. But on the other hand, however, a high range of angles permits only a small working distance (the distance between the objective lens and sample).

2. Air, water, or oil:

   The other factor in the equation for NA is the refractive index n. Because the refractive index of air is unity, using water ($n = 1.33$) or oil immersion ($n = 1.52$) increases resolution (*see* **Note 3**).

3. Transmission:

   For SHG or THG imaging, it is necessary to use an objective lens (and the other optical components) with a wide spectral range for light transmission up to infrared radiation.

In our experience, the intensity of SHG signal from myosin is higher in forward scattering direction. Therefore, using two objective lenses with the same (or similar) features (one lens for laser focusing and one lens for signal collection) in thin samples enables images with higher contrast. In addition to this fact, some investigations show that the forward and the backward scattered images are significantly different from each other [8].

2.2.3 Half-Wave Plate

The intensity of SHG signal of myosin is dependent on the polarization of the excitation laser beam related to the myofibril orientation [9]. To ensure the highest possible contrast, it is advantageous to install a rotatable lambda half-wave plate into the beam path. In addition to this, a rotatable lambda half-wave plate enables investigations of polarization dependency of SHG signal.

2.2.4 Filter/Dichroic Mirrors

The usage of a short-pulsed laser allows the simultaneous excitation of a broad range of fluorescent dyes. Therefore, we need filters and dichroic mirrors to separate the SHG signal from other signals like autofluorescence. Because the SHG signal is exactly half the wavelength of the incident laser light, we can use a bandpass filter with a small transmission range (e.g. 20 nm). The choice of additional dichroic mirrors or filters is dependent on the available number of photomultipliers and fluorescent dyes (*see* **Note 4**).

2.3 Image Processing with Fiji

Most microscopes have a software with numerous analytical algorithms. In addition to this, we need special applications for our analyses.

Fiji is an open-source software for image processing based on ImageJ. There are many plugins available that enable visualization and analysis of images from different origins. Fiji can also be used in combination with other image processing applications (e.g. Imaris) and is capable of reading out metadata from images.

3 Methods

3.1 Muscle Preparation

For single fiber preparation we need a binocular preparation microscope (*see* **Note 5**).

3.1.1 Enzymatic Dissociation to Obtain Living Single Fibers

An enzymatic dissociation is a gentle method to separate single fibers from the muscle for investigations on living cells. For the analysis of ultrastructure in muscle fibers, it is the preferred method and recommended for small muscle fibers like *Mm. interossei.*

1. Prepare your chosen muscle in cold Ringer's solution (4 °C or on ice) and remove as much as possible the connective tissue.
2. Transfer the muscle to a small petri dish with Ringer's solution containing collagenase. The preferred concentrations are 1–3 mg/ml and with an incubation time of 30–45 min at 37 °C (*see* **Note 6**).
3. Stop the enzyme action with three times washing with cold Ringer solution.

4. Triturate the muscle with a fire-polished Pasteur pipette until no large muscle fragments are visible. The result of a correct enzymatic treatment are many almost transparent single fibers on the bottom of the petri dish (*see* **Note 7**).
5. Transfer any number of single fibers together with Ringer solution to a recording chamber with a cover slip. To prevent drying out, cover the chamber with a second cover slip with a small spacer in between.

3.1.2 Mechanic Isolation of Living Single Fibers

A new method in muscle research is the combined investigation of muscle force compared to structure within the muscle fiber. For this one needs long muscle fibers like *soleus* or *extensor digitorum longus* muscles, which one can clamp in a measurement device including a force transducer. The preparing of single fiber needs experience.

1. Prepare your chosen muscle in cold Ringer solution (4 °C or on ice) and remove as much as possible the connective tissue.
2. Pin the muscle on a silicon coated petri dish with HPS solution.
3. Remove carefully single fibers or fiber bundles with tweezers from muscle.
4. Transfer the fibers to a recording chamber with Ringer solution. Fix the ends of muscle fibers with some Vaseline strips.

3.1.3 Mechanic Isolation of Fixed Single Fibers

At the end of the preparation of living cells (Subheadings 3.1.1 and 3.1.2) one can add an aqueous fixing solution like formaldehyde (1–3.7%) or glutaraldehyde (0.1–1%) in Ringer's solution.

It is also possible to fix the whole muscle first. The fixation leads to a higher stiffness of muscle fibers, so one may be able to separate single fibers or fiber bundles (e.g., from muscles with much fibrosis).

1. Prepare your chosen muscle in cold Ringer solution (4 °C or on ice) and remove as much as possible the connective tissue.
2. Pin the muscle on an appropriately large cork or polystyrene plate under slight stretch.
3. Put the pinned muscle in a tube with fixing solution, like formaldehyde (1–3.7%) or glutaraldehyde (0.1–1%) in Ringer's solution, at least overnight at 4 °C.
4. Store the fixed muscle in fixing solution at room temperature.
5. Remove and transfer carefully with tweezers single fibers or fiber bundles from muscle to the recording chamber.

3.1.4 Cryosection of Muscle Tissue

One can also investigate tissue sections with the SHG microscopy. However, the slicing procedure leads to small ruptures on the cutting edges, which is counterproductive for structural investigations. Therefore, I recommend to cut thick slices at least 30 μm. Additionally, it is difficult to find the right cutting plane to obtain longitudinal sections of muscle tissue. For cryosections, one can use fixed or unfixed tissue.

- Sliced unfixed muscle tissue has to be fixed immediately with cooled fixing solution before the sample thaws. Otherwise the muscle tissue shrinks and loses its structure. The fixation solution attaches the sample to the slide or cover slip. Cover the wet sample with a cover slip. For a longer storage, seal the borders of cover slip with Entellan® or similar.
- Of course, it is not necessary to fix sliced fixed muscle tissue, but they are not well attached to the slide. If you want to perform a stain with fluorescence dyes, you must make sure that the sections do not float away from slide (*see* **Note 8**).

3.1.5 Muscle Tissue Sections by Vibratome

Because cryosections are only possible up to 50 μm, three-dimensional imaging of frozen muscle tissues is limited. Thicker samples up to 300 μm can be obtained with a vibratome.

1. Prepare your chosen muscle in cold Ringer solution (4 °C or on ice) and remove as much as possible the connective tissue.
2. Pin the muscle on the tendons under slight stretch to a piece of cork or similar.
3. Fix the muscle 24 h (or longer) in 1% formaldehyde at 4 °C.
4. Dissolve a 2.5% Low Melt Agarose in physiological saline.
5. Remove the muscle from cork and transfer it into a small well or dish with warm and liquid agarose solution.
6. The agarose solidifies at 4 °C in a refrigerator (overnight).
7. Remove the agarose muscle from well and trim it if necessary.
8. Note the specification of your vibratome.
9. Clamp the agarose block in the vibratome with the orientation of muscle fibers parallel to the blade and cool the sample with ice.
10. For good results, I recommend a cut thickness of 75–100 μm, a small amplitude (0.6–0.8 mm), and high frequency of vibration with small forward speed (0.1–0.2 mm/s). The settings are also dependent on the kind of blade and blade angle.
11. Transfer the sections onto a slide or cover slip and cover it with a second cover slip.
12. For a longer storage, seal the borders of cover slip with Entellan® or similar.

3.2 SHG Imaging

For best results and the highest contrast in your images, you need the right settings. Many parameters determine the success of the image capture and especially the quantification based on it.

3.2.1 Alignment of Optical Components

Dependent on your microscope setup, you have to check the different optical components for their alignment.

If you use two objective lenses or an additional condenser for collecting light, it is important that they have the same focal point and optical axis. In general, a high end microscope for SHG imaging should have a Köhler illumination with an adjustment for condenser or a second objective lens. Bring a test sample in the focus of the first (fixed) objective lens and partially close the iris diaphragm, so you can see a small spot of the microscope light. Adjust the z-direction of your second lens, so the edges of your iris diaphragm are recognizable and focused. Move the light spot with the x-y-adjustment to the middle of your field. For an improvement of adjustment, it is helpful to open the iris diaphragm up to the borders of your field of view (Fig. 1).

For a homogenous illumination of your sample, it is necessary that the direction of the laser beam is identical with the optical axis of your objective lens. Use an autofluorescent diagnostic slide or a recording chamber filled with a fluorescent dye like fluorescein and determine the consistency and evenness of illumination. If you were not satisfied with the illumination, then the beam path must be corrected. This work should be performed only by an experienced user.

3.2.2 Scan Frequency, Resolution, Laser Power

The scan frequency, the laser power, and the image resolution are inextricably linked to each other. It is your choice which parameter is more important. Using examples, I explain the recommended setting.

The choice of image resolution is dependent on the used objective lens as well as the scale of distinguishable structures. For example, the physical resolution of an objective lens with 1.1 NA is, depending on the wavelength, about 0.3 μm. That means, this objective lens is suitable for imaging individual sarcomeres in skeletal muscle, because the sarcomere length is about 2 μm. An image

Fig. 1 Performing an adjustment of optical pathway with Köhler illumination (*left* to *right*). Close and focus the iris diaphragm. Adjust the light spot to the *middle* of field of view

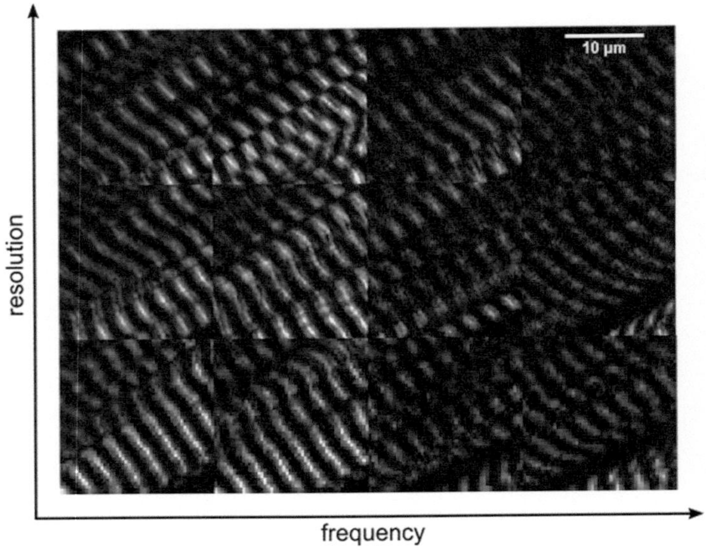

Fig. 2 Effect of frequency and resolution to a SHG signal from skeletal muscle. The distance between M-lines is 2.9 μm (sarcomere length, stretched). The resolution is between 0.1 (*top*) and 0.4 (*below*). Because the physical resolution is limited (about 0.3 μm, 1.1 NA), there are no differences in the image quality of *middle* and *top* column. A higher scan frequency leads to deficits in brightness or contrast

resolution of 0.2 μm per pixel (middle columns in Fig. 2) is also suitable for this sort of imaging, and increasing of the image resolution entails no improvement.

The scan frequency and the image resolution result in the frame time. A scan frequency of 1400 Hz means that 1400 lines per seconds are recorded. The frame time of an image with 256 × 256 px with this frequency is about 200 ms and a long time for dynamic processes. On the other hand, an image with 1024 × 1024 px with a scan frequency of 200 Hz needs about 5 s but provides more details and a higher contrast. Therefore, fast imaging is only possible with a loss of image details, but it enables imaging of dynamic processes. The loss of brightness and contrast can be compensated for by increasing laser power. High resolution imaging requires patience (Fig. 2).

3.3 Measurement of Polarization Dependency

Myosin is a motor protein. The head of myosin slides along the actin filament to generate force. This walk is possible because the head has two different states, relaxed and rigor, which have an effect on the polarization dependence of SHG signal.

1. Record SHG image series from myofibrils or muscle fibers under rotation of the lambda half-wave plate. The rotation range of 180° with steps of 2° (91 images) is enough, because a rotation angle $\alpha = 1°$ of the lambda half-wave plate changes the polarization angle by $\varphi = 2°$.
2. Load the image series in Fiji (based on ImageJ).

3. Define a region of interest (ROI) with a selection tool in Fiji. This region should include structures with the same orientation.
4. Add this ROI to the *ROI Manager*.
5. Select in the *Set Measurements* of *Analyze* menu *Mean Gray Value*.
6. Open *Multi Measure* in the ROI Manager and activate *measure all 91 slices* and press *OK*.
7. Plot the results (possibly normalize your data for comparison with other measurements). You should get a diagram like Fig. 3. Fit your data with Eq. (1).

$$I_{SHG} = B\left(\left(\sin^2(\varphi - A) + \gamma\cos^2(\varphi - A)\right)^2 + \left(\sin\left(2(\varphi - A)\right)\right)^2\right) + C \quad (1)$$

The parameter A and C are shift parameters in horizontal and vertical direction, and B is a scale factor. These parameters are dependent on microscope settings like laser power, PMT voltage, and offset as well as the starting point of half-wave plate. The interesting parameter is γ, a ratio of tensor components of the second-order susceptibility $\chi^{(2)}$ [10] (Fig. 3).

3.4 Fourier Component Analysis of Muscle Cell Damage and Prediction of Force Loss

The smallest functional and morphological repetitive unit in a muscle fiber and the origin of force generation is the sarcomere. A chain of sarcomeres builds a myofibril. A muscle fiber contains hundreds or thousands of myofibrils. In a healthy muscle fiber, all myofibrils have the same direction (main direction) and are interconnected at the z-disks of sarcomeres. This arrangement of myofibrils leads to

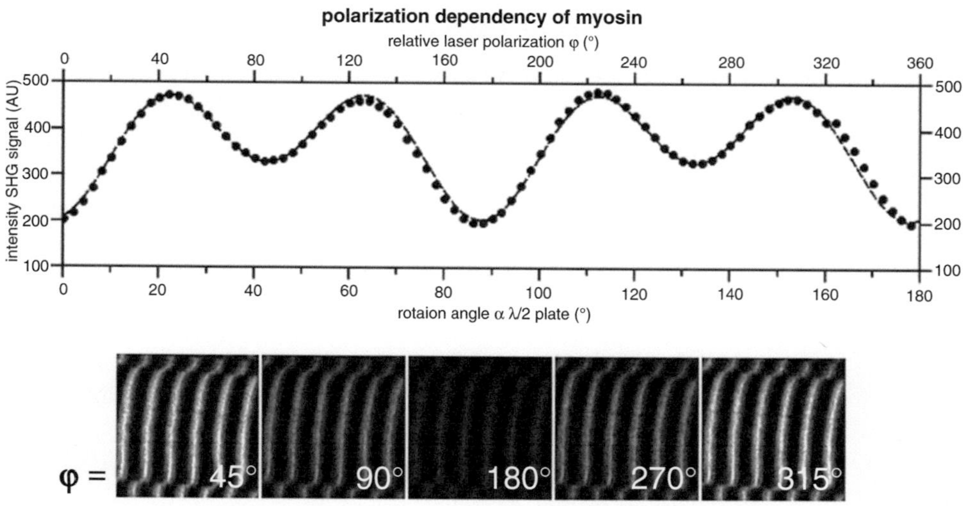

Fig. 3 Plot of polarization-dependent intensity of the SHG signal. Data points were fitted with Eq. (1). The overall minimum is found at φ about 0° and 180°, and a local minimum at about 90° and 270°. The intensity of SHG signal is maximal at φ about 50°, 130°, 230°, and 310°

the visible striation pattern in the conventional microscopy and an optimized force generation in fiber direction.

The high resolution visualization of myosin and different image processing tools allows the quantification of abnormalities. Sarcomeres or myofibrils with deviations from the main direction contribute less to the force generation. Therefore, the analysis of deviations allows a prediction of force loss.

1. Perform 2D or 3D recordings of SHG signal of single fibers.
2. Load the images in Fiji.
3. Open in the *Analyze* menu the *Directionality* plugin [11].
4. Choose *Fourier components*, the number of bin (180°/*n*bin = bin width) and the start angle. Activate *display table* for exporting the data and press *OK*.
5. We obtain a histogram with a fit of normal distribution and a table with fit parameters.
6. The parameter *Dispersion* is the standard deviation σ of normal distribution and describes the deviation of myofibrils to the maximum x_c.

$$y = y_0 + Ae^{\left(-\frac{(x-x_c)}{2\sigma^2}\right)} \quad (2)$$

7. Note: The maximum of normal distribution is not necessarily the main direction of the fiber.
8. For three-dimensional analyses, one can add up all histograms of all slices of a 3D recording and determine the standard deviation of the resulting plot (Fig. 4).

3.5 Biomotoric Efficiency

The myonuclei in a skeletal muscle fiber regulate its metabolic processes. Therefore, the relation between nuclear and myosin volume can be an estimate of biomotoric efficiency of a muscle fiber [12].

1. After preparation of single fibers described in Subheading 3.1, add a nucleus dye to the recording chamber. I recommend ethidium bromide (EtBr) for living cells and propidium iodide (PI) for fixed cells, because the emission spectrum is perfect to be separated from the SHG signal. Hoechst or DAPI is also possible, but it is difficult to find a good combination of laser wavelength, filters, and dichroic mirrors.
2. Incubate the dye at room temperature for at least 30 min.
3. Perform 3D recordings and simultaneously record the multiphoton fluorescence signal of nuclear staining and the SHG signal from myosin with two photomultipliers.
4. Activate *Area fraction* in the *Set Measurements* of *Analyze* menu in Fiji.
5. Open the *ROI Manager* in the *Tools* of *Analyze* menu.

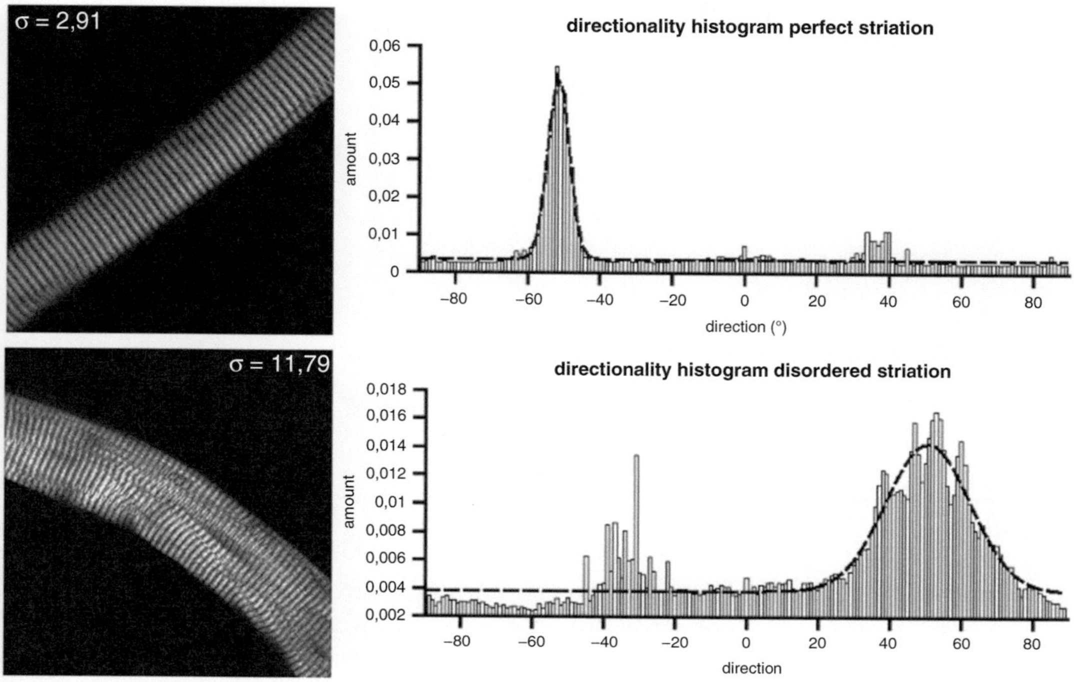

Fig. 4 Analysis of striation directions in muscle fibers with normal distribution fit (*dashed line*). A perfect pattern results in a small peak with low standard deviation ($\sigma = 2.91$). Disordered pattern in the muscle fiber below increases the standard deviation ($\sigma = 11.79$)

6. Load the two channel image stack in Fiji. Split channels in the *Image Color* menu.
7. For one channel, set a threshold in the *Image Adjust* menu to separate the background from the signal. Use the stack histogram and if necessary, an automatic thresholding like *otsu*.
8. Select the ROI in your Image (in the case of one fiber per stack select the whole image) and press Add in the ROI manager. (If necessary delete old ROIs.)
9. Press *More* and *Multi Measure* in the *ROI Manager*, and we obtain a table with the results.
10. Press *Summarize* in the *Results* menu of table and take the mean.
11. Repeat the points 7–10 for the other channel.
12. The means are the percentage of nuclear (NV) or myosin volume (MV) in relation to the total volume of stack.
13. Calculate the biomotoric efficiency with Eq. (3) (Fig. 5).

$$\eta_{mot} = \frac{MV}{MV + NV} \tag{3}$$

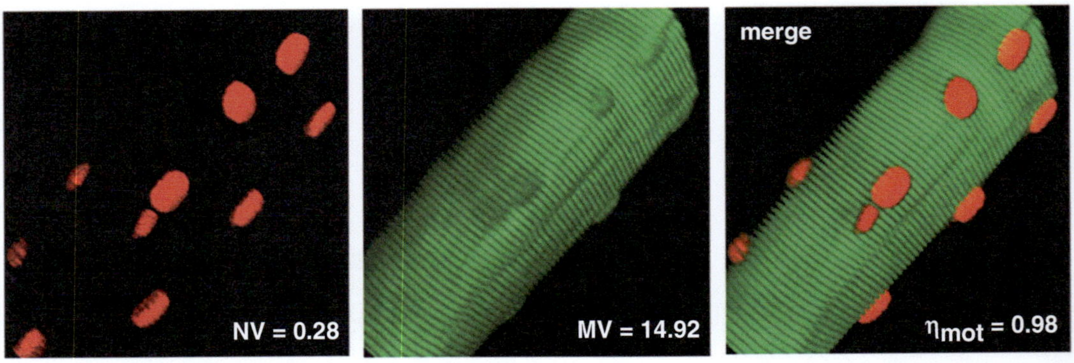

Fig. 5 3D-reconstruction from a muscle fiber performed with the 3D Viewer plugin in Fiji. The volume of nuclei (*red*) and myosin (*green*) was measured by thresholding

4 Notes

1. The real parameter for nonlinear optical effects is the intensity I or the photon flux Φ. The relation between these parameters is $\Phi \cdot hf = I$, with the photon energy $E = hf$. Therefore, the choice of a laser source with a smaller pulse width or the using of an objective lens with a higher NA increases the photon flux in the focus and results in a higher contrast.
2. In most cases, objective lenses with a high NA have a small free working distance (FWD). Because of that, in an inverted microscope, one cannot image samples on a slide or in a normal petri dish or multiwell plate. Use cover slips!
3. It is important in which medium the biological sample is embedded. Different refraction indices in the beam path lead to aberrations. In most cases, our samples are embedded in water or solutions with a similar refraction index; therefore, it is best to use an objective lens which is designed for water immersion.
4. There is another detection method without filters and mirrors. The isolation of fractions of emission spectrum is also possible with prism or grating.
5. In addition to the conventional illumination of a preparation microscope from below or above it is useful to have a lateral illumination to increase the contrast.
6. The type of collagenase, the concentration, and the incubation time for the enzymatic isolation of muscle fibers is dependent on the producer and the activity of enzyme. The change of a batch requires the modification of the incubation time or collagenase concentration. Orientate to the enzyme activity, if given.
7. One can also perform the enzymatic treatment in Tyrode or in phosphate buffered saline (PBS) but not in HPS. The composition of solutions and the pH is important for the success of enzymatic isolation.

8. It is easy to perform cryosections on fixed muscles. Because these samples are not well attached on the slide, one can put the fixed cryosections directly in a staining solution. After the staining procedure gently place the sample on a cover slip or a slide.

References

1. Davidovits P, Egger MD (1969) Scanning laser microscope. Nature 223:831
2. Davidovits P, Egger MD (1971) Scanning laser microscope for biological investigations. Appl Opt 10:1615–1619
3. Denk W, Strickler JH, Webb WW (1990) Two-photon laser scanning fluorescence microscopy. Science 248:73–76
4. Freund I, Deutsch M, Sprecher A (1986) Connective tissue polarity. Optical second-harmonic microscopy, crossed-beam summation, and small-angle scattering in rat-tail tendon. Biophys J 50:693–712
5. Campagnola PJ, Millard AC, Terasaki M, Hoppe PE, Malone CJ, Mohler WA (2002) Three-dimensional high-resolution second-harmonic generation imaging of endogenous structural proteins in biological tissues. Biophys J 82:493–508
6. Friedrich O, Both M, Weber C, Schurmann S, Teichmann MD, von Wegner F, Fink RH, Vogel M, Chamberlain JS, Garbe C (2010) Microarchitecture is severely compromised but motor protein function is preserved in dystrophic mdx skeletal muscle. Biophys J 98:606–616
7. Buttgereit A, Weber C, Garbe CS, Friedrich O (2013) From chaos to split-ups—SHG microscopy reveals a specific remodelling mechanism in ageing dystrophic muscle. J Pathol 229:477–485
8. Légaré F, Pfeffer C, Olsen BR (2007) The role of backscattering in SHG tissue imaging. Biophys J 93:1312–1320
9. Schürmann S, von Wegner F, Fink RH, Friedrich O, Vogel M (2010) Second harmonic generation microscopy probes different states of motor protein interaction in myofibrils. Biophys J 99:1842–1851
10. Both M, Vogel M, Friedrich O, von Wegner F, Kunsting T, Fink RH, Uttenweiler D (2004) Second harmonic imaging of intrinsic signals in muscle fibers in situ. J Biomed Opt 9:882–892
11. Liu ZQ (1991) Scale space approach to directional analysis of images. Appl Opt 30: 1369–1373
12. Buttgereit A, Weber C, Friedrich O (2014) A novel quantitative morphometry approach to assess regeneration in dystrophic skeletal muscle. Neuromuscul Disord 24:596–603

Chapter 19

Assessment of Population and ECM Production Using Multiphoton Microscopy as an Indicator of Cell Viability

Martin Vielreicher and Oliver Friedrich

Abstract

Multiphoton microscopy allows continuous depth-resolved, nondestructive imaging of scaffold-seeded cells during cell or tissue culture. Spectrally separated images in high resolution can be provided while cells are conserved in their native state. Here we describe the seeding of mesenchymal stem cells to bacterial nanocellulose hydropolymer scaffolds followed by 2-channel imaging of cellular autofluorescence (AF) and collagen-I formation using second harmonic generation (SHG) signals. With this approach the simultaneous observation of the progression of cell morphology and production of extracellular matrix as hallmarks of viability and cell fitness is possible.

Key words Multiphoton imaging, Cell viability, Extracellular matrix formation, Collagen-I, SHG, Cellular autofluorescence, Scaffold, Bacterial nanocellulose

1 Introduction

Cell viability assessment has been performed in multiple ways mostly using photometric analysis based on colorimetry, fluorescence, or luminescence and detection of enzymatic activity (mitochondrial and dehydrogenase activity, e.g., MTT assay) [1–3]. For the purpose of imaging researchers relied on results from cell staining and fluorescence microscopy. Classically, live-dead screenings were carried out using a staining protocol with two viability markers that discriminate between live and dead cells [4, 5]. Live cells, for example, take up Calcein AM and transform it into fluorescing Calcein (green) present in the cytoplasm whereas a second marker passively enters dead cells through ruptured plasma membranes and specifically binds to DNA (blue or red fluorescence depending on the selected marker).

Multiphoton imaging (MPI) became increasingly important through the last decade after its power was demonstrated in the imaging not only of tissue sections but also of living cells [6–10]. Apart from the fact that MPI allows to perform the full spectrum

Daniel F. Gilbert and Oliver Friedrich (eds.), *Cell Viability Assays: Methods and Protocols*, Methods in Molecular Biology, vol. 1601, DOI 10.1007/978-1-4939-6960-9_19, © Springer Science+Business Media LLC 2017

of applications of classical (confocal) fluorescence microscopy it adds on the benefits of noninvasiveness, depth penetration, and higher sensitivity [11, 12] (Fig. 1). Moreover, additional applications are possible like the imaging of certain biological macromolecules like collagen-I or -III, myosin-II, or tubulin using SHG [14, 15]. MPI builds upon excitation by two simultaneously incoming near-infrared photons from a pulsed, mode-locked laser.

In the field of tissue engineering the population of scaffolds is a key parameter for successful construct development. Here, on

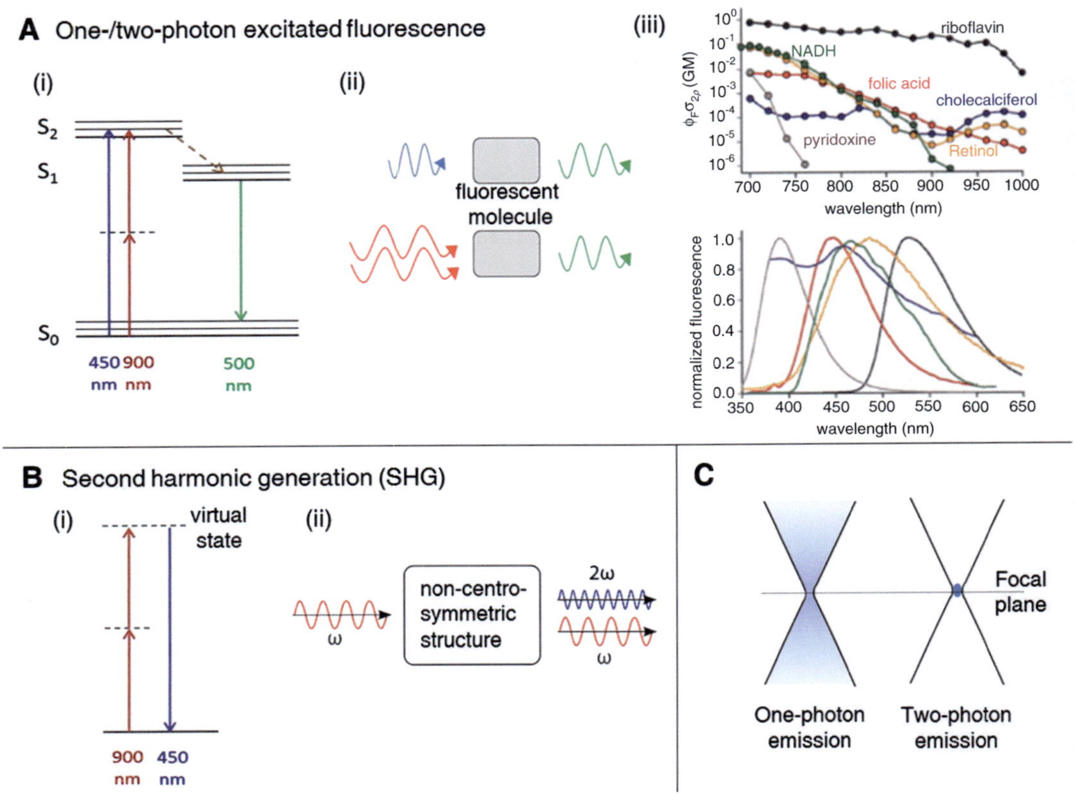

Fig. 1 Principle of multiphoton microscopy. In (**a**) the principle of 2-photon fluorescence is explained. The Jablonski diagrams (*i*) show the energy levels involved and the excitation and emission processes. Other than with classic 1-photon excitation with, for example, 450 nm (*blue*), 2-photon excitation requires 2 near-simultaneous (*ii*) photons of 900 nm (half the energy, *red*). Emission (*green*) is independent on the excitation process. (*iii*) Sources of cellular AF. Two-photon action cross sections of various AF cellular molecules are displayed (*top*) together with the normalized emission spectra (*bottom*) [13] (copyright 2003 National Academy of Sciences). (**b**) Energy diagram of SHG, a nonlinear scattering effect (*i*). No photons are absorbed and no molecular/quantum states are involved in this process. (*ii*) Two photons of 900 nm simultaneously striking a highly ordered, non-centrosymmetric material may induce a virtual energy state transition and the release of a photon of exactly twice the energy of the incoming photons (450 nm). (**c**) In classic 1-photon excitation, fluorescence is generated all along the light path with maximum intensity in the focal plane (*left*). A pinhole is required for blocking out-of-focus fluorescence. Two-photon excitation occurs exclusively in a tiny focal volume (~1 μm^3) causing much less photo-bleaching

the one hand MPI can visualize cell populations by their natural AF [16]. In addition, as tissue formation goes along with extracellular matrix (ECM) formation like the production of collagen-I, ECM imaging is of great benefit for the validation of cell viability (as compared to Fig. 2). Collagen-I is ubiquitously present and the most abundant protein matrix marker in animal tissue [17]. Here, we provide a step-by-step protocol for the preparation, seeding, and growth of mesenchymal stem cells followed by simultaneous high-resolution cell and collagen-I MPI over time (Fig. 3). The described technique is a sophisticated, but highly informative method for cell viability testing of artificial tissues that produce collagen-I (e.g., bone, tendon, ligament, skin, connective tissue, and others).

2 Materials

2.1 Cell Culture

1. Mesenchymal stem cells (MSCs), isolated from rat femur marrow (*see* **Note 1**).
2. Growth medium: DMEM-Ham's F12 (basal medium) containing 20% FBS (fetal bovine serum), 2 mM L-glutamine, and 1× penicillin-streptomycin.
3. Phosphate-buffered saline (PBS), no Ca^{2+} and Mg^{2+}.
4. Trypsin solution, 0.05% trypsin, 0.02% EDTA.

2.2 Bacterial Nanocellulose (BNC) Scaffolds

1. Production: biotechnologically from *Komagataeibacter xylinus* strain DSM 14666 (*see* **Note 2**), synthesis by static cultivation (~7 days) at the interface between air and culture medium resulting in fleeces with a layered structure [18].

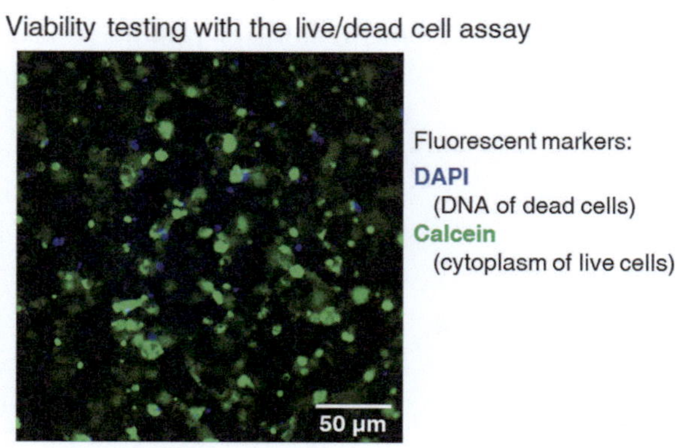

Fig. 2 Live-dead fluorescence imaging assay. Caco-2 cells were seeded on a PET membrane and grown for 2 days, then co-stained with DAPI/Calcein AM and imaged using fluorescence microscopy. Almost all cells are viable (*green*), only a low number of dead cells (*blue*) is found at this early time-point and low cell density

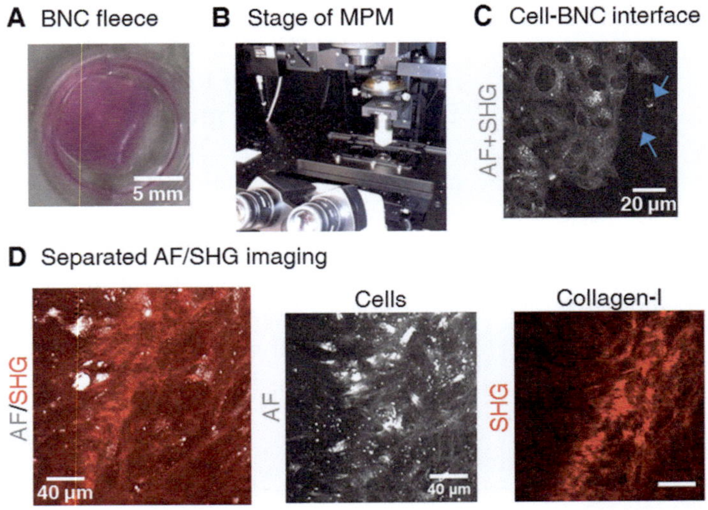

Fig. 3 MPI of MSCs seeded on BNC fleece biomaterial. Biocellic+ hydro biomaterial fleeces (**a**) were equilibrated in medium before MSCs were seeded. (**b**) Stage of the multiphoton microscope where the sample is positioned. Excitation and fluorescence detection occurs from below (*backward direction*). (**c**) An unseparated multiphoton image (AF + SHG) is displayed from the interface between cells and the BNC surface. Cells are visualized by their AF. Weak BNC-specific SHG signals become visible as dots and fibers (*blue arrows*). (**d**) Imaging of collagen-I formation as an indicator of cell viability. Emission is recorded in two channels (AF and SHG). AF (*grey*) and SHG signals (*red*) can be arranged in a two-color merged image (*top*). Cells were cultured over 10 days in a low serum medium with elevated ascorbic acid content

2. Macroscopic features: robust (tensile strength: 0.5–0.6 MPa [19]) and almost transparent material (water content >99%) that can be thermally sterilized.
3. Microscopic constitution: basic units are characteristic cellulose micro-fibrils of uniform thickness and low fiber length variations. Intermolecular hydrogen bonds between cellulose monomers enable the formation of nanostructured fibers (diameter 5–20 nm) and fiber bundles which organize into anisotropic networks with an average pore size of 1 × 3.5 μm and low size variations [20]. BNC consists of both amorphous and highly crystalline regions and shows a SHG pattern [20–22].
4. Biological properties: good biocompatibility, due to its biological nature highly biomimetic.

2.3 Multiphoton Microscopy

1. Glass bottom dishes: sterile, 35 mm diameter. Glass properties: thickness: ≤170 μm (*see* **Note 3**), superior optical quality (low AF, high spectral transmission), excellent flatness, suitable for high-resolution microscopy applications.
2. Microscopic system: high-performance multifocal multiphoton microscope including a multiphoton laser, a beam shaper

(for conditioning the laser pulse shape), a high-speed galvanometric XY-scanner, and a XYZ-stage operated with stepper motors enabling micromechanical movement (x, y, and z drive) (*see* **Note 4**).

3. Laser: femtosecond-pulsed titan-sapphire (Ti:Sa) laser with dispersion pre-compensation and tunable wavelengths of λ = 710–980 nm (near infrared), pulse length <150 fs, repetition rate: 80 MHz (*see* **Note 5**), delivers spatially confined excitation resulting in very thin optical sections (~1 μm) (focal volume: ~1 μm^3, Fig. 1).
4. Optical components:
   (a) Objective lens: 40×, water immersion objective, working distance: 0.62 mm (*see* **Note 6**), high numerical aperture (N.A.: 1.1), high transparency in the UV-vis and IR range.
   (b) Filters: Long-pass dichroic mirror (cut-off wavelength: 460 nm) for separation of wavelengths <460 nm (reflected) from >460 nm (pass through), reflected wavelengths are narrowed to 395–415 nm by a 405/20 bandpass filter (*see* **Note 7**).
   (c) All components optimized for use with near-infrared pulsed light.
5. Detection: signal is exclusively collected in backward direction by high-sensitivity GaAsP photomultiplier detectors (PMTs) mounted in non-descanned configuration close to the back aperture of the objective (*see* **Note 8**).

2.4 Software and Analysis

1. Image acquisition: microscope supplier's software.
2. Data analysis and processing: Fiji software ([23], www.fiji.sc).

3 Methods

3.1 Scaffold Preparation

1. Cut BNC sheets to ~8 × 8 mm patches with a sterilized scalpel (Fig. 3a). Some skill is required as fleeces are robust with a thickness of ~2 mm (highly hydrated material).
2. Add patches to a sterile 50 ml cell culture plastic tube filled with growth medium and equilibrate overnight by slight agitation, e.g., using a test tube rotator (*see* **Note 9**).
3. Place patches in a multi-well plate (24-well format), overlay with growth medium, and store for 15 min in the cell incubator to equilibrate for temperature and pH.

3.2 Cell Preparation for MPI

1. Culture MSCs in a CO_2 incubator at 37 °C, 5% CO_2, and 90–95% relative humidity. MSCs usually have a doubling time of ~24 h.

2. Culture cells in growth medium for a maximum of 3 weeks (requirement for primary cells like MSCs) in polystyrene flasks (tissue culture grade, T-75 and T-25 format). Media must be exchanged every 2–3 days.
3. For splitting and preparation for seeding, wash cells with PBS, detach in trypsin solution (3 ml, 3 min at 37 °C), dilute with 9 ml medium (to stop the action of trypsin) and collect by centrifugation (1000 ×*g*, 3 min). After resuspension of the pellet (2 ml growth medium) add 70 μl of the cell suspension to a cell counting chamber (Neubauer improved, depth: 0.1 mm). Follow your chamber supplier recommendations for determination of cell number and density.

3.3 Seeding and Culture of Cell-BNC Constructs

1. Remove medium from the incubated BNC patches and add cell suspensions (3×10^5 cells/ml). Immediately transfer the plate to the incubator to enable cell adhesion and growth (*see* **Note 10**).
2. When reaching 100% confluence (usually after ~2–3 days) add fresh growth medium or specific reference medium for stimulating enhanced collagen fibril formation (*see* **Note 11**).
3. Imaging at days 1, 2, 3, 5, 7, 10, and 14 is well-chosen time-points to track the course of collagen formation (*see* **Note 12**). Formation of detectable collagen-I fibers, however, is most likely beginning from day 3.
4. Change the medium in advance of imaging if too many floating dead cells are found (*see* **Note 13**).
5. For imaging, flip over the patches carefully in order to access the cells from below for MPI (inverted system). This is an ultimately required step for cells to be located within the objective's working distance.
6. Keep imaging time as short as possible as physical conditions (room temperature, low atmospheric CO_2 content) negatively affect cells resulting in delayed proliferation and growth in continued culture (*see* **Note 14**).

3.4 Multiphoton Imaging

1. For optimized imaging results (low background signal) protect the microscope stage (Fig. 3b) from any diffraction light (Fig. 1a, b, *see* **Note 15**).
2. Use a wavelength λ = 810 nm which is ideal for collagen-I 2-photon excitation. Wavelengths >460 nm are detected in channel 1 (cellular AF); shorter wavelengths get reflected and then filtered with a 405/20 bandpass filter to image SHG in channel 2 (collagen-I). Detection of SHG requires a higher laser power at the sample position than detection of AF (P = ~30 vs. 15–20 mW). Ultra-high-sensitivity PMT's are recommended for its detection.

3. Place a drop of water on top of the objective and position the sample. Focus the sample by using transmission white light or low intensity light from a fluorescence lamp (epi-fluorescence mode).

4. Fine-tune the z-position in scanning mode until signal appears in the acquisition software. Adapt the laser power to signal intensity and optimize PMT voltage enhancement to obtain best signal-to-noise ratios in both detection channels.

5. For fast cell screening in scanning mode select a low image size (e.g., 150 × 150 μm, 512 × 512 pixels). For recording high-quality images larger image sizes (200 × 200 μm, 1024 × 1024 pixels) can be chosen.

6. BNC produces both a characteristic AF and SHG signal. For imaging adapt the z-position until the material surface is reached and the typical signal pattern of adherent cells appears (*see* Subheading 2.1, **Item 1**, Fig. 3c, *see* **Note 16**).

7. In channel 2 SHG signal appears in the shape of fibrous networks. As signals are usually relatively weak at least 4× averaging is recommended to get high-quality images. Collagen SHG signal from cultured MSCs is structured finely (Fig. 3d) and generally weaker than large and thick collagen fiber bundles found in native tissue. As controls, sections from rat tail, skin, or bone tissue can be used. These will give excellent SHG signal. Alternatively, cell spheroids [17] or collagen-I hydrogels [24] can be applied.

8. The SHG signal intensities will change over time and will be most prominent between days 6 and 12. During the progression of cell culture the cell layer will increase in thickness (up to ~60–70 μm possible).

9. Record 3D image stacks for getting an impression on collagen fiber networks in space. Depending on the imaged volume the z-distance between individual images should be set between 0.5 and 1 μm for stacks up to ~20 μm and to ~5 μm for depths <100 μm. This will provide a compromise between acquisition speed, 3D resolution, and data volume (*see* **Note 17**).

10. For higher resolved images of collagen fibers we recommend fluorescent staining with labeled CNA35 (collagen adhesion) [25]. A possible way is to select labels that fluoresce in the green, orange, or red part of the spectrum and couple them to CNA35 using commercial labeling kits. SHG imaging in parallel (405 nm, blue) allows comparisons between both imaging modalities (*see* **Note 18**).

11. Statements on viability based on MPI are valid as cells can be followed over weeks due to the noninvasive nature of this technology (*see* **Note 19**).

12. Changes in cell morphology and collagen-I will become obvious in AF and SHG images (Fig. 3d). Increased AF of cells is frequently observed and very often a sign of lowered cell fitness, cell stress, or cell death. De-adherent, rounded cells appear much brighter in AF images as fluorescent molecules are concentrated in a much smaller area. For distinguishing cell- from BNC-derived signals we recommend to run a control sample (BNC without cells) in parallel.
13. This protocol is compatible with cell seeding to alternative materials including polymeric, ceramic, hydrogel scaffolds or metal surfaces (*see* **Note 20**).
14. A large number of parameters can be analyzed in the study of collagen-I formation (*see* **Note 21**).

3.5 Analysis

1. To enable Fiji to gather all required information from the acquired images these should be imported by the Plugins-BioFormats Importer (*see* **Note 22**).
2. Adjust images for brightness and contrast by selecting Image-Adjust-Brightness-Contrast (*see* **Note 23**).
3. Introduce scale bars to the images by using Analyze-Tools-Scale Bar.
4. Color-code images with the Image-Lookup Tables tool.
5. To create overlays from AF and SHG images (as in Fig. 3d) apply the "Merge Channels" function (Image-Color-Merge Channels) which allows the selection of a large variety of colors on demand.
6. Arrange 3D image stacks in the form of videos/video collections (*see* **Note 24**).
7. Collagen networks can be analyzed by orientation analysis with the Orientation J plugin in Fiji (only possible with single images [17, 26]).

4 Notes

1. MSCs are multipotent stromal cells that can be differentiated into a variety of cell types including osteoblasts (bone cells), chondrocytes (cartilage cells), adipocytes (fat cells), and others. These cells were isolated using density gradient centrifugation followed by FACS analysis (tested for $CD90^+$, $CD29^+$, $CD54^+$, $CD106^+$ (endothelial cells), and $CD45^+$ (leucocytes)). MSCs exhibit good growth, morphology, and osteogenic differentiation according to alkaline phosphatase (AP) test [17]. The long and thin MSCs are characterized morphologically by a small cell body (which contains a large and round nucleus)

with long cell extensions. MSCs are known to produce large amounts of collagen-I.

2. Supplier of BNC: JeNaCell GmbH (Jena, Germany). Remnants of microorganisms were removed from the fleeces by chemical treatment. The material was routinely tested for the presence of pro-inflammatory bacterial lipopolysaccharides (LPS).
3. Thin glass surfaces are ultimately required for diffraction of the pulsed photons and focusing them to the sample. Sufficiently high photon densities are a prerequisite for 2-photon excitation to occur (Fig. 1a).
4. The microscope is inverted, i.e., the objective is positioned below the sample. The multiphoton system features excellent sectioning capabilities even in thick and almost nontransparent samples.
5. Multiphoton excitation is generally possible in a broad wavelength range and depends on the 2-photon action cross section of the respective fluorescent molecule (Fig. 1a–iii, top).
6. Such upscale and high-quality multiphoton microscopy objectives are also available in other magnifications (e.g., 32×).
7. Reflected photons contain SHG signals at exactly half the excitation wavelength of 810 nm (SHG is also referred to as frequency doubling, Fig. 1b).
8. This means that the detectors have the shortest possible light path (with no pinhole and fewer optical elements unlike in classic confocal detection) which helps to overcome signal loss. This dramatically increases the collection of emitted light and sample depth penetration.
9. This step is carried out to exchange the internal liquid volume (pure water, >99%) for growth medium. Deionized water left within the material will have an inhibitory effect on cell adhesion.
10. Initial seeding density should be high enough to yield fast monolayer formation (adaptations in seeded cell number may be required). Cells adhere to the BNC surface efficient and fast. Cell coverage could be followed by trans-illumination microscopy during culture. However, some experience is required to differentiate the cells from lacunae present all over the BNC surface. The phenol red (contained in the DMEM-F12 basal medium) indicates when the medium is used up by a color shift from red to orange.
11. Reference media compositions: Media with reduced serum content (1% FBS) and supplementation of 100 μg/ml ascorbic acid are useful for stimulating elevated collagen-I fibril formation. In addition to growth medium osteogenic medium contains 25 μg/ml L-ascorbic acid, 100 nM dexamethasone, and

10 mM ß-glycerol-phosphate. This medium stimulates the differentiation of MSCs into bone-forming osteoblasts that produce elevated collagen-I [17].

12. Intensity and abundance of SHG signal will be time dependent as collagen-I fiber formation follows a biological time course. MPI of the samples is an ECM-based cell fitness assay. It may be compared to classic live-dead staining running in parallel (e.g., based on Calcein AM/DAPI, Fig. 2) and correlated accordingly. Calcein AM is a nonfluorescent, hydrophobic compound that easily permeates intact, live cells. Its intracellular hydrolysis by esterases produces Calcein, a hydrophilic, strongly fluorescent (green) compound that is well retained in the cell cytoplasm. DAPI (4″,6-diamidino-2-phenylindole) is a fluorescent stain (blue) that specifically binds to DNA. As DAPI passes through intact cell membranes only very slowly, it can be used to differentiate between live and dead cells.
13. We did not experience better imaging quality when changing to phenol red-free PBS buffer although this component exhibits some fluorescence.
14. For intended long-term imaging an incubation chamber placed on the microscope stage is a good choice. Such chambers provide comparable physical conditions like in a cell incubator.
15. Photo-bleaching (and its cyto- and genotoxic effects) is not a major problem in MPI. In particular, generation of SHG causes no bleaching as no energy loss occurs in this nonlinear scattering process (unlike fluorescence processes).
16. Signal appears within a few μm depending on the depth of the cell layer. The nucleus appears black. The majority of AF is located around the nucleus and derives from NADH and flavins (Fig. 1a–iii, bottom) in mitochondria and other organelles. Cell extensions and the plasma membrane also show AF.
17. The surface of BNC is partially ripped with cracks of various depths into which MSCs could settle, grow, and form collagen-I ECM. MPI with its large penetration depth is the ideal technology for the imaging of such 3D structured surfaces.
18. Fluorescence imaging based on CNA35 collagen-I binding domain provides more details on collagen fiber networks and brighter images. However, the required labeling procedure negatively impacts cell viability and continued culture is not possible. In addition CNA35 binds to all the collagen-I present including intracellular monomers which are accessible in damaged or dead cells, so only live cell staining will provide good images of collagen networks.
19. In a study by Dittmar et al. (2012, [8]) cell viability was assessed by spectral analysis of 2-photon excited fluorescence.

Viable cells showed predominantly blue fluorescence with a peak emission around 470 nm, whereas dead cells mainly emitted green fluorescence (peak intensity around 560 nm).

20. Imaging of cell viability on reference materials: All scaffolds need to be sterilized in advance (e.g., 2 h in 70% ethanol) and equilibrated in serum-free MSC medium. Depending on hydrophobicity and other material features cell adhesion to some materials might be low. In this case MSCs require more time for attachment. To overcome poor adhesion materials can be coated with ECM proteins (e.g., collagen, laminin, fibronectin) or matrigel. Materials need to be tested thoroughly for their AF and SHG signatures before imaging seeded cells and collagen-I in order to separate signals specific for the cells. Many biomaterials are auto-fluorescent [27] and some (like BNC or silk [28]) in addition generate SHG signals. Knowledge of the micro-topography of material surfaces is also very helpful for imaging as it may strongly impact cell viability [29].
21. These include time dependence of collagen-I formation, influence of different media and supplements/growth factors, analyses of fiber networks (fiber orientation using the Fiji plugins "Orientation J" or "Directionality"), analyses of cell and ECM layer thickness, and influence of the scaffold on collagen formation (analysis by relating cell seeding results to other biomaterials or to biomaterial-free culture systems like 3D spheroids).
22. All image recording parameters can be found in Image-Show Info. Comprehensive information on Fiji software and its tools and plugins are available online.
23. Images suffering from high background signal can be processed by defining a background ROI (region of interest) followed by mean intensity calculation (using Analyze-Histogram) and subtracting the mean value from the image (choose image, then select Process-Math-Subtract). To enhance image quality various filtering algorithms (e.g., Process-Filters-Median) and many more useful analysis and processing tools are available in Fiji.
24. In addition image stacks can be presented spatially using the plugins "3D Viewer" or "Volume Viewer". A minimum overlap between images should exist for the software to calculate a continuum of structures, i.e., the z-distance between images should be sufficiently low. For the calculation of 3D representations the signal intensity levels have to be relatively high. Unfortunately, unlike in fluorescence images, in SHG images from collagen networks from cultured MSCs provide only relatively low intensities. Therefore SHG image quality has to be optimized.

Acknowledgments

This work was supported by the Emerging Fields Initiative (EFI) of the University of Erlangen-Nürnberg (project TOPbiomat) and the Erlangen Graduate School in Advanced Optical Technologies (SAOT) within the German Excellence Initiative. We want to thank Dana Kralisch and Nadine Hessler (JeNaCell GmbH) for providing BNC fleeces.

References

1. Gerhardt L-C, Widdows KL, Erol MM et al (2011) The pro-angiogenic properties of multi-functional bioactive glass composite scaffolds. Biomaterials 32:4096–4108. doi:10.1016/j.biomaterials.2011.02.032
2. Musson DS, Naot D, Chhana A et al (2015) In vitro evaluation of a novel non-mulberry silk scaffold for use in tendon regeneration. Tissue Eng Part A 21:1539–1551. doi:10.1089/ten.tea.2014.0128
3. Ramaswamy Y, Wu C, Van Hummel A et al (2008) The responses of osteoblasts, osteoclasts and endothelial cells to zirconium modified calcium-silicate-based ceramic. Biomaterials 29:4392–4402. doi:10.1016/j.biomaterials.2008.08.006
4. Papadopoulos N, Dedoussis G, Spanakos G et al (1994) An improved fluorescence assay for the determination of lymphocyte-mediated cytotoxicity using flow cytometry. J Immunol Methods 177:101–111
5. Yao J, Korotkova T, Smith RL (2011) Viability and proliferation of pluripotential cells delivered to tendon repair sites using bioactive sutures—an in vitro study. J Hand Surg [Am] 36:252–258. doi:10.1016/j.jhsa.2010.10.004
6. Rice WL, Kaplan DL, Georgakoudi I (2010) Two-photon microscopy for non-invasive, quantitative monitoring of stem cell differentiation. PLoS One 5:e10075. doi:10.1371/journal.pone.0010075
7. Lee H, Teng S, Chen H et al (2006) Imaging human bone marrow stem cell morphogenesis in polyglycolic acid scaffold by multiphoton microscopy. Tissue Eng 12:2835–2841
8. Dittmar R, Potier E, Van Zandvoort M et al (2012) Assessment of cell viability in three-dimensional scaffolds using cellular auto-fluorescence. Tissue Eng Part C Methods 18:198–204. doi:10.1089/ten.tec.2011.0334
9. Chen W-L, Huang C-H, Chiou L-L et al (2010) Multiphoton imaging and quantitative analysis of collagen production by chondrogenic human mesenchymal stem cells cultured in chitosan scaffold. Tissue Eng Part C Methods 16:913–920. doi:10.1089/ten.TEC.2009.0596
10. Zoumi A, Yeh A, Tromberg BJ (2002) Imaging cells and extracellular matrix in vivo by using second-harmonic generation and two-photon excited fluorescence. Proc Natl Acad Sci U S A 99:11014–11019. doi:10.1073/pnas.172368799
11. Helmchen F, Denk W (2005) Deep tissue two-photon microscopy. Nat Methods 2:932–940. doi:10.1038/nmeth818
12. Ustione A, Piston DW (2011) A simple introduction to multiphoton microscopy. J Microsc 243:221–226. doi:10.1111/j.1365-2818.2011.03532.x
13. Zipfel WR, Williams RM, Christie R et al (2003) Live tissue intrinsic emission microscopy using multiphoton-excited native fluorescence and second harmonic generation. Proc Natl Acad Sci U S A 100:7075–7080. doi:10.1073/pnas.0832308100
14. Van der Rest M, Garrone R (1991) Collagen family of proteins. FASEB J 5:2814–2823
15. Orgel JPRO, Miller A, Irving TC et al (2001) The in situ supermolecular structure of type I collagen. Structure 9:1061–1069
16. Vielreicher M, Schürmann S, Detsch R et al (2013) Taking a deep look: modern microscopy technologies to optimize the design and functionality of biocompatible scaffolds for tissue engineering in regenerative medicine. J R Soc Interface 10:20130263. doi:10.1098/rsif.2013.0263
17. Vielreicher M, Gellner M, Rottensteiner U et al (2015) Multiphoton microscopy analysis of extracellular collagen I network formation by mesenchymal stem cells. J Tissue Eng Regen Med. doi:10.1002/term.2107
18. Iguchi M, Yamanaka S, Budhiono A (2000) Bacterial cellulose—a masterpiece of nature's arts. J Mater Sci 35:261–270
19. Wiegand C, Moritz S, Hessler N et al (2015) Antimicrobial functionalization of bacterial nanocellulose by loading with polihexanide

and povidone-iodine. J Mater Sci Mater Med 26:245. doi:10.1007/s10856-015-5571-7
20. Kralisch D, Hessler N, Klemm D et al (2010) White biotechnology for cellulose manufacturing—the HoLiR concept. Biotechnol Bioeng 105:740–747. doi:10.1002/bit.22579
21. Nadiarnykh O, Lacomb RB, Campagnola PJ et al (2007) Coherent and incoherent SHG in fibrillar cellulose matrices. Opt Express 15:3348–3360
22. Brackmann C, Zaborowska M, Sundberg J et al (2012) In situ imaging of collagen synthesis by osteoprogenitor cells in microporous bacterial cellulose scaffolds. Tissue Eng Part C Methods 18:227–234. doi:10.1089/ten.tec.2011.0211
23. Schindelin J, Arganda-Carreras I, Frise E et al (2012) Fiji: an open-source platform for biological-image analysis. Nat Methods 9:676–682. doi:10.1038/nmeth.2019
24. Lee P-F, Yeh AT, Bayless KJ (2009) Nonlinear optical microscopy reveals invading endothelial cells anisotropically alter three-dimensional collagen matrices. Exp Cell Res 315:396–410. doi:10.1016/j.yexcr.2008.10.040
25. Boerboom RA, Krahn KN, Megens RTA et al (2007) High resolution imaging of collagen organisation and synthesis using a versatile collagen specific probe. J Struct Biol 159:392–399. doi:10.1016/j.jsb.2007.04.008
26. Rezakhaniha R, Agianniotis A, Schrauwen JTC et al (2011) Experimental investigation of collagen waviness and orientation in the arterial adventitia using confocal laser scanning microscopy. Biomech Model Mechanobiol 11(3–4):461–473. doi:10.1007/s10237-011-0325-z
27. Sun YEN, Tan H, Lin S et al (2008) Imaging tissue engineering scaffolds using multiphoton microscopy. Microsc Res Tech 71:140–145. doi:10.1002/jemt.20537
28. Rice WL, Firdous S, Gupta S et al (2008) Noninvasive characterization of structure and morphology of silk fibroin biomaterials using non-linear microscopy. Biomaterials 29:2015–2024. doi:10.1016/j.biomaterials.2007.12.049
29. Bottan S, Robotti F, Jayathissa P et al (2015) Surface-structured bacterial cellulose with guided assembly-based biolithography (GAB). ACS Nano 9:206–219. doi:10.1021/nn5036125

Chapter 20

Average Rheological Quantities of Cells in Monolayers

Haider Dakhil and Andreas Wierschem

Abstract

Measuring rheological properties of cells in monolayers enables quantifying average cell properties in single experimental runs despite large cell-to-cell variations. Here, we describe how to modify a commercial rotational rheometer to accomplish the necessary precision for a monolayer rheometer and we delineate the steps for setting up experiments detecting average viscoelastic cell properties.

Key words Cell rheology, Linear viscoelasticity, Narrow-gap rheometry, Cell monolayer

1 Introduction

Rheology is the study of deformation of matter under the influence of stress. Rheological studies enable quantification of material properties such as viscosity of fluids or viscoelasticity of solids and fluids [1, 2]. Rheological properties of biological cells are closely related to their physiological activities and have vital functional implications such as mechanical stability, adjustment to environmental load, migration, proliferation, phagocytosis, or contraction. For single cell studies, different techniques like optical and magnetic tweezers, atomic force microscopy, magnetic twisting cytometry, micropipettes, microplates, cell poking, and particle tracking micro-rheology have been employed [3, 4]. Depending for instance on cell identity, life cycle, shape, structure, and level of proteins [5], cells show variations in stiffness and dynamic moduli by an order of magnitude [6]. Hence, to quantify the impact of drugs, aging, or diseases on the viscoelastic cellular properties, it is of great importance to determine average viscoelastic cell properties. Averaging over a large number of cells in a single experimental run can be achieved by studying cells in a monolayer between rheometer disks in the parallel-disk configuration [7] and detecting the cell coverage [6].

To study average rheological properties of cells in a monolayer, the cells have to be adhered between two transparent plates with a

Daniel F. Gilbert and Oliver Friedrich (eds.), *Cell Viability Assays: Methods and Protocols*, Methods in Molecular Biology, vol. 1601, DOI 10.1007/978-1-4939-6960-9_20, © Springer Science+Business Media LLC 2017

gap-width variation to be a fraction of the average cell height. This is beyond the scope of commercial rheometers that usually have a systematic error in the gap width of about ±25 μm and more [8–10]. To overcome this limitation and meet the requirements for cell monolayers, we developed a setup for commercial rotational rheometers significantly reducing the variation to less than ±1 μm [4, 6].

2 Materials

Cells in suspension like, for instance, HeLa cells or fibroblasts that can be adhered to solid substrates with proteins like, for instance, fibroblasts [11]. Use the following materials:

1. Cultural medium for 3 T6 Swiss Albino murine fibroblasts: Dulbecco's modified Eagle medium (DMEM, Invitrogen), 25 mM HEPES and 0.5 mM $NaHCO_3$. Add 10% (v/v) fetal bovine serum (FBS) to the cultural medium (*see* **Note 5**).
2. Cultural medium for primary human dermal fibroblasts: Dulbecco's modified Eagle medium (DMEM (1×) + GlutaMAX-I) supplemented with 10% (v/v) fetal calf serum (FCS), and an antibiotic combination of 1% Pen Strep, at pH 7.4, 310.16 K, and 5% CO_2 (*see* **Note 5**).
3. Cultural medium for HeLa (ACC 57) cells: RPMI 1640 medium supplemented with 10% (v/v) fetal bovine serum (FBS) and 4 mM L-Glutamine at pH 7.4, 310.16 K, and 5% CO_2 (*see* **Note 5**).
4. Phosphate buffer solution (PBS).
5. 9 vol.% fibronectin, an adhesion promoting protein, dissolved in PBS for 1 h prior to coating procedure (100 μl fibronectin in 1 ml PBS).
6. 1 vol.% Calcein AM solution (20 μl Calcein Am in 2 ml fresh cultural medium).
7. 1 vol.% propidium iodide solution (20 μl propidium iodide in 2 ml fresh cultural medium).
8. Digital b/w camera equipped with a filter block for fluorescence imaging and with a 5× objective lens. The filter block (C-FL Epi-FL FITC, EX 465–495, DM 505, BA 515–555, Olympus) is incorporated in the lens tube for fluorescence imaging with Calcein AM (*see* **Note 6**).
9. Confocal interferometric sensor: Working distance: 20 mm, measuring range: 90 μm.
10. UDS 200 rotational rheometer from Physica.

3 Methods

All procedures may be carried out at room temperature unless otherwise specified (*see* **Note 1**).

3.1 Cell Preparation

1. Cell detachment from inner surface of the culture flask: Remove the culture medium from the culture flask (*see* **Note 2**).
2. Rinse the flask with 5 ml of phosphate buffered solution (PBS).
3. After gently shaking the flask, remove and discard the PBS.
4. Add 5 ml Accutase solution to the culture flask. Incubate the flask (at room temperature) for 30 min to let the cells detach from its inner surface.
5. Check the cells under an inverted microscope to determine if at least 90% of the cells are detached and floating, which is identified by their round shape. If the cells are not detached after 30 min, incubate them for another 1–2 min.
6. As soon as cells are floating, add 10 ml fresh culture medium to the culturing flask.
7. Rinse down several times the sidewalls of the flask with new culture medium to wash all cells off the wall and into the solution.
8. Remove and transfer the liquid that contains the cells into a 50 ml conical tube. Centrifuge the suspension for 8 min at $180 \times g$ to separate the cell pellet from the culture suspension.
9. Discard the supernatant and count the cells under the microscope to quantify the amount of cells needed in an individual rheological experiment.

3.2 Cell Staining

Cell staining with fluorescent agents like, for instance, Calcein AM to optically detect cells covering the area in the gap between the rheometer plates.

1. Stain the cells with 1 vol.% Calcein AM.
2. Incubate the cells in the incubator at a temperature of 37 °C for about 1 h to let the agent pass through the cell membrane.
3. Remove and transfer the liquid that contains the cells into a 50 ml conical tube. Centrifuge the suspension for 8 min at $180 \times g$ to separate the cell pellet from the residual Calcein AM in the culture suspension (*see* **Note 4**).
4. Discard the supernatant and add 2 ml fresh medium. Mix it gently with the cell pellet.
5. In case dead cells should be indicated optically in the rheometer gap, one can add 1 vol.% propidium iodide.

3.3 Adhesive Protein Solution

1. Add 100 μl fibronectin protein stock solution to 1 ml phosphate buffer solution (PBS) to prepare a final concentration of 9 vol.% of fibronectin solution, as a promoting adhesive solution.
2. Leave the prepared solution 1 h prior to the coating procedure to let the protein dissolve well in the PBS.

3.4 The Narrow-Gap Rheometer

To study rheological properties of cells in a monolayer, the cells have to be adhered between two transparent plates, *see* Fig. 1. One plate is brought into azimuthal oscillation at controlled torque or controlled strain. This load is exerted with a commercial rotational rheometer, which also detects the mechanical response of the cell monolayer to this load. To obtain reproducible results, the gap-width variation between the two plates needs to be a fraction of the average cell height. Since in commercial rheometers the gap-width variation is about ±25–70 μm, a modification is necessary to obtain the high degree of precision, *see* Fig. 1 [8–10]:

1. Stationary lower glass plate with a diameter of 75 mm and evenness of at least 1λ and rotation upper glass plate with a diameter of 50 mm and evenness of $\lambda/10$ where λ is the testing wavelength (633 nm) (*see* **Note 3**).
2. The rotating plate is attached to a measurement head of the rheometer. The head has a diameter with 25 mm.
3. To align the stationary plate perpendicular to the rotation axis, it is fixed to a tripod, which is mounted on the rheometer table. The tripod is aligned by 3 μm screws with a fine resolution of 1 μm and fixed to the rheometer with three screws after adjustment.

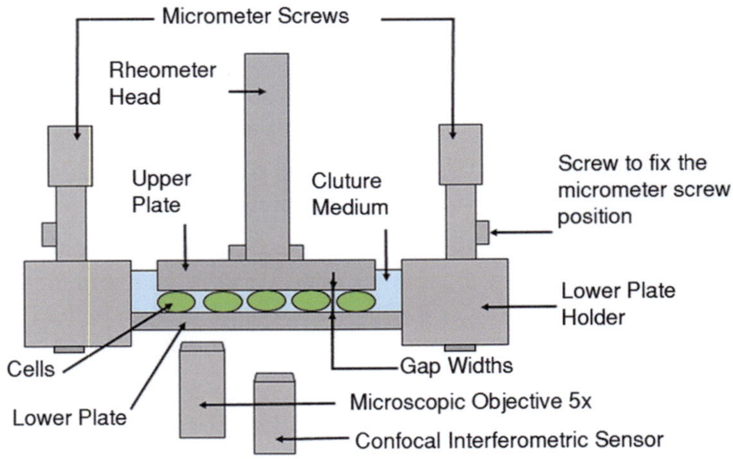

Fig. 1 Sketch of the setup in the customized rheometer. The gap width between the rheometer plates is measured with a confocal interferometric sensor; the cells are viewed with a camera

3.5 Alignment of the Narrow-Gap Rheometer

Procedure to align the glass plates of the rheometer with the desired precision of about ±1 μm [12]:

1. After carefully cleaning the glass plates with ethanol or acetone, place the top glass plate in contact with the top of the bottom glass plate and glue it to the rheometer head using a dissolvable adhesive like Vitralit [12].
2. Lift the upper plate by about 50 μm.
3. Use a confocal interferometric sensor to measure the gap width between the plates near the edge of the smaller plate at three different locations, forming a triangle. During these measurements, the upper plate rotates at a constant speed of 0.01 s^{-1} to measure the variation in gap width in time.
4. Due to the inclination of the upper plate with respect to the rotating rheometer axis, the gap width varies periodically during rotation. Since the lower plate is not perfectly perpendicular with respect to the rheometer axis, the mean value of the gap width is different at each position of the sensor. Fig. 2a shows the measurement at beginning of this procedure.

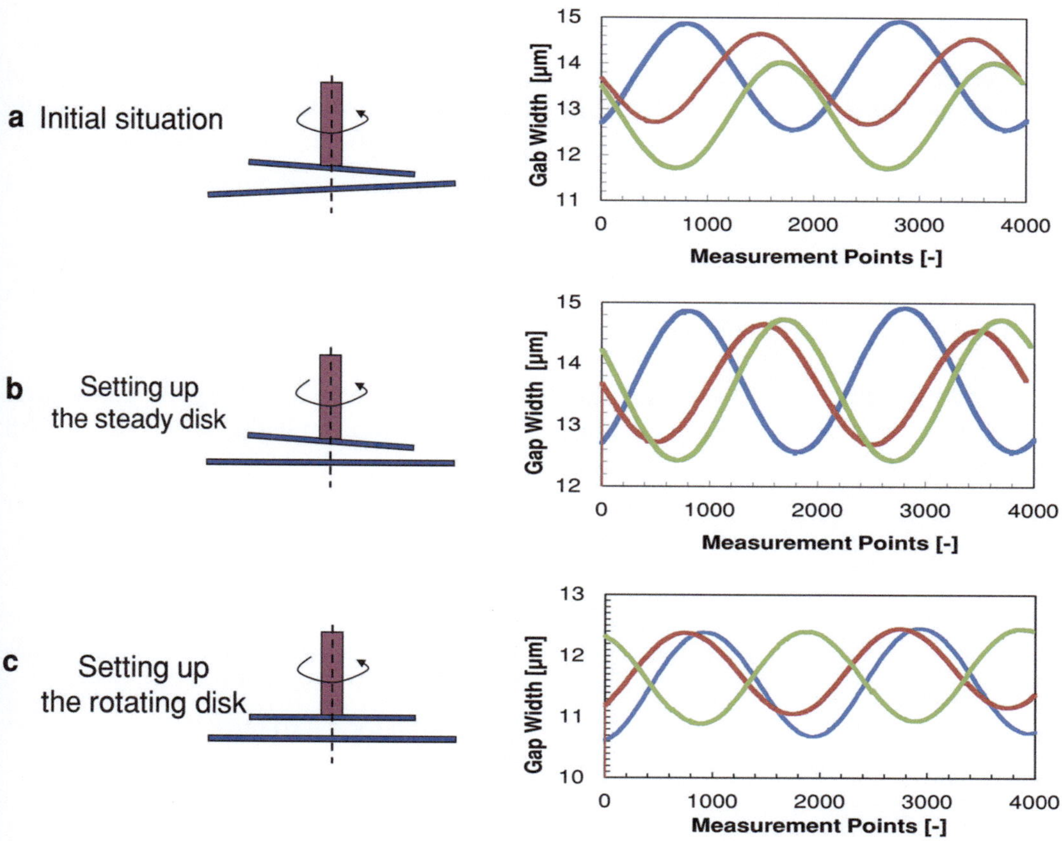

Fig. 2 Setting up the rheometer disks. The upper disk rotates slowly while the gap width is detected with a sensor at three different locations: before adjustment (**a**), after adjusting the lower plate (**b**), after final adjustment of the upper disk (**c**)

5. Adjust the lower plate with micrometer screws that is shown in Fig. 1a to minimize deviations in the mean gap width. Fig 2b shows an example after this step.
6. Dismantle the upper disk from the rheometer head by leaving it in an adequate solvent like acetone for at least 1 h and subsequent exposure to hot air for a short period of time.
7. Place the upper disk on the leveled lower plate and glue it to the rheometer head.
8. Check parallelism again with the sensor to achieve undulations at each location with an amplitude of about 1 μm, *see* Fig. 2c.
9. Rheometer software: Press zero-gap position to bring down the upper glass plate in contact with the lower glass plates at minimum velocity.
10. Measure the zero-gap prior to each experiment with the sensor as a three point average and then adjust the lower plate to make sure the rheometer zero-gap is identical with sensor reading. This drastically reduces the error in zeroing.

If the upper plate is already properly aligned, one may start directly with **step 8**.

3.6 Coating the Rheometer Plates

1. Coat both rheometer glass plates with the adhesive protein solution by introducing the solution into the gap between the glass plates at a gap width of 200 μm with micropipette.
2. Leave the adhesive protein solution between the glass plates for 1 h at a gap width of 50 μm.
3. Rinse the plates three times with PBS. Therefore, the upper plate can be lifted up and down. Suck the solution with a pipette.

3.7 Introducing Cells Between the Rheometer Plates

1. Introduce the cell suspension between the glass plates after lifting the upper plate to 200 μm gap width via sucking it with capillary forces.
2. Leave the suspension on the lower glass plate of the rheometer for 20 min to let the cells settle down and adhere to the lower plate.
3. Lower the upper rheometer plate at minimum speed of the rheometer to confine the cells at a gap width that is sufficiently smaller than the average height of the cells.
4. Leave the cells in this position for 1 h to promote perfect adhesion between both plates.
5. Lift the rheometer plate up to the desired gap width at which the cell monolayer is to be studied.

3.8 Detecting the Cell Coverage

To determine average rheological quantities of the cells, one needs to quantify the cell coverage in the gap between the rheometer plates, i.e., the relative area between the plates that is covered by the cells.

1. 40–50 fluorescence images are taken near the edge of the upper plate with the fluorescence microscope. Since deformation is strongest at maximum distance from the rheometer's turning axis, the rheometer signal is mainly due to the cells at the outer rim [11]. Therefore, the cell coverage in the gap determined from pictures at the outer rim (*see* **Note 6**).
2. The relative area covered by the cells can be determined, for instance, with software DetecTIFF [13] or with the software ImageJ using the Analyze-Analyze Particle window (*see* **Note 7**).

3.9 Measuring Viscoelastic Properties

To study the viscoelastic properties of the cell monolayer, the upper rheometer plate carries out oscillatory motions around its axis. In amplitude sweeps, the oscillation frequency is fixed and the oscillation amplitude, hence the strain amplitude of the sample, is changed; in frequency sweeps the strain amplitude is fixed and the frequency is changed. The linear viscoelastic range is determined from amplitude sweeps. It is identified as the range of low amplitudes at which the storage and loss moduli do not change with amplitude. The storage modulus G' quantifies the elastic response of the sample, and the loss modulus G" the dissipated part. In Fig. 3, the linear viscoelastic range of the medium and of the cell monolayer is below a strain amplitude of 1% (*see* **Note 8**).

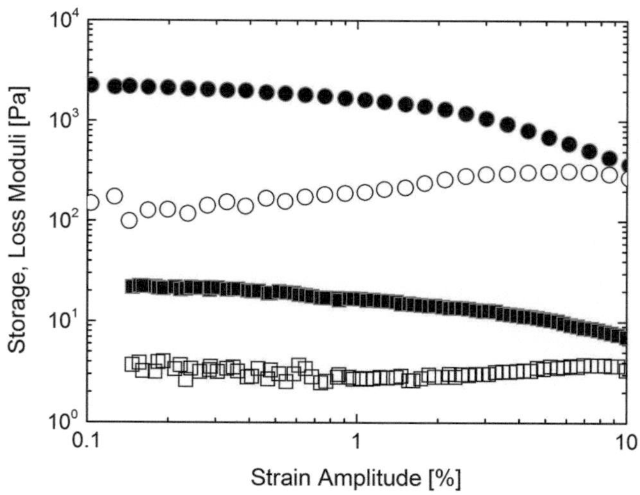

Fig. 3 Amplitude sweep of mouse fibroblasts (*circles*) and of the cell culture medium (*squares*). *Closed* and *open* symbols indicate the storage and the loss modulus, respectively

1. Amplitude sweep: Carry out the sweep in the controlled strain mode at strain amplitudes between 0.1% and a strain amplitude where storage and loss moduli start deviating from plateau values (typically between 1% and 10%). Fix the frequency (typically 1 Hz).
2. Frequency sweep: Carry out the sweep in a controlled strain mode at frequencies that typically range between 0.1 Hz and 30 Hz. Fix the strain amplitude at a value well within the linear viscoelastic range determined with the amplitude sweeps (a typical value is 0.2%).

3.10 Calculation of the Average Cell Moduli

Figure 3 shows the storage and the loss modulus of the medium, G'_{medium} and G''_{medium}, respectively, and those of the cell monolayer, G' and G". The moduli of the pure medium are much smaller than those of the cell monolayer with medium. Average cell moduli G'_{cell} and G''_{cell} can be determined from the measured moduli and the mean cell coverage close to the gap rim, c:

$$G'_{cell} = \frac{G' - (1-c)G'_{medium}}{c}. \quad (1)$$

Figure 4 shows the average storage modulus of different types of cells in the linear viscoelastic regime. Standard deviation is about 20%.

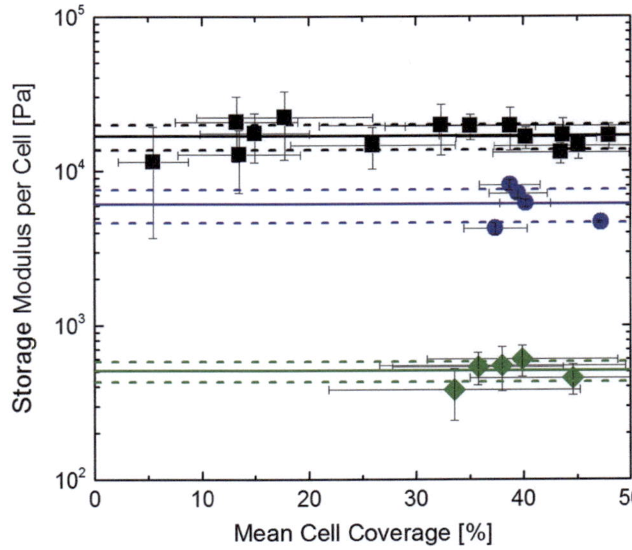

Fig. 4 Storage modulus per cell of mouse fibroblasts (*black squares*), human fibroblasts (*blue circles*), and HeLa cells (*green diamonds*). *Solid lines* show the mean modulus; *thin dashed lines* indicate the range within standard deviation between the measurements [11] (License no. 3898181445518, Springer, Rheologica Acta)

4 Notes

1. Before using the safety cabinet, be sure it is clean and sterile.
2. While preparing the cells for experiments wear powder-free Nitrile gloves to avoid any infections.
3. After each experiment, clean the glass plates with 70 vol.% ethanol to kill any microorganisms. Then, use 2 vol.% SDS to clean the glass plate from any remaining protein residues.
4. All wastes like conical tubes, pipette tips, culture flasks, gloves etc. have to be autoclaved.
5. Lid and neck of medium bottles must be kept clean and dry, especially after warming-up in the water bath, to avoid contamination.
6. Camera magnification: Compromise between imaging large areas to obtain good statistics for the cell coverage and sufficient resolution for precisely determining the cell size.
7. Optimal cell coverage in the monolayer is between 30 and 50%. At lower concentrations, the rheological signal may be rather weak and the scatter in local cell coverage rather large. At higher cell concentrations, cell agglomeration becomes significant making it difficult to obtain a monolayer.
8. To study cells at minimum pre-stress, two methods can be used:
   - Setting the final gap width to the average height of the cells.
   - Zeroing the normal force of the rheometer with the measurement head of the rheometer lifted up in air. When bringing down the measurement head, compressed cells yield a positive normal force. Minimum average pre-stress is obtained with the cells at a gap width where the normal force is zero again.

Acknowledgments

We gratefully acknowledge financial support by Deutscher Akademischer Austauschdienst (DAAD), Germany, and the Ministry of Higher Education & Scientific Research, Iraq.

References

1. Chhabra RP, Richardson JF (2008) Non-Newtonian flow and applied rheology, 2nd edn. Butterworth-Heinemann, Oxford, pp 1–55
2. Steffe JF (1996) Rheological methods in food process engineering. Freeman Press, East Lansing
3. Verdier C, Etienne J, Duperray A et al (2009) Review: rheological properties of biological materials. Comptes Rendus Physique 10:790–811
4. Kollmannsberger P, Fabry B (2011) Linear and nonlinear rheology of living cells. Annu Rev Mater Res 41:75–97

5. Weissman-Shomer P, Fry M (1975) Chick embryo fibroblasts senescence in vitro: Pattern of cell division and life span as a function of cell density. Mech Ageing Dev 4:159–166
6. Cai P, Mizutani Y, Tsuchiya M et al (2013) Quantifying cell-to-cell variation in power-law rheology. Biophys J 105:1093–1102
7. Fernández P, Heymann L, Ott A et al (2007) Shear rheology of a cell monolayer. New J Phys 9:419–447
8. Davies GA, Stokes JR (2005) On the gap error in parallel plate rheometry that arises from the presence of air when zeroing the gap. J Rheol 49:919–922
9. Davies GA, Stokes JR (2008) Thin film and high shear rheology of multiphase complex fluids. J Non-Newtonian Fluid Mech 148:73–87
10. Pipe CJ, Majmudar TS, Mckinley GH (2008) High shear rate viscometry. Rheol Acta 47:621–642
11. Dakhil H, Gilbert DF, Malhotra D et al (2016) Measuring average rheological quantities of cell monolayers in the linear viscoelastic regime. Rheol Acta 55:527–536
12. Dakhil H, Wierschem A (2014) Measuring low viscosities and high shear rates with a rotational rheometer in a thin-gap parallel-disk configuration. Appl Rheol 24:63795–63801
13. Gilbert DF, Meinhof T, Pepperkok R et al (2009) DetecTiff©: a novel image analysis routine for high-content screening microscopy. J Biomol Screen 14:944–955

Chapter 21

Measurement of Cellular Behavior by Electrochemical Impedance Sensing

Simin Öz, Achim Breiling, and Christian Maercker

Abstract

There is a great demand for label-free in vitro assays in a high-throughput context, in order to measure cell viability and analyze cellular functions like cell migration or cell differentiation under noninvasive conditions. Here, we describe impedance measurement to quantify dynamic changes on cell morphology in real time. In order to monitor physiological changes, cells are grown in tissue culture vessels where gold electrodes are incorporated at the bottom. An alternating current signal of several kHz is applied to the electrodes and the resulting voltage is measured to calculate the cellular impedance. Since impedance is closely related to the area of the electrodes covered by the growing cells, parameters such as cell number, size of the cells attached to the electrodes, and cell-cell and cell-substrate/extracellular matrix interactions contribute to the overall impedance values.

Key words In vitro experiments, Cell based assays, Cell viability, Cell differentiation, Impedance sensing

1 Introduction

Electrochemical measurements in tissue culture to generate dielectric resistance profiles were first reported by the pioneering work of Giaever and Keese [1] and were further developed into a new field of biosensors now called electric cell-substrate impedance sensing (ECIS) [2]. Based on this principle, several systems evolved each equipped with its own specific features including sensors for pH, temperature, glucose, or oxygen-consumption [3, 4]. To monitor cell behavior electrically, cells are grown on special tissue culture vessels in which gold electrodes are incorporated at the bottom of the cultureware. An alternating current (AC) signal of several kHz is applied on the electrodes and the resulting voltage is measured to calculate the impedance; the AC equivalent to electric resistance, where the resulting output unit is in Ohms. As cells start to attach to the electrodes they serve mainly as insulators and restrict the current flow, thus leading to an increase in the impedance (Fig. 1).

Daniel F. Gilbert and Oliver Friedrich (eds.), *Cell Viability Assays: Methods and Protocols*, Methods in Molecular Biology, vol. 1601, DOI 10.1007/978-1-4939-6960-9_21, © Springer Science+Business Media LLC 2017

In general, impedance is closely related to the covered area of the electrodes and is thus mainly influenced by the cell number and size of the cells attached to the electrodes, but also cell-cell and cell-substrate/ECM interactions further contribute to the overall impedance values. Therefore, the impedance profiles, varying in magnitude and initial slopes depending on cell morphology and attachment of the cells, may reflect the cell type, cell state, and cell growth and this method may be used as a valuable tool to characterize the dynamics of cells in real time in a noninvasive manner [2].

A mathematical model was developed which determines three parameters describing impedance data (for details *see* [5, 6]). Therein, the current flows beneath the cells are defined by the cell-substrate interaction, which mainly depends on the cell size and the space between the cells and the substratum. The spaces and formed junctions between the cells determine the current flow through the cell layer and define the barrier resistance (Rb). Additionally, the membrane capacitance (Cm) can be included as well. Further interpretation of impedance data can be obtained if the phase of voltage is considered. The method discussed in this

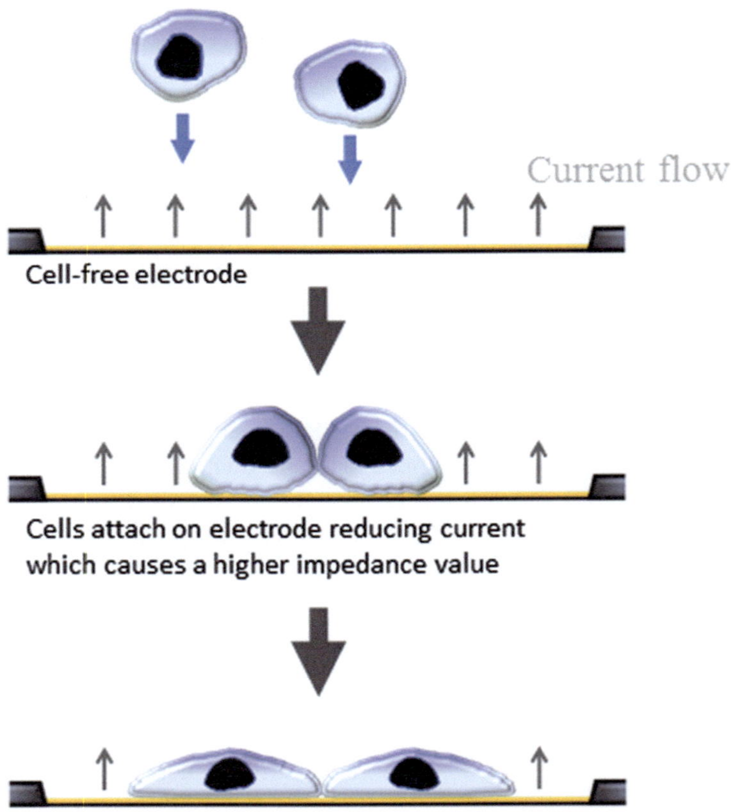

Fig. 1 Principle of electrochemical measurement (kindly provided by Michael Angstmann)

chapter focuses on the measurement of impedance (Z) without regard to other parameters.

In order to monitor the dynamics of cell behavior, we use the ECIS system from Applied Biophysics where we describe measurements in 8-well plates (Fig. 2), which is suitable for functional analyses on a small scale, in particular to follow cell behavior, e.g., upon small molecule treatment or siRNA-mediated depletion of specific genes. Moreover, cells on the wells can be observed under the microscope. For high-throughput applications, e.g., viability tests in drug target validation assays, 96-well plates are recommended.

The ECIS system also allows more specialized assays, such as cell migration after electroporation ("wound healing"), or testing the endothelial barrier functions (trans-epithelial electrical resistance, TEER) with the TransFilter Adapter. However we will not be going over these techniques here [7, 8].

Another system which we tested but is not shown here is the xCELLigence real-time cell analysis (RTCA) system by ACEA Biosciences. The wells of the biochips of this system are not accessible by microscopy; however, this device is very convenient for high-throughput applications with up to five 96-well plates in one experiment. The RTCA system also is applicable for transmigration assays on specialized xCELLigence transwell plates [9].

2 Materials

2.1 Cell Culture

1. Cell line: For most experiments, we used the human cell line NTERA 2 D1 (NT2, kind gift from Peter W. Andrews, University of Sheffield). The culturing conditions and medium mentioned below are specific for this cell line and have to be adjusted for each cell line.

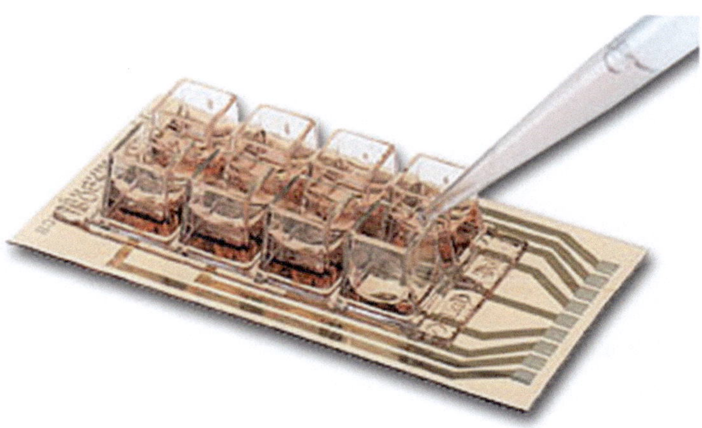

Fig. 2 ECIS 8W10E+ array (Applied Biophysics)

2. Cell culture medium: Cells were maintained in Dulbecco's Modified Eagle Medium supplemented with 10% fetal bovine serum, 200 U/ml penicillin, and 200 μg/ml streptomycin.
3. Cell culture flasks: Tissue Culture Flasks.
4. Cell cultivation: Cell culture was performed under sterile hoods to avoid contamination. Cells were kept at 37 °C in an incubator with a humidified 5% CO_2 atmosphere. Cells were grown to 70% and maintained by scraping and splitting in 1:4 dilution every 3–4 days. With this cell line we did not use trypsin for splitting as it induces differentiation of the cells. Cell density was determined using Bio-Rad TC20. Cells were mixed with Trypan Blue Dye at a 1:1 ratio prior to counting to assess viability as viable cells exclude trypan blue.

2.2 Measuring Instrument

1. ECIS instrument: With the ECIS Model 1600 and appropriate software, the resistance of the cells can be analyzed. For measurement, an oscillator sets an AC signal of 1 μV amplitude at a frequency of 45 kHz in series with a 1 M resistor resulting in an approximately constant current of 1 μA.
2. Arrays: To record changes in electrical resistance, the ECIS Model 1600 (Applied Biophysics) is equipped with 8-well slides with impedance sensing gold electrodes implemented at the bottom (ECIS 8W10E+ arrays, Fig. 2) [10].

3 Methods

3.1 Preparation of Cells in ECIS Arrays

1. Work under the cell culture hood under sterile conditions. Pre-incubate 8W10E+ ECIS Cultureware arrays (which contain 40,250-μm gold electrodes per well) with complete cell culture medium for 4 h at 37 °C (*see* **Note 1**).
2. Remove cell culture medium and add 400 μl cell suspension containing 2×10^4 cells to each well (*see* **Note 2**). In order to equally distribute the cells over the wells, gently pipet the cell suspension up and down. In case the impedance will be measured over 2–3 days, it is recommended to use medium not any less than 400 μl, as by time the medium will evaporate. If there are any wells left empty, fill with 400 μl of medium, as this can serve as a no-cell control. In order to improve reproducibility, we recommend to seed cells in duplicates, triplicates, or quadruplicates.

3.2 Measurement of Impedance

1. Place the arrays on the appropriate ECIS model that is in an incubator with suitable temperature and CO_2 settings, according to the demands of the respective cell line (*see* Subheading 2.1).
2. Record impedance at 45 kHz in 5-min intervals for up to 4–5 days, until cells reach complete confluency (*see* **Notes 3** and **4**).

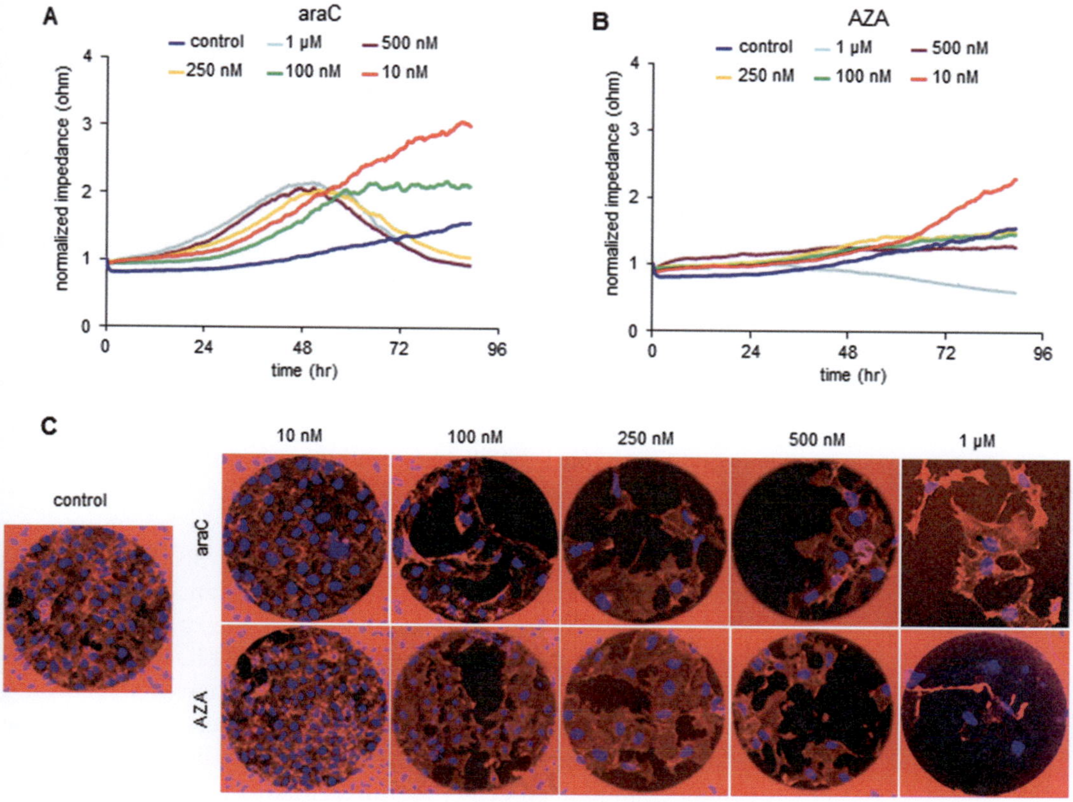

Fig. 3 Typical ECIS result and control by light microscopy. (**a**) Impedance profiles of NT2 cells treated with different concentrations of 1β-arabinofuranosylcytosine (araC) during a 4 day period. Concentrations above 100 nM are severely cytotoxic, which leads to a drastic drop in impedance values after 48 h. Measurements were executed at 45 kHz in 5-min intervals for 96 h. One representative experiment is shown. Single diagrams show the mean of at least three experiments including standard deviations and statistical tests. (**b**) Impedance profiles of NT2 cells treated with different concentrations of 5-azacytidine (AZA) during a 4 day period. Concentrations above 100 nM strongly induce proliferative defects, which prevents the increase of impedance values. Measurements were executed at 45 kHz in 5-min intervals for 96 h. (**c**) Phalloidin staining of growing cultures on ECIS electrodes. Fluorescence images (10× magnification) of NT2 control cells and NT2 cells treated with the indicated concentrations of AZA and araC. The *circular dark* region is the electrode measuring area covered by the cells. Cells were stained with Phalloidin TRITC (*red*) and DAPI (*blue*) [11]

3. For comparative analysis, impedance data are normalized to the values measured at time point zero. Defining these reference values as 1.0, the later time points indicate changes in the relative impedance which is dimensionless. A typical readout is shown in Fig. 3a (*see* **Note 5**).

4 Notes

1. This step is not mandatory for all cell lines. However, since many media contain components of the extracellular matrix, pre-incubation may help to coat the well and therefore improve cell attachment.

2. Cell density: The cell density and even coverage of the electrodes are critical for reliability and reproducibility of data. Therefore, different cell densities have been tested and checked by microscopy (*see* Fig. 3b). Optimal cell density also is cell-type specific. We applied cell numbers between 5×10^4 cells/well (0.6125×10^5 cells/cm^2) and 2×10^4 cells per well. If you work with 96-well plates or other systems (e.g., xCELLigence), again, the optimal cell density has to be experimentally defined first. With the cell types we tested, generally, 2×10^4 cells per well was appropriate.
3. Duration of the experiment: The duration of the experiments will vary depending on the cell type of interest, applied treatments etc., however, we recommend measuring the impedance until cells reach complete confluency. Moreover, the purity of the cell type or the differentiation state should be checked in parallel, if it is critical, for example, when working with primary cells. This can be done by seeding cells on microscopy slides in parallel and staining by antibodies specific for certain markers. The attachment of the cells to the gold electrodes can be observed within the first 24 h. In case cellular differentiation is to be measured, generally, recording impedance over 96–120 h can be recommended. Again, the optimal time course has to be determined individually, because it is dependent on cell type, differentiation lineage, treatment of cells etc. With our system, we measured cell differentiation of the embryonic cancer stem cell line NT2 after treatment with nucleoside analogues and upon siRNA-mediated depletion of stem cell markers up to 4 days [11] as well as after osteogenic or adipogenic induction of mesenchymal stem cells isolated from bone marrow or fat tissue [12]. In case cells are to be treated during the run, we recommend adding the necessary compounds directly to the arrays without displacing them off the ECIS system.
4. The variability between in vitro experiments with living cells always has to be taken into consideration. Therefore, performing experiments in at least duplicates is strongly recommended. Moreover, in addition to experimental replicates, biological replicates are important, especially when working with primary cells (e.g., mesenchymal stem cells isolated from bone marrow).
5. If the readout needs to be compared with data from the xCELLigence system, the corresponding values from the supplier-defined readout have to be recalculated first.

References

1. Giaever I, Keese CR (1984) Monitoring fibroblast behavior in tissue culture with an applied electric field. Proc Natl Acad Sci U S A 81:3761–3764
2. Giaever I, Keese CR (1993) A morphological biosensor for mammalian cells. Nature 366:591–592
3. Kirstein SL, Atienza JM, Xi B et al (2006) Live cell quality control and utility of real-time cell electronic sensing for assay development. Assay Drug Dev Technol 4:545–553
4. Thedinga E, Kob A, Holst H et al (2007) Online monitoring of cell metabolism for studying pharmacodynamic effects. Toxicol Appl Pharmacol 220:33–44
5. Giaever I, Keese C (1991) Micromotion of mammalian cells measured electrically. Proc Natl Acad Sci U S A 88:7896–7900
6. Lo CM, Keese CR, Giaever I (1995) Impedance analysis of MDK cells measured by electric cell-substrate impedance sensing. Biophys J 69:2800–2807
7. Keese CR, Wegener J, Walker SR et al (2004) Electrical wound-healing in vitro. Proc Natl Acad Sci U S A 101:1554–1559
8. Clark PR, Kim RK, Pober JS et al (2015) Tumor necrosis factor disrupts claudin-5 endothelial tight junction barriers in two distinct NF-kappaB-dependent phases. PLoS One 10(3):e0120075
9. Rees MD, Thomas, SR (2015) Using cell-substrate impedance and live cell imaging to measure real-time changes in cellular adhesion and de-adhesion induced by matrix modification. Vis Exp (96):e52423
10. Bilandzic M, Stenvers KL (2014) Assessment of ovarian cancer spheroid attachment and invasion of mesothelial cells in real time. J Vis Exp (87). doi:10.3791/51655
11. Öz S, Maercker C, Breiling A (2013) Embryonic carcinoma cells show specific dielectric resistance profiles during induced differentiation. PLoS One 8:e59895
12. Angstmann M, Brinkmann I, Bieback K et al (2011) Monitoring human mesenchymal stromal cell differentiation by electrochemical impedance sensing. Cytotherapy 13: 1074–1089

Chapter 22

Nano-QSAR Model for Predicting Cell Viability of Human Embryonic Kidney Cells

Serena Manganelli and Emilio Benfenati

Abstract

Traditional Quantitative Structure-Activity Relationships (QSAR) models based on molecular descriptors as translators of chemical information show some drawbacks in predicting toxicity of nanomaterials due to their unique properties and to their nonhomogeneous structure.

This chapter provides instructions on how to use CORAL, freely available software for building nano-QSAR models. CORAL makes use of descriptors based on "quasi-SMILES" representing physicochemical features and/or experimental conditions as an alternative to traditional SMILES encoding chemical structure to build up predictive nano-QSAR models for cytotoxicity.

Key words Nanoparticles, Cell viability, Nano-QSAR, CORAL, Quasi-SMILES

1 Introduction

Nanoparticles (NPs), nano-objects with all three dimensions at the nanoscale, are commonly included in a larger matrix or substrate referred to as nanomaterial (NM) that measures less than 100 nm at least one dimension [1, 2].

The past decade has been characterized by an increase of NMs applications in several areas of science and technology [3]. These materials are commonly employed in consumer products (cosmetics and sunscreens, stain-resistant clothing, self-cleaning windows), in biomedical fields (drug delivery agents, biosensors, or imaging contrast agents), and also in various electronics systems and space technology [3, 4].

However, information concerning the possible toxicological implications derived from the use of these materials for the human and environmental health is still few [5]. Considering their wide range of applications, the impact of NMs on human health and the environment is of great interest. Nanotoxicology has been proposed as a new field of toxicology necessary to fill the gap in knowledge

Daniel F. Gilbert and Oliver Friedrich (eds.), *Cell Viability Assays: Methods and Protocols*, Methods in Molecular Biology, vol. 1601, DOI 10.1007/978-1-4939-6960-9_22, © Springer Science+Business Media LLC 2017

related to the adverse health effects likely to be caused by exposure to NMs [4].

As already mentioned a very small dimension characterizes NPs. Thanks to the small sizes, these substances have highly desirable properties from mechanical, electrical, and chemical points of view. Interestingly this same property is also one of the main reasons of toxicological reactivity [6]. However, the NPs size is not the only factor that can lead to adverse effects for human health. In particular, nanoscale materials have far larger surface area than larger-scale materials of similar masses: this is an important feature affecting bio-reactivity. As surface area per mass of a material increases, a greater amount of the material can be exposed to surrounding environment, thus affecting reactivity [7]

Also size distribution, agglomeration state, shape, crystal structure, chemical composition, surface area, surface chemistry, surface charge, and porosity may influence the biological activity of these substances [8].

The human body can be exposed to NPs through different routes such as inhalation, ingestion, injection, and dermal exposure. After absorption, these substances can be transported to blood leading to the possibility to cause adverse reactions in several organs. Among the organs involved, the kidney could be a target since it receives high blood supply from the total organism and it has an active role in elimination of xenobiotics [9].

Silicon dioxide, also known as silica (SiO_2), is one of the most commonly used NMs [10, 11]. SiO_2 can be divided into types: crystalline and noncrystalline (amorphous) silica. Amorphous silica can be further divided into naturally occurring amorphous silica and synthetic forms. The latter represents the SiO_2 intentionally manufactured which has found widespread applications [12]. Silica NPs are employed in chemical mechanical polishing and as additives to drugs, cosmetics, printer toners, varnishes, and food stuffs; in biomedical and biotechnological fields, such as biosensors for the simultaneous assay of glucose, lactate, l-glutamate, and hypoxanthine in rat striatum; and biomarkers for leukemia cell identification using optical microscopy imaging, cancer therapy, DNA delivery, drug delivery, and enzyme immobilization [13]. Since SiO_2 found many applications in the everyday life, special attention should be paid to their potential toxic effects [10, 11]. Traditionally, toxicity studies on NPs often focused on lung damage, since these substances can easily disperse into the air due to their very low density and so can be inhaled. However, SiO_2 (45 nm) was found also in liver, urinary bladder, and kidney after intravenous injection in mice [14]. Moreover, it was demonstrated that SiO_2 NPs with a size ranging from of 50 and 100 nm were eliminated with the urine, proving the possibility of exposure to these substances also for other organs than lungs [15]. The mechanism of SiO_2 toxicity is still poorly understood. However, recent studies on SiO_2 NPs toxicity demonstrated that these substances induce oxidative stress and pro-inflammatory responses in the rodent model and in several types of cultured mammalian cell lines [12, 15]. Oxidative stress, mediated by reactive oxygen species (ROS) production, is a key event in the mechanism of toxicity for many NPs. A healthy human organism is able to develop a series of cellular defense mechanism in order to decrease the ROS level. However, this is not always the case. Indeed, when the ROS production is high, many events such as lipid peroxidation occur that can finally lead to cellular dysfunction or death [15, 16].

Since the production of synthetic NPs is increasing, the exposure to these substances is supposed to increase consequently in the coming years. In this context, the challenge for scientists is to develop new knowledge and approaches for the safety evaluation of NPs to be used for the assessment of the safety of NMs [6, 17].

In silico methods can be used for toxicological characterization of NPs since they have many advantages in terms of cost, time-effectiveness, and ethical implications [18]. Among in silico approaches, quantitative structure-activity relationships (QSAR) seem to be the most promising. Basically, QSAR are predictive computational models that aim at defining a mathematical equation between the variance in molecular structures (molecular descriptors) and the variance in a given property (endpoint) for a set of compounds [19]. Aside from toxicological activity, this approach can be employed for predicting various physical–chemical properties as well, and, in such cases, it is named as QSPR (quantitative structure–property relationship) [20].

Besides in silico models, a number of cell-based assays have been widely employed in toxicological studies to determine the viability and/or cytotoxicity after exposure to NPs [21]. The lactate dehydrogenase (LDH) leakage assay, the neutral red, and the 3-[4, 5-dimethylthiazol-2-yl]-2, 5 diphenyl tetrazolium bromide (MTT) assay are the most commonly employed tests and they use colorimetric or fluorescent dyes as markers to determine cell viability through assessment of membrane integrity or cell metabolism [21, 22].

In particular, the MTT assay is based on the protocol described for the first time by Mossmann [23]. MTT is a yellow water-soluble tetrazolium dye that succinate dehydrogenase, present within the mitochondria, can convert into insoluble purple formazan by cleavage of the tetrazolium ring. Since formazan production can only occur in metabolically active cells, the level of activity is a measure of the cells viability (Fig. 1) [22]. The amount of MTT-formazan produced is measured spectrophotometrically. However, a limitation of MTT assay is that it is not able to discriminate between a cytotoxic (cell-killing effects) and a cytostatic (reduced growth rate) effect [24].

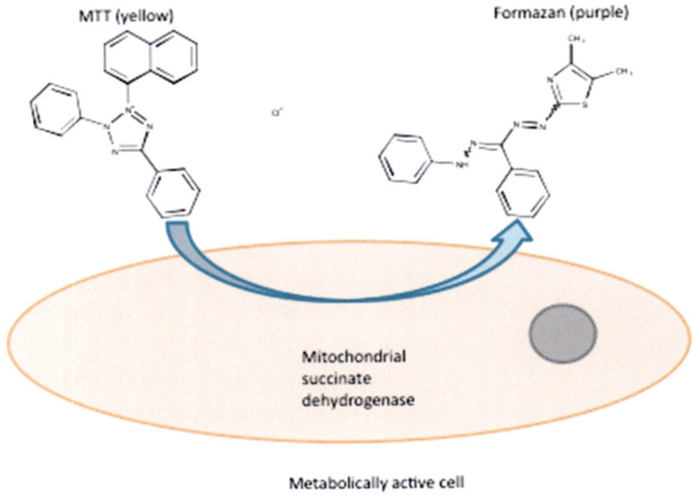

Fig. 1 Principles of MTT assay: MTT assay aims to measure the reduction of yellow 3-(4,5-dimethylthiazol-2-yl)-2,5-diphenyl tetrazolium bromide (MTT) by mitochondrial succinate dehydrogenase

The MTT enters the cells and passes into the mitochondria where it is reduced to an insoluble, purple colored formazan product. Since reduction of MTT can only occur in metabolically active cells, the level of this activity is a measure of the viability of the cells.

In order to assess in vitro renal toxicity, the human embryonic kidney cell line (HEK293) is a widely used model [25]. This cell line is well characterized for its relevance to the toxicity model in human and it has been previously used for toxicological assessments [5, 26–28].

To provide an overview of one of recently introduced strategy of QSAR modeling for NPs (nano-QSAR), in the below subsections of this chapter we propose to illustrate the different steps involved in model building using the CORAL software. Specifically, we describe how we used this software to construct a QSAR model for SiO_2 NPs [29] on a dataset of in vitro data derived from cytotoxicity studies (MTT assay) conducted on HEK293 cell line [5]. The algorithm we propose makes use of descriptors, which are mathematical function of size, concentration, and exposure time. The so-called optimal descriptors are based on "quasi-SMILES". Differently from "Simplified Molecular Input Line Entry specification" (SMILES), which are strings of characters encoding molecular structures, quasi-SMILES are strings representing particular conditions, not of the molecular structure. Numbers and/or alphabet letters can specify each of these features [29].

2 Materials

Encode and combine each physicochemical property (e.g., size) and experimental condition (e.g., concentration and exposure time) using different items (e.g., numbers or letters) into strings (quasi-SMILES). Assign an identifier (ID number) to each quasi-SMILES and associate it with the corresponding experimental cell viability value. The outcomes of these procedures are enclosed in Tables 1 and 2 (*see* **Note 1**).

3 Methods

CORAL [30] is standalone application software for constructing regression and classification QSAR/QSPR for classical chemicals and for nanomaterials based on the Monte Carlo technique. Download the software and the user manual from the CORAL website [30]. In the zip folder "CORALSEA," you will find two subfolders:

- "DataBases," containing many pre-built QSAR/QSPR models for different endpoints.
- "MyCORALSEA," containing two subfolders (REGRESSION and CLASSIFICATION), each one with an executable file to build QSAR/QSPR models.

Table 1
Codes for different size, concentration, and exposure time values of SiO_2-Nps

Eclectic information	Value	Code
Physico-chemical property		
Np size (nm)	20 50	X Y
Experimental condition		
Concentration (μg mL^{-1})	25 50 100 200	1 2 3 4
Exposure time (h)	0 12 24 36 48	A B C D E

Table 2
Values of cell viability (%) for 20 and 50 nm SiO_2- NPs at different concentrations (25–200 μg mL^{-1}) and exposure times

Particle size, nm	Concentration, μg mL^{-1}	Exposure time, h	Quasi-SMILES	Cell viability (%)
X	1	A	X1A	100.04
X	2	A	X2A	99.78
X	3	A	X3A	100.04
X	4	A	X4A	100.04
X	1	B	X1B	101.73
X	2	B	X2B	98.35
X	3	B	X3B	96.15
…	…	…	…	…
…	…	…	…	…
X	4	D	X4D	27.06
X	1	E	X1E	75.01
X	2	E	X2E	62.75
X	3	E	X3E	26.72
X	4	E	X4E	20.30
Y	1	A	Y1A	99.95
Y	2	A	Y2A	99.95
Y	3	A	Y3A	100.11
Y	4	A	Y4A	99.86
Y	1	B	Y1B	101.74
Y	2	B	Y2B	102.07
…	…	…	…	…
…	…	…	…	…
Y	3	E	Y3E	53.01
Y	4	E	Y4E	26.68

After choosing the kind of model to build (regression or classification), you can run the executable file CORALSEA.exe.

In this subsection, we outline the basic steps to create a regression nano-QSAR model using the CORAL software, which are the following:

1. Preparation of the input file.
2. Definition of the method.
3. Search for the best threshold (T^*) and number of epochs (N^*).
4. Building and validation of the model.

3.1 Preparation of the Input File

CORAL requires a text input file, which contains a list of chemicals indicated by their identifiers (any ID number such as CAS number), the SMILES strings, and experimental activities (*see* **Note 2**). The original version of the software has been improved and adapted to the use of nanomaterials (*see* **Note 3**). In this case, CORAL needs a list of alphanumeric strings, named “quasi-SMILES,” encoding the relevant topological features (but also other encoded parameters, relevant for nanoparticles), and the endpoint values. A definite item (e.g., number or alphabet letter) specifies each of these features. Use data contained in Table 2 to prepare *n* different sets of quasi-SMILES randomly split in training (≈60%), calibration (≈20%), and validation sets (≈20%). The first two “visible” sets are involved in the model's building while the third one is called “invisible” and is used for its validation. Follow the instructions below to prepare the input file:

1. Indicate “+” for the training set; “# “for the calibration set; and for the”*” for the validation set items.
2. Identifier, i.e., a sequential number, or CAS number.
3. Quasi-SMILES encoding relevant parameters.
4. Endpoint value.
   You will obtain an input file organized as follows:

```
#1 X1A    100.04
+2 X2A    99.78
#3 X3A    100.04
+4 X4A    100.04
#5 X1B    101.73
+6 X2B    98.35
```

3.2 Definition of the Method

After loading the input file containing the list of quasi-SMILES of training and calibration sets and the endpoint values, the selection of method for the calculation of descriptors can be performed (Fig. 2).

For modeling toxicity of bulk materials CORAL allows calculating graph or SMILES based descriptors [31]. In the case of nano-QSAR, the choice of descriptors depends on the type of nanomaterial. For fullerene derivatives the user can apply both

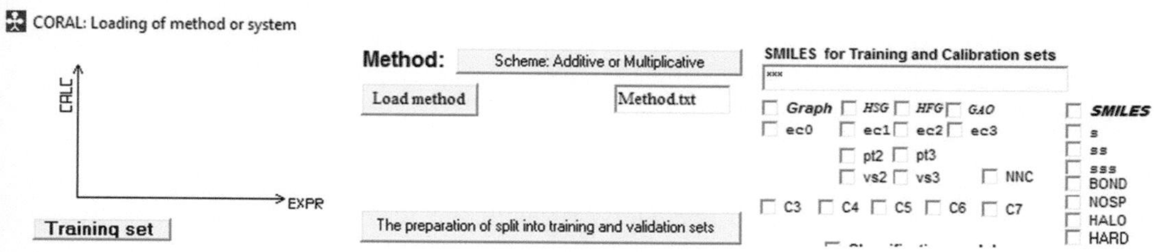

Fig. 2 Method selection for calculating descriptors

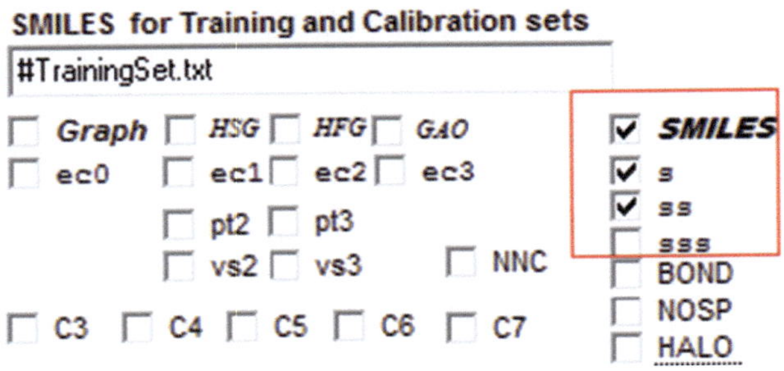

Fig. 3 Attributes available for quasi-SMILES

SMILES and graph [32], but for most nanoparticles (e.g., inorganic oxides) the description of the only chemical structure is not sufficient, or inappropriate. Hence apply options related to SMILES (not related to graph) and the attributes will represent the features encoded by quasi-SMILES (Fig. 3). For modeling cell viability of SiO_2 nanoparticles use the *Sk* local attributes.

Then select the additive scheme of the CORAL software (Fig. 2), which calculates the "Descriptor of Correlation Weights" (*DCW*) as the sum of correlation weights associated with each component of the quasi-SMILES [29, 33], according to Eq. (1):

$$DCW\left(Threshold, N_{epoch}\right) = £CW\left(Ck\right) \quad (1)$$

Where *CW(Ck)* are the correlation weights for codes of size, concentration, and exposure time, extracted from *k*-th quasi-SMILES. The so-called optimal descriptor (*DCW*) is used to calculate the predicted value for the endpoint of interest using a one variable linear equation, as follows:

$$Endpoint = C_0 + C_1 \times DCW\left(Threshold, N_{epoch}\right) \quad (2)$$

When the method is loaded, you can choose different options according to your task. Define the parameters (D_{start}, $D_{precision}$, N_{epoch}) described in the pdf manual "ReadMe." D_{start} and $D_{precision}$ refer to the incremental value to be assigned to the correlation weights. Finally, save the defined method.

3.3 Search for the Best Threshold (T*) and Number of Epochs (N*)

Threshold (*T*) and N_{epoch} (*N*) are parameters of the Monte Carlo optimization method. In order to obtain a model with a good predictive potential, search for the preferred threshold (*T**) and number of epochs (*N**), which give the best statistics for the calibration set. The threshold classifies codes as either rare (less reliable features) or not rare features, which are used by the model and labeled as active. The *DCW* is calculated with the *CWs* only of active features *Ck*, excluding those related to rare ones.

The N_{epoch} is the number of cycles (sequence of modifications of *CWs* for all codes involved in model development) for the optimization.

In order to select the preferable T^* and N^* evaluate the contents of the following output files generated by the software:

1. File Search/#a.txt containing the average statistics of the models for the selected *T* and *N*.
2. File Search/#r.txt containing statistics of the models for each probe of the Monte Carlo optimization.
3. File Search/#BestMDL.txt containing statistical results of the best models for calibration set.

3.4 Building and Validation of the Model

Having data on the preferred T^* and N^* build the model based on their numerical values (Fig. 4). Confirm that the files in the folder "Model" can be replaced by those containing information on the newly built model (Fig. 5) and "Save system" (Fig. 6).

Repeat the calculations several times for various splits (e.g., five) into training, calibration, and external validation sets. The applicability domain for the models can be defined according to the distribution of the attributes and their role [29].

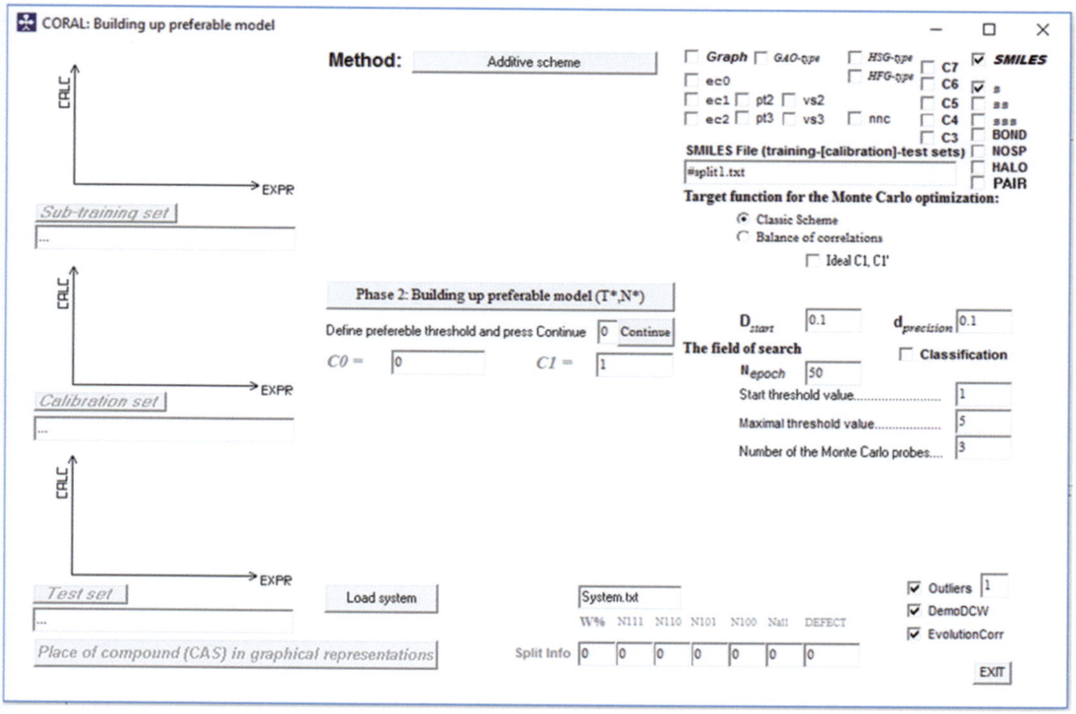

Fig. 4 Building of the preferred model by inserting values for the preferred T^* and N

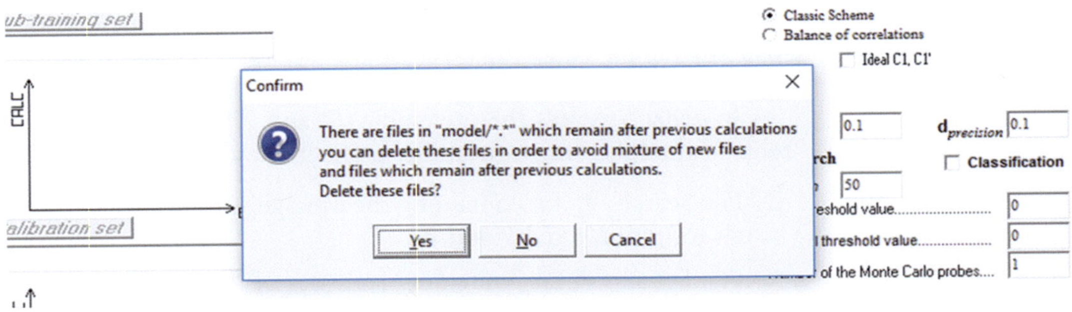

Fig. 5 Message displayed before the new model building to confirm the replacement of the existing files in the Model folder

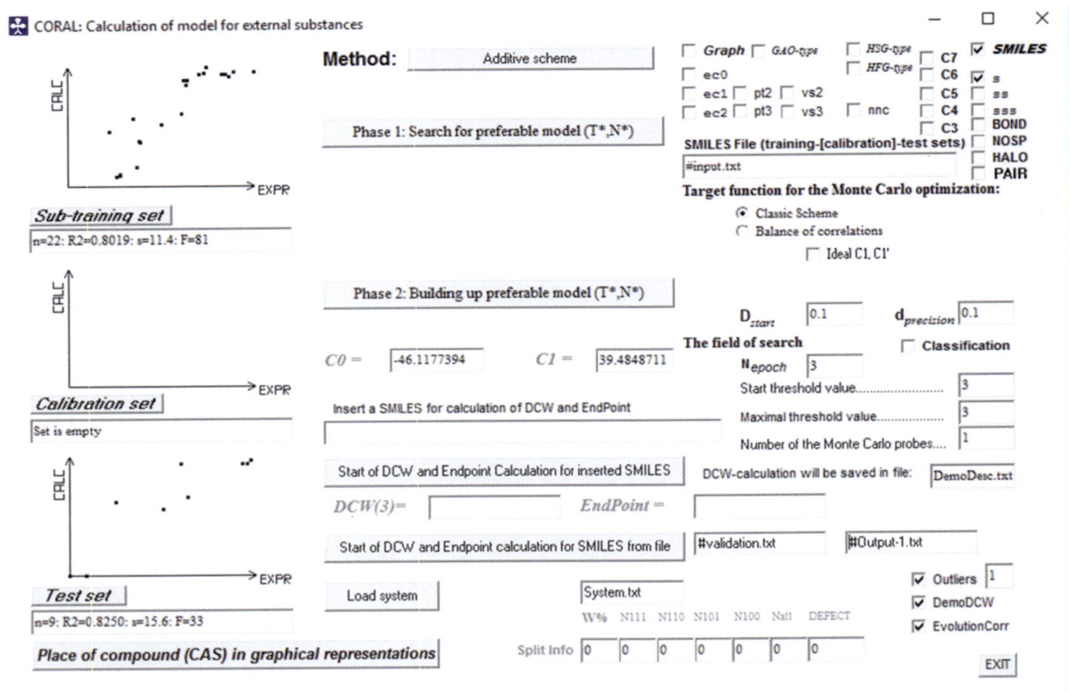

Fig. 6 Built model: the built models must be saved before being used for predicting the external validation set

3.5 Interpretation of Results

CORAL calculates a number of statistical parameters for the internal and external validation of newly built regression models: the squared correlation coefficient (R^2), the leave-one-out (*LOO*) cross-validated correlation coefficient (Q^2), the standard error of the estimation (s), and the variance ratio (F) [34]. These are computed as follows:

$$R^2 = \frac{\left[\sum\left(y_{obs} - \overline{y_{obs}}\right)\left(y_{pred} - \overline{y_{pred}}\right)\right]^2}{\sum\left(y_{pred} - \overline{y_{pred}}\right)^2 \sum\left(y_{obs} - \overline{y_{obs}}\right)^2} \quad (3)$$

where y_{obs} and y_{pred} are the observed and predictive values while $\overline{y_{obs}}$ and $\overline{y_{pred}}$ are the mean values of the observed and predictive values respectively;

$$Q^2 = 1 - \frac{\sum \left(y_{obs} - y_{pred} \right)^2}{\sum \left(y_{obs} - \bar{y} \right)^2} \tag{4}$$

where $\bar{y}$ means average activity value of the training set, whereas y_{obs} and y_{pred} represent observed and *LOO*-predicted activity values of the training set.

$$F = \frac{\dfrac{\sum \left(y_{calc} - \overline{y_{obs}} \right)^2}{p}}{\dfrac{\sum \left(y_{obs} - y_{calc} \right)^2}{n - p - 1}} \tag{5}$$

where n is the number of compounds and p is the number of predictor variables.

$$s = \sqrt{\frac{PRESS}{n}} \tag{6}$$

where *PRESS* represents the predicted residual sum of squares and n is number of compounds. *PRESS* is given by the expression:

$$PRESS = \sum \left(y_{obs} - y_{pred} \right)^2 \tag{7}$$

According to the criteria indicated by Golbraikh and Tropsha [35] revised by Tropsha [36], a model has high predictive power if the following conditions are fulfilled:

1. $Q^2 > 0.5$.
2. $R^2 > 0.6$.
3. $(R^2 - R_0^2)/R^2 < 0.1$ or $(R^2 - R_0^{'2})/R^2 < 0.1$.
4. $0.85 \leq k \leq 1.15$ or $0.85 \leq k' \leq 1.15$.
5. $|R_0^2 - R_0^{'2}| < 0.3$.

where R_0^2 and $R_0^{'2}$ are squared correlation coefficients for regression through the origin, calculated between predicted versus experimental values and between experimental versus predicted values, and k and k' are the slopes in the former and later cases respectively; Q^2 is calculated for the training sets, while all other criteria are calculated for the validation sets.

$$R_0^2 = 1 - \frac{\sum \left(y_{obs} - k \times y_{pred} \right)^2}{\sum \left(y_{obs} - \overline{y_{obs}} \right)^2} \tag{8}$$

$$R'^2_0 = 1 - \frac{\sum \left(y_{pred} - k' \times y_{obs} \right)^2}{\sum \left(y_{pred} - \overline{y_{pred}} \right)^2} \tag{9}$$

$$k = \frac{\sum \left(y_{pred} \times y_{obs} \right)^2}{\sum \left(y_{pred} \right)^2} \tag{10}$$

$$k' = \frac{\sum \left(y_{pred} \times y_{obs} \right)^2}{\sum \left(y_{obs} \right)^2} \tag{11}$$

Check the model performance fulfills these criteria. CORAL also calculates the validation parameters R_m^2, average R_m^2, and ΔR_m^2 on the calibration sets related, as proposed in the literature [34, 37]. The R_m^2 metric is calculated based on the correlations between observed and predicted values with (R^2) and without $\left(R_0^2 \right)$ intercept for the least squares regression lines as shown in the following equations:

$$R^2 \times \left(1 - \sqrt{\left(R^2 - R_0^2 \right)} \right) \tag{12}$$

R'^2_m metric calculation is based on the correlations between predicted and observed values with (R^2) and without $\left(R'^2_0 \right)$ for the least squares regression lines, as follows:

$$R'^2_m = R^2 \times \left(1 - \sqrt{\left(R^2 - R'^2_0 \right)} \right) \tag{13}$$

Average R_m^2 and ΔR_m^2 are calculated as follows:

$$\overline{R_m^2} = \frac{\left(R_m^2 + R'^2_m \right)}{2} \tag{14}$$

$$R_m^2 = \left| R_m^2 - R'^2_m \right| \tag{15}$$

According to the literature, a model has predictive potential if $R_m^2 > 0.5$, average $R_m^2 > 0.5$, and $\Delta R_m^2 < 0.2$ [38]. Check the values of these parameters for the evaluation of the predictability (*see* **Note 4**).

4 Notes

1. In our previous work [29] the endpoint considered for nano-QSAR analysis was cell viability (%) of HEK293 cells, measured by the MTT assay. The MTT results, expressed as percentage of cell viability (%), were taken from the literature [5]. HEK293 cells were exposed to 20 and 50 nm silica nanoparticles

at increasing concentrations of 25, 50, 100, and 200 μg mL^{-1} for 12, 24, 36, and 48 h. These physicochemical property (size) and experimental condition (concentration and exposure time) were encoded and combined to obtain "quasi-SMILES."

2. The nature of the molecular graph can be hydrogen suppressed graph (*HSG*), hydrogen filled graph (*HFG*), and graph of atomic orbitals (*GAO*). In the case of *HSG* and *HFG*, vertices are representations of the chemical elements (such as carbon, nitrogen, oxygen, etc.), while in the case of *GAO*, vertices are representations of electronic structure. SMILES attributes include the local and the global ones. Local SMILES attributes are defined by a sequence of atoms and bonds present in the SMILES string, while global SMILES attributes look for the presence of certain atoms and bonds which may be contained in the molecule, as detailed below. *Sk, SSk*, and *SSSk* are local SMILES attributes that are representations of molecular fragments, composed of one, two, or three elements, respectively. An example can be provided by the SMILES code Clc1ccccc1, whose local attributes can be represented as *Sk* = (Cl,c,1,c,c,c,c,c,1), *SSk* = (Clc,c1,cc,cc,cc,cc,cc,c1), and *SSSk* = (Clc1,c1c,ccc,ccc,ccc,ccc,cc1). *BOND*, *PAIR*, *NOSP*, and *HALO* are global SMILES attributes. These involve the presence of the following elements in the target SMILES: *BOND* refers to the presence/absence of double (=), triple (#), and stereochemical (@) bonds; *PAIR* indicates the co-occurrence of two elements among the followings: F, Cl, Br, I, N, O, S, P, #, =, @; *NOSP* indicates the presence/absence of nitrogen, oxygen, sulfur, and phosphorus; *HALO* refers to the presence/absence of halogens.

3. The peculiarities of nanoparticles described in the introduction affect their toxicological behavior. Thus the modeling scheme for toxicological endpoints related to nanoparticles can be structured differently from that used for classical chemicals, as follows: the endpoint is a mathematical function of all available information, not only of molecular structure. The heterogeneous so-called eclectic information may consist of: (a) chemical composition of the NPs, (b) conditions of synthesis/preparation of the NPs, (c) physicochemical features of NPs (size, surface area, porosity, symmetry, electromechanical properties, etc.), (d) experimental conditions of the cytotoxicity assay used to determine the endpoint (exposure time, presence/absence of serum, etc.). In order to define a predictive model for an endpoint related to cytotoxicity of nanomaterials the traditional paradigm for QSAR modeling "Endpoint = F (molecular structure)" can be replaced by a new one: "Endpoint = F (eclectic information)."

As possible limitation, it should be considered that the use of information associated with the properties of NPs (such as size, shape, etc.) greatly depends on the measurements of experimental values. This has two important consequences: (a) the developed model needs experimental values to describe the property of the material; (b) the procedure to generate the experimental value may influence such a value, and so the model.

4. Cell viability as a function of the optimal descriptors, used for building the nano-QSAR model was calculated on five splits of data according to Eq. (2) as follows:

Split1

$$\textit{Cell viability}(\%) = -24.5778(\pm 2.5421) + 29.364(\pm 0.7808) \times DCW(2,2) \quad (6)$$

$n = 13$, $R^2 = 0.9276$, $Q^2 = 0.8992$, $s = 8.12\%$, $F = 141$ (training set)
$n = 13$, $R^2 = 0.7971$, $s = 14.4\%$ (calibration set)
$n = 14$, $R^2 = 0.7288$, $s = 12.7\%$ (validation set)

Split2

$$\textit{Cell viability}(\%) = -135.37(\pm 8.6778) + 60.0345(\pm 2.4084) \times DCW(1,15) \quad (7)$$

$n = 9$, $R^2 = 0.9268$, $Q^2 = 0.8829$, $s = 7.51\%$, $F = 89$ (training set)
$n = 16$, $R^2 = 0.6878$, $s = 18.6\%$ (calibration set)
$n = 15$, $R^2 = 0.7204$, $s = 14.7\%$ (validation set)

Split3

$$\textit{Cell viability}(\%) = -7.555(\pm 1.7333) + 20.3850(\pm 0.3811) \times DCW(2,3) \quad (8)$$

$n = 12$, $R^2 = 0.9521$, $Q^2 = 0.9354$, $s = 6.89\%$, $F = 19$ (training set)
$n = 14$, $R^2 = 0.8205$, $s = 13.9\%$ (calibration set)
$n = 14$, $R^2 = 0.6838$, $s = 13.9\%$ (validation set)

Split4

$$\textit{Cell viability}(\%) = 27.3716(\pm 5.4223) + 22.5382(\pm 1.875) \times DCW(2,9) \quad (9)$$

$n = 11$, $R^2 = 0.7963$, $Q^2 = 0.6505$, $s = 13.1\%$, $F = 35$ (training set)
$n = 15$, $R^2 = 0.7538$, $s = 15.9\%$ (calibration set)
$n = 14$, $R^2 = 0.7996$, $s = 13.5\%$ (validation set)

Split5

$$\textit{Cell viability}(\%) = -47.8716(\pm 4.2635) + 31.733(\pm 1.1206) \times DCW(1,3) \quad (10)$$

$n = 10$, $R^2 = 0.9075$, $Q^2 = 0.8648$, $s = 8.73\%$, $F = 78$ (training set)
$n = 14$, $R^2 = 0.8155$, $s = 16.9\%$ (calibration set)
$n = 16$, $R^2 = 0.7948$, $s = 21.2\%$ (validation set)

where n is the number of nanoparticles system in each set; R^2 is the square correlation coefficient, Q is leave-one-out cross-validated correlation coefficient, s is the standard error of the estimation, F is the variance ratio (*see* above). You can find these results in the output file of the software.

Based on the statistical results, the five models have good predictability because they fulfill the criteria described above.

Acknowledgments

We kindly acknowledge the EC Project PEPTICAPS (Project Reference 686141).

References

1. Buzea C, Pacheco Blandino II, Robbie K (2007) Nanomaterials and nanoparticles: sources and toxicity. Biointerphases 2:17–172
2. National Institute for Occupational Safety and Health (2009) Approaches to safe nanotechnology: managing the health and safety concerns associated with engineered nanomaterials. Publication Number 2009–125
3. Puzyn T, Leszczynska D, Leszczynski J (2009) Toward the development of "Nano-QSARs": advances and challenges. Small 22:2494–2509
4. Arora S, Rajwade JM, Paknikar KM (2012) Nanotoxicology and in vitro studies: the need of the hour. Toxicol Appl Pharmacol 258: 151–165
5. Wang F, Gao F, Lan M, Yuan H, Huang Y, Liu J (2009) Oxidative stress contributes to silica nanoparticle-induced cytotoxicity in human embryonic kidney cells. Toxicol in Vitro 23: 808–815
6. Oberdorster G (2009) Safety assessment for nanotechnology and nanomedicine: concepts of nanotoxicology. J Intern Med 267:89–105
7. Nel AE, Lutz M, Darrell V, Tian X, Hoek Eric MV, Somasundaran P, Klaessig F, Castranova V, Thompson M (2009) Nat Mater 8:543–557
8. Oksel C, Ma CY, Liu JJ, Wilkins T, Wang XZ (2015) (Q)SAR modelling of nanomaterial toxicity: a critical review. Particuology 21:1–19
9. Pujalté I, Passagne I, Brouillaud B, Tréguer M, Durand E, Ohayon-Courtès C, L'Azou B (2011) Cytotoxicity and oxidative stress induced by different metallic nanoparticles on human kidney cells. Part Fibre Toxico 8:10
10. Chen JF, Ding HM, Wang JX, Shao L (2004) Preparation and characterization of porous hollow silica nanoparticles for drug delivery application. Biomaterials 25:723–727
11. Fuller JE, Zugates GT, Ferreira LS, Ow HS, Nguyen NN, Wiesner UB, Langer RS (2008) Intracellular delivery of core-shell fluorescent silica nanoparticles. Biomaterials 29:1526–1532
12. Park EJ, Park K (2009) Oxidative stress and pro-inflammatory responses induced by silica nanoparticles in vivo and in vitro. Toxicol Lett 184:18–25
13. Eom HJ, Choi J (2009) Oxidative stress of silica nanoparticles in human bronchial epithelial cell, Beas-2B. Toxicol in Vitro 23:1326–1332
14. He Q, Zhang Z, Gao F, Li Y, Shi J (2011) In vivo biodistribution and urinary excretion of mesoporous silica nanoparticles: effects of particle size and PEGylation. Small 7:271–280
15. Passagne I, Morille M, Rousset M, Pujalté I, L'Azou B (2012) Implication of oxidative stress in size-dependent toxicity of silica nanoparticles in kidney cells. Toxicology 299: 112–124
16. Manke A, Wang L, Rojanasakul Y (2013) Mechanisms of nanoparticle-induced oxidative stress and toxicity. BioMed Res Int 2013: 942916. doi:10.1155/2013/942916
17. Kahru A, Dubourguier HC (2010) From ecotoxicology to nanoecotoxicology. Toxicology 269:105–119
18. Winkler DA, Mombelli E, Pietroiusti A, Tran L, Worth A, Fadeel B, McCallh MJ (2013) Applying quantitative structure-activity relationship approaches to nanotoxicology: current status and future potential. Toxicology 313:15–23
19. Roy K, Kar S, Das RN (2015) A primer on QSAR/QSPR modeling: fundamental concepts, SpringerBriefs in Molecular Science. Springer, New York
20. Todeschini R, Consonni V, Gramatica P (2009) Chemometrics in QSAR. In: Brown S, Tauler R, Walczak R (eds) Comprehensive Chemometrics, vol 4. Elsevier, Oxford, pp 129–172
21. Monteiro-Riviere NA, Inman LW, Zhang AO (2009) Limitations and relative utility of screening assays to assess engineered nanoparticle toxicity in a human cell line. Toxicol Appl Pharmacol 234:222–235
22. Fotakis G, Timbrell JA (2006) In vitro cytotoxicity assays: comparison of LDH, neutral red, MTT and protein assay in hepatoma cell lines following exposure to cadmium chloride. Toxicol Lett 160:171–177
23. Mossmann T (1983) Rapid colorimetric assay for cellular growth and survival: application to proliferation and cytotoxicity assays. J Immunol Methods 65:55–63
24. Plumb JA (2004) Cell sensitivity assays: the MTT assay. Methods Mol Med 88:165–169
25. Thomas P, Smart TG (2005) HEK293 cell line: a vehicle for the expression of recombinant proteins. J Pharmacol Methods 51: 187–200

26. Florea AM, Splettstoesser F, Büsselberg D (2007) Arsenic trioxide (As_2O_3) induced calcium signals and cytotoxicity in two human cell lines: SY-5Y neuroblastoma and 293 embryonic kidney (HEK). Toxicol Appl Pharmacol 220:292–301
27. Reddy ARN, Reddy YN, Krishna DR, Himabindu V (2010) Multi wall carbon nanotubes induce oxidative stress and cytotoxicity in human embryonic kidney (HEK293) cells. Toxicology 272:11–16
28. Ji LL, Chen Y, Wang ZT (2008) The toxic effect of pyrrolizidine alkaloid clivorine on the human embryonic kidney 293 cells and its primary mechanism. Exp Toxicol Pathol 60:87–93
29. Manganelli S, Leone C, Toropov AA, Toropova AP, Benfenati E (2016) QSAR model for predicting cell viability of human embryonic kidney cells exposed to SiO_2 nanoparticles. Chemosphere 144:995–1001
30. CORAL-QSAR/QSPR, http://www.insilico.eu/coral/. Accessed October 2016
31. Manganelli S, Benfenati E, Manganaro A, Kulkarni S, Barton-Maclaren TS, Honma M (2016) New quantitative structure-activity relationship models improve predictability of Ames mutagenicity for aromatic azo compounds. Toxicol Sci 153:316–326
32. Toropova AP, Toropov AA (2015) Quasi-SMILES and nano-QFAR: united model for mutagenicity of fullerene and MWCNT under different conditions. Chemosphere 139:18–22
33. Toropov AA, Toropova AP, Benfenati E, Gini G, PuzynT LD, Leszczynski J (2012) Novel application of the CORAL software to model cytotoxicity of metal oxide nanoparticles to bacteria *Escherichia coli*. Chemosphere 89: 1098–1102
34. Roy K, Roy PP (2008) Comparative QSAR studies of CYP1A2 inhibitor flavonoids Using 2D and 3D descriptors. Chem Biol Drug Des 72:370–382
35. Golbraikh A, Tropsha A (2002) Beware of q^2! J Mol Graphics Model 20:269–276
36. Tropsha A (2010) Best practices for QSAR Model development, validation, and exploitation. Mol Inf 29:476–488
37. Roy K, Mitra I, Kar S, Ojha PK, Das RN, Kabir H (2012) Comparative studies on some metrics for external validation of QSPR models. J Chem Inf Model 52:396–408
38. Roy K, Chakraborty P, Mitra I, Ojha PK, Kar S, Das RN (2013) Some case studies on application of "Rm^2" metrics for judging quality of quantitative structure-activity relationship predictions: emphasis on scaling of response data. J Comput Chem 34: 1071–1082

Erratum to: Functional Viability: Measurement of Synaptic Vesicle Pool Sizes

Jana K. Wrosch and Teja W. Groemer

Erratum to:
Chapter 15 in: Daniel F. Gilbert and Oliver Friedrich (eds.),
***Cell Viability Assays: Methods and Protocols*, Methods in Molecular Biology,**
vol. 1601, DOI 10.1007/978-1-4939-6960-9_15

When the volume published the chapter coauthor, Jana K. Wrosch, did not include the acknowledgment that this work was performed in fulfillment of the requirements for obtaining the degree "Dr. rer. biol. hum." at the University of Erlangen-Nuremberg.

The updated online version of this chapter can be found at
http://dx.doi.org/10.1007/978-1-4939-6960-9_15

Daniel F. Gilbert and Oliver Friedrich (eds.), *Cell Viability Assays: Methods and Protocols*, Methods in Molecular Biology, vol. 1601, DOI 10.1007/978-1-4939-6960-9_23, © Springer Science+Business Media LLC 2017

Erratum to: Phenotyping Cellular Viability by Functional Analysis of Ion Channels: GlyR-targeted Screening in NT2-N Cells

Katharina Kuenzel, Sepideh Abolpour Mofrad and Daniel F. Gilbert

Erratum to:
Chapter 16 in: Daniel F. Gilbert and Oliver Friedrich (eds.),
***Cell Viability Assays: Methods and Protocols*, Methods in Molecular Biology, vol. 1601, DOI 10.1007/978-1-4939-6960-9_16**

The volume was published with partial acknowledgement. The current version of the volume is updated with the complete acknowledgement as follows:

The authors gratefully acknowledge funding of the Staedtler Stiftung and Bavarian Equal Opportunities Sponsorship—Förderung von Frauen in Forschung und Lehre (FFL)—Promoting Equal Opportunities for Women in Research and Teaching. Mofrad SA performed the present work in fulfillment of the requirements for obtaining the degree Dr. rer. biol. hum. at the Friedrich- Alexander University of Erlangen-Nürnberg (FAU).

The updated online version of this chapter can be found at
http://dx.doi.org/10.1007/978-1-4939-6960-9_16

Daniel F. Gilbert and Oliver Friedrich (eds.), *Cell Viability Assays: Methods and Protocols*, Methods in Molecular Biology, vol. 1601, DOI 10.1007/978-1-4939-6960-9_24,

Index

A

B

C

Daniel F. Gilbert and Oliver Friedrich (eds.), *Cell Viability Assays: Methods and Protocols*, Methods in Molecular Biology, vol. 1601, DOI 10.1007/978-1-4939-6960-9, © Springer Science+Business Media LLC 2017

D

E

F

N

O

P

Q

R

S

T

U

V

W

X

Y

MIX
Papier aus verantwortungsvollen Quellen
Paper from responsible sources
FSC® C105338

If you have any concerns about our products,
you can contact us on
ProductSafety@springernature.com

In case Publisher is established outside the EU,
the EU authorized representative is:
Springer Nature Customer Service Center GmbH
Europaplatz 3, 69115 Heidelberg, Germany

Printed by Libri Plureos GmbH
in Hamburg, Germany